Assessing infiltration and exfiltration on the Performance of Urban Sewer Systems (APUSS)

Assessing infiltration and exfiltration on the Performance of Urban Sewer Systems (APUSS)

Edited by

J. Bryan Ellis* and Jean-Luc Bertrand-Krajewski**

***Urban Pollution Research Centre, Middlesex University,**
The Burroughs, Hendon, NW4 4BT, London, UK.
email: B.Ellis@mdx.ac.uk
****LGCIE, INSA de Lyon – Université de Lyon,**
34 Avenue des Arts, 69621 Villeurbanne Cedex, France.
email: jean-luc.bertrand-krajewski@insa-lyon.fr

IWA Publishing
London • New York

Published by **IWA Publishing**
Alliance House
12 Caxton Street
London SW1H 0QS, UK
Telephone: +44 (0)20 7654 5500
Fax: +44 (0)20 7654 5555
Email: publications@iwap.co.uk
Web: www.iwapublishing.com

First published 2010

Typeset in India by OKS Prepress Services.

British Library Cataloguing in Publication Data
A CIP catalogue record for this book is available from the British Library

Library of Congress Cataloging-in-Publication Data
A catalog record for this book is available from the Library of Congress

ISBN 10: 184339149X
ISBN 13: 9781843391494

Acknowledgement
The results obtained in this publication have been obtained within the framework of the EC funded research project "*Assessing infiltration and exfiltration on the Performance of Urban Sewer Systems (APUSS)*", contract no. EVK1-CT-2000-00072, coordinated by LGCIE, INSA de Lyon – Universite de Lyon, France (FR) and including Urban Pollution Research Centre, Middlesex University, London (UK); EAWAG, Dubendorf, Switzerland (CH); Institute of Water Management, Dresden University of Technology, Germany (DE) DHI-Hydroinform a.s, Prague, Czech Republic (CZ) IRSA-CNR, Rome, Italy (I); LNEC, Lisbon, Portugal (PT); Department of Sanitary & Ecological Engineering, Technical University, Prague, Czech Republic (CZ); DHI a.s. Prague, Czech Republic (CZ); Hydroprojekt, Prague, Czech Republic (CZ) and Emschergen-ossenschaft, Essen, Germany (DE).

This project was organised within the "Energy, Environment and Sustainable Development Programme" in the 5th Framework Programme for "Science Research and Technological Development" of the European Commission and was part of the CityNet cluster; the network of European research projects on integrated urban water management.

Contents

Edited by Bryan Ellis and Jean-Luc Bertrand-Krajewski. ISBN: 9781843391494. Published by IWA Publishing, London, UK.

Foreword

Wolfgang Rauch and Zizzimos Vergos

Sewer systems are among the most valuable and cost-intensive structures in the urban environment representing a considerable national capital asset for most countries. Typically these sewer pipes are designed with a life expectancy of approximately 50 years but some of them have already lasted for more than 100 years. Thus it is clear that the majority of the existing pipe network has been constructed with rules, guidelines and principles which differ from the ones generally applied today. Most notably, until three decades ago, the tightness of joints between pipes was not considered to be an issue and sewer infiltration has been seen as unavoidable (and sometimes even as a desired feature for drainage). A second aspect of the rather long life expectancy of sewers is their proneness to degradation, defects and subsequent leakage. This has led to increasing local flooding, groundwater pollution and reductions in sewage treatment plant performance.

Following the above, a typical sewer system has to be considered as being an inherently leaky system that does not comply with the EU Standard EN 752 which requires a "watertight" system. The water industry is now beginning to realise that even small exfiltration losses ($<2\%$) can be potentially prejudicial to long term environmental quality and sustainable urban water resource management. Wastewater can also exfiltrate from the sewer network, thus contributing to groundwater pollution in the urban environment. In cases where the groundwater table is above the sewer, water can infiltrate which leads to huge amounts of parasitic water and unnecessary costs to the treatment plant. Such infiltration has long been recognized by water authorities, but the APUSS project has also highlighted the potential problems that might be associated with exfiltration leakage.

Following early awareness and publications in the 1990's, a group of universities, consultants, companies and municipalities applied for funding under the 5th Framework programme of the EU to assess this problem of sewer infiltration and exfiltration. This book summarizes the achievements made during the course of the project APUSS. A significant effort of the group has been devoted towards the development of innovative tracer measurements in sewers to assess the amount of exfiltration and infiltration given the inherent difficulties and high costs associated with traditional pressure testing and CCTV survey. Although not a simple and straightforward technique, the APUSS tracer methodology has contributed significantly to scientific innovation in the field. The methodology and its field application are described in several chapters and illustrate the difficulties of identifying and quantifying losses from sewer defects by applying conventional techniques. Following major achievements towards a better understanding of the role of sewer sediments and the colmation layer to the exfiltration process, the book depicts conceptual modelling of the phenomena on both catchment and sewer reach scale. The final chapters deal with performance indicators and economic issues. The development of cost-structure approaches for typical sewer systems also provides a basis for benchmark sewer asset surveys.

Overall, the APUSS project has led to significant new insights to the problem of leaky sewer systems and drawn attention to the urgent need to safeguard the capacity and efficiency of both existing and future sewer infrastructure. It has promoted a better scientific understanding of sewer processes under conditions of sediment deposition, with proposals for new definitions for the growth of "blocking" zones in open sewer joints and fractures. The role of sediment in achieving quasi-sealing conditions in sewers is now much better understood as well as the circumstances

Edited by Bryan Ellis and Jean-Luc Bertrand-Krajewski. ISBN: 9781843391494. Published by IWA Publishing, London, UK.

under which equilibrium breakthrough can occur to cause disruption of the sealed joint or opening. This book is a most valuable source of information for scientists and practitioners interested in the phenomena.

Professor Dip-Eng, Dr Wolfgang Rauch, Past member and Chairman IWA/IAHR Joint Committee on Urban Drainage (JCUD), Austrian Water & Waste Association Board member, Editor "Water Research": Institute of Environmental Engineering, University of Innsbruck, Innsbruck. Austria.

Zizzimos Vergos, EU Scientific Officer for the APUSS project, Head of Infrastructure, EU Delegation to Kenya, Nairobi, Kenya.

Authors

Baer Emilie, LGCIE, INSA de Lyon – Université de Lyon, 34 Avenue des Arts. 69621 Villeurbanne Cedex, France, emiliebaer@yahoo.fr .... chapter 11

Barraud Sylvie, LGCIE, INSA de Lyon – Université de Lyon, 34 Avenue des Arts. 69621 Villeurbanne Cedex, France, Sylvie.barraud@insa-lyon.fr .... chapter 11

Bertrand-Krajewski Jean-Luc, LGCIE, INSA de Lyon – Université de Lyon, 34 Avenue des Arts. 69621 Villeurbanne Cedex, France, jean-luc.bertrand-krajewski@insa-lyon.fr .... chapters 1, 7, 11, 13

Cardoso Adriana, LNEC, Av.do Brasil 101, 1700-066 Lisbon, Portugal, macardoso@lnec.pt .... chapter 12

de Bénédittis Jérôme, LGCIE, INSA de Lyon – Université de Lyon, 34 Avenue des Arts. 69621 Villeurbanne Cedex, France, jerome.debenedittis@veoliaeau.fr .... chapter 7

Ellis John Bryan, Urban Pollution Research Centre, Middlesex University, The Burroughs, Hendon, London, NW4 4BT, UK, B.Ellis@mdx.ac.uk .... chapters 1, 5, 6, 9, 13

Franz Torsten, Institute for Urban Water Management, University of Technology, George-Bahrstrasse 1-3, 01069 Dresden, Germany, Torsten.franz2@mailbox.tu-dresden.de .... chapter 10

Frehmann Torsten, Emschergenussenschaft, Kronprinzenstrasse 24, 45128 Essen, Germany, Frehmann.torsten@eglv.de .... chapter 12

Giulianelli Mario, Water Management, IRSA-CNR, Via Salaria, CP 10 00016 Monterotondo St., Rome, Italy, m.giulianellu@irsa.rm.cnr.it .... chapter 8

Guyer Willi, Institute of Environmental Engineering, Urban Water Management, ETHZ, Wolfgang-Pauuli-Strasse 15, 8093 Zurich, Switzerland, guyer@ifu.baug.ethz.ch .... chapter 2

Joannis Claude, LCPC, Division Eau et Environnement, BP4129, 44341 Bouguenais Cedex, France, claude.joannis@lcpc.fr .... chapter 3

Karpf Christian Institute for Urban Water Management, University of Technology, George-Bahrstrasse 1-3, 01069 Dresden, Germany, christian.karpf@tu-dresden.de .... chapter 6

Edited by Bryan Ellis and Jean-Luc Bertrand-Krajewski. ISBN: 9781843391494. Published by IWA Publishing, London, UK.

KOHOUT David, Department of Sanitary & Ecological Engineering, Technical University, Pod Pat'ankou 30/5, 16612 Prague 6. Czech Republic, kohoutd@mat.fsv.cvut.cz .... chapter 4

KRACHT Oliver, Institute for Geology, University of Bern, Baltzerstrasse 1/3, 3012 Bern, Switzerland, oliver.kracht@geo.unibe.ch .... chapters 1, 2, 13

KREBS Peter, Institute for Urban Water Management, University of Technology, George-Bahrstrasse 1-3, 01069 Dresden, Germany, Peter.Krebs@tu-dresden.de .... chapters 6, 7, 10

LE GAUFFRE Pascal, LGCIE, INSA de Lyon – Université de Lyon, 34 Avenue des Arts. 69621 Villeurbanne Cedex, France, pascal.le-gauffre@insa-lyon.fr .... chapter 11

METELKA Tomas, DHI a.s., SW Department, Na Vrsich 1490/5. Prague 10, Czech Republic, t.metelka@dhi.czx .... chapter 10

PATERAKIS Nick, Urban Pollution Research Centre, Middlesex University, The Burroughs, Hendon, London, NW4 4BT. UK, n.paterakis@mdx.ac.uk .... chapter 9

PLISKA Zdenik, Hydroprojekt, Taborska 31, 14000 Prague 4, Czech Republic, zplis@hydroproject.cz .... chapter 10

POLLERT Jaraslov, Department of Sanitary & Ecological Engineering, Technical University, Pod Pat'ankou 30/5, 16612 Prague 6, Czech Republic, pollert@lermo.cz .... chapter 4

PRIGIOBBE Valentina, Institute for Process Engineering, ETHZ, Sonneggstrasse 3, 8092 Zurich. Switzerland, valentina.prigiobba@ipe.mavt.ethz.ch .... chapter 8

PRINC Ivan, Department of Sanitary & Ecological Engineering, Technical University, Pod Pat'ankou 30/5, 16612 Prague 6. Czech Republic, princ@mat.fsv.cvut.cz .... chapter 4

PRYL Karol, DHI-Hydroinform, Na Vrisch 5, Prague 10, Czech Republi, k.pryl@dhi.cz .... chapter 10

REVITT David Michael, Urban Pollution Research Centre, Middlesex University, The Burroughs, Hendon, London, NW4 4BT, UK, M.Revitt@mdx.ac.uk .... chapters 1, 5, 6, 9

RIECKERMANN Jörg, Urban Water Management, EAWAG, Ueberlandstrasse 133, CH-8600 Dubendorf, Zurich, Switzerland, joerg.rieckermann@eawag.ch .... chapters 1, 2, 13

RUTSCH Mandy, Institute for Urban Water Management, University of Technology, George-Bahrstrasse 1-3, 01069 Dresden, Germany, mandy.rutsch@mailbox.tu-dresden.de .... chapters 7, 10

Chapter 1

APUSS: Assessing the significance of infiltration and exfiltration on the performance of urban sewer systems

J. Bryan Ellis, Jean-Luc Bertrand-Krajewski, D. Michael Revitt and Jörg Rieckermann

INTRODUCTION

Urban sewer systems constitute a very significant infrastructure heritage in most European cities. Many working sections of the Roman urban drainage system still remain in use today, notably the sixth century great sewer (cloaca maxima) draining the Forum in Rome. The first beginnings of modern drainage practices were intended for stormwater (impervious surface runoff) control and were initiated in European cities such as Paris, London and Hamburg in the 1840–1870 period. Sewer design perspectives also changed in the late nineteenth century with the emergence of the Rational Formula which converted rainfall to runoff volumes. By the late nineteenth century, engineers possessed various design concepts and methods for wastewater collection and disposal systems and, for the next 100 years, these would become the standard tools used in urban drainage design and operation throughout the world.

Since the 1960s, rapid developments have occurred in urban drainage practice which can be largely linked to the introduction of enhanced computing facilities, such that it is now possible to calculate flows and fluxes in sewers and drainage networks with relatively high precision and resolution to support cost-effective design, analysis and operation. Whilst these advances have certainly helped in coping with urban flooding and have substantially improved the health of urban citizens, progress in water quality considerations, and particularly those addressing the aquatic impacts of increased urban population growth and their activities, upon both surface and groundwaters have been much slower. More recently, major changes in drainage design and operation philosophy have been introduced in many developed countries as a result of:

- the introduction and adoption of the concepts of ecological integrity and sustainable approaches to environmental and water resource management set within a catchment-wide framework,
- acceptance of the need to consider urban drainage, wastewater systems and receiving waters in a holistic, integrated manner,
- the continued development of computer power and an associated range of new analytical and real-time control techniques as well as dynamic spatial software capabilities.
- data availability, models and management systems (GIS, on-line sensors, discharge monitors, 1D/2D modelling approaches etc.,)
- administrative and organisational reforms (privitization, cost-sharing etc.,)

Edited by Bryan Ellis and Jean-Luc Bertrand-Krajewski. ISBN: 9781843391494. Published by IWA Publishing, London, UK.

However, throughout Europe, urban wastewater and drainage systems suffer from a combination of insufficient or over-designed capacity, institutional weaknesses, infrastructure deterioration and frequently inappropriate and/or reactive remedial measures. These problems have been compounded by an ageing infrastructure, lack of maintenance, increased urbanisation and more recently, the effects of climate change. The major consequences of this include structural and operational failures, temporary service breakdowns, receiving water quality deterioration, local flooding and excessive capital and operational costs. According to the UK water regulator (OFWAT) for example, 45% of sewage derived flooding is generated from hydraulic overloading and a further 40% (>5000 incidences per year) by sewer blockages (OFWAT, 2002) with annual sewer repair rates varying between 14–211incidences per 100 km sewer length. However, more recent research (Arthur *et al.*, 2008) has suggested that 84% of sewage-derived flooding incidents in the UK (>26,000 per annum), fall into "other causes" than hydraulic overloading, with some 90% attributable to blockages. If this is the case, then blockages probably represent the prime cause of losses to sewer serviceability (and hence flooding) across the world for both dry and wet weather flow conditions. Urban drainage and water systems now comprise some 40% of the replacement value of all European municipal-owned infrastructure assets, with the total European water sector representing over 1% of EU GDP (with a 5% annual growth rate) and a spend in the order of €50 billion per annum for water and wastewater services. Some 20% of EU-funded infrastructure investments are for urban water/drainage systems and the eastward extension of EU member states has sharply increased the investment need due to a poor or even a lack of basic sewer systems.

It is within the context of a sustainable operational function, with negligible (or minimal) infiltration gains or exfiltration loss, that the structural quality and functional efficiency of urban sewers can be considered as key parameters to guarantee the conveyance of domestic and trade wastewater to sewage treatment works (STWs). Infiltration of groundwater is particularly detrimental to treatment plant efficiency; hydraulic overloading due to infiltrated volumes of "clean" groundwater can reach up to and even exceed 100% of the wastewater volume in some cities, whilst dilution of pollutant concentrations leads to a lower pollutant removal efficiency. Exfiltration of wastewater, on the other hand, can lead to groundwater contamination which is of particular significance where the groundwater resource is locally abstracted for drinking water production. Both problems are critical on a long term basis for sustainable urban water management and have important economic consequences for cities and sewer system operators through the European Union as reported for example in the case studies of the companion EU AISUWRS project (Wolf *et al.*, 2009). The European standard EN 752-2 (CEN, 1996) indicates basic performance criteria applicable to any sewer system. Among these criteria, the following two are especially relevant: i) receiving waters should be protected against pollution; ii) the structural integrity of urban sewer systems, including water tightness, should be guaranteed. In order to accurately evaluate the performance of urban sewer systems, public and private operators need appropriate methods and techniques to locate and measure sewer gains and losses. This objective provided the driving focus for the APUSS project and the basis for the collaborative multi-disciplinary work undertaken in urban hydrology, environmental engineering science, risk assessment, social science and information technology.

APUSS STRUCTURE AND DEVELOPMENT

During the period 2001–2004, the APUSS project (Assessing Infiltration and Exfiltration on the Performance of Urban Sewer Systems), financed by the European Commission under the 5th R&D Framework Programme, brought together universities, SMEs (engineering consultants and wastewater companies) and municipalities in seven European countries to address some of the questions and issues relating to sewer infiltration and exfiltration (I/E). Four main Work Areas (WA) and associated work package (WP) sub-study areas were established within the project programme each with identified tasks and deliverables:

WA1) development of new measurement methods

- (WP1); exfiltration and infiltration
- (WP2); sediment interactions and effects
- (WP3); house connections

WA2) tests and field applications of the new methods under various field and laboratory contexts

- (WP4); identification and development of a conceptual model
- (WP5); field test studies in European cities
  - (WP5.1) Switzerland: I/E development and testing in live sewers within the city of Zurich
  - (WP5.2) Germany: I/E testing within the cities of Berlin and Dresden
  - (WP5.3) France: I/E testing within the city of Greater Lyon
  - (WP5.4) UK: clean/live experimental sewer rig studies; including field work with (and funded by) a consortium of UK wastewater utility companies (lead by Scottish Water) and I/E testing with Thames Water Utilities Ltd in Greater London
  - (WP5.5) Italy: I/E testing with the city of Rome

(WA3) implementation of associated models and tools
- (WP6); Aquabase software development and testing
- (WP7); similarity approaches
- (WP8); leakage approaches

WA4) socio-economic aspects linked to infiltration and exfiltration (I/E)
- (WP9); Development of performance indicators (PIs) and assessment
- (WP10); economic assessment of I/E performance indicators (PIs) and testing with the cities of Bottrop and Gladbeck in Germany.

Table 1 shows the initial work flow chart envisaged for the various work areas and sub-study work packages with the success of Work Area 1 (WA1) being seen as critical to the implementation and progress of all other work areas, being particularly dependent on the methods developed within Work Package 1 (WP1). The innovative tracer methodology for the identification and quantification of exfiltration loss developed by EAWAG within WP1 was essentially based on the application of sodium chloride (NaCl) as the tracer. This was because one explicit goal of the APUSS project was to avoid complicated analytical techniques which could not generally be available to end-users. Salt tracing formed the basis for all the exfiltration case studies undertaken in the test cities with the exception of those undertaken in the UK. For the London and Dundee exfiltration work conducted within WP5.4, the tracer used was the fluorescent dye, Rhodamine WT.

Table 1: APUSS work flow chart.

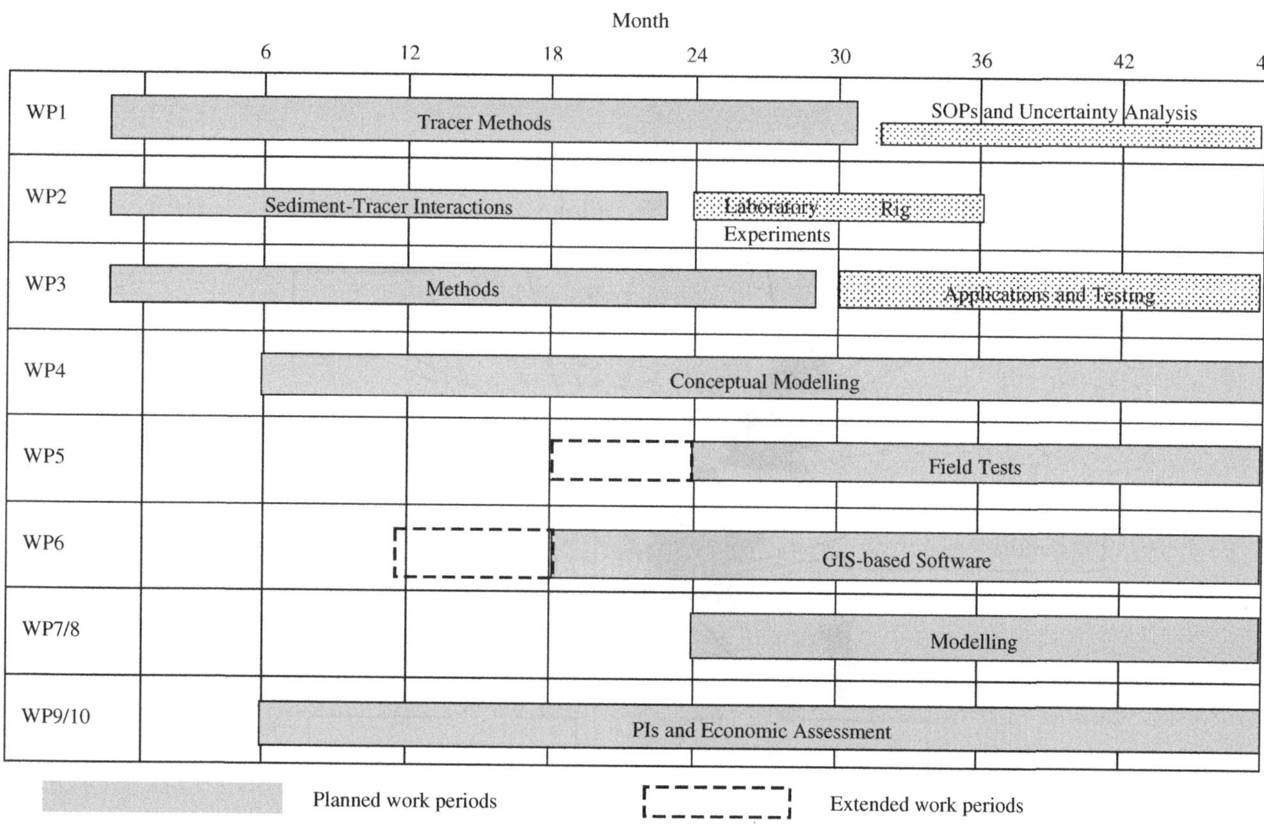

After a brief introduction to the APUSS partners, the following chapters outline and discuss the objectives, methods and main results obtained within each Work Area. References are given to APUSS reports and documents, as well as scientific and technical papers for readers who are interested in further detailed information. All documents and methods produced during the project are publicly available on the APUSS website at http://www.insa-lyon.fr/Laboratoires/URGC-HU/apuss.

Ten scientific partners from seven European countries have contributed to the APUSS project as indicated in Figure 1a, which also shows the group leaders appointed for each Work Package (WP). All partners were associated with end-users (sewer and/or wastewater operators) concerned with I/E, who contributed to the project by providing assistance, access to experimental sites, operational data, and additional funding (2). A total of 14 end-users drawn from 9 European cities collaborated in testing the APUSS methodological procedures (Figure 1b), with most of the Work Packages involving several partners and end users. This supported cross-study integration and end-user requirements as well as developing a common understanding of the goals and methods involved in each task.

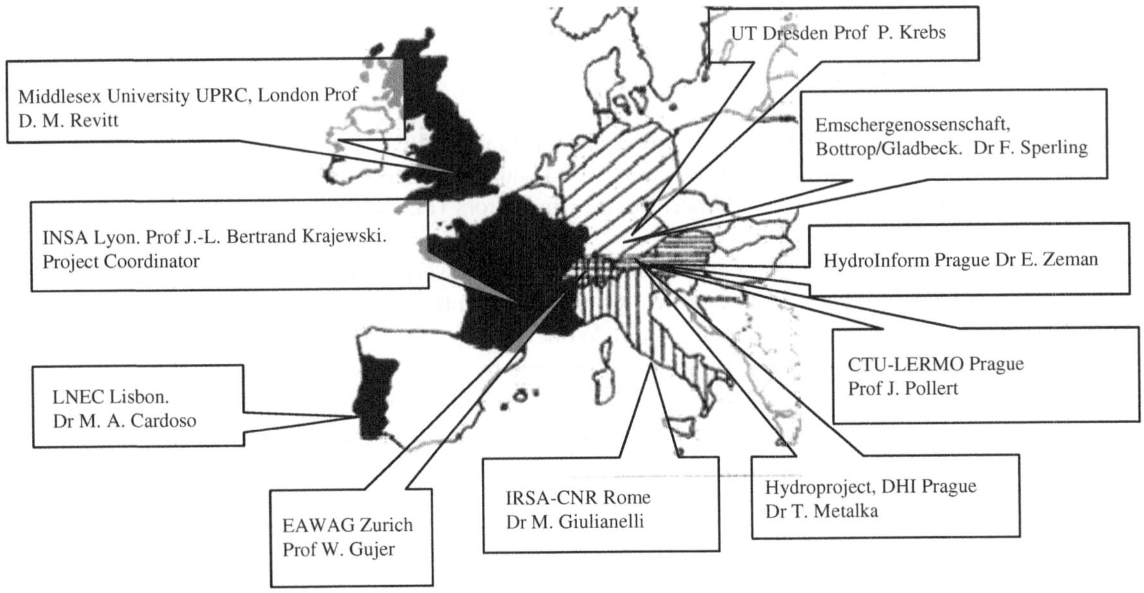

Figure 1a: Location of APUSS partners and Work Package (WP) Leaders.

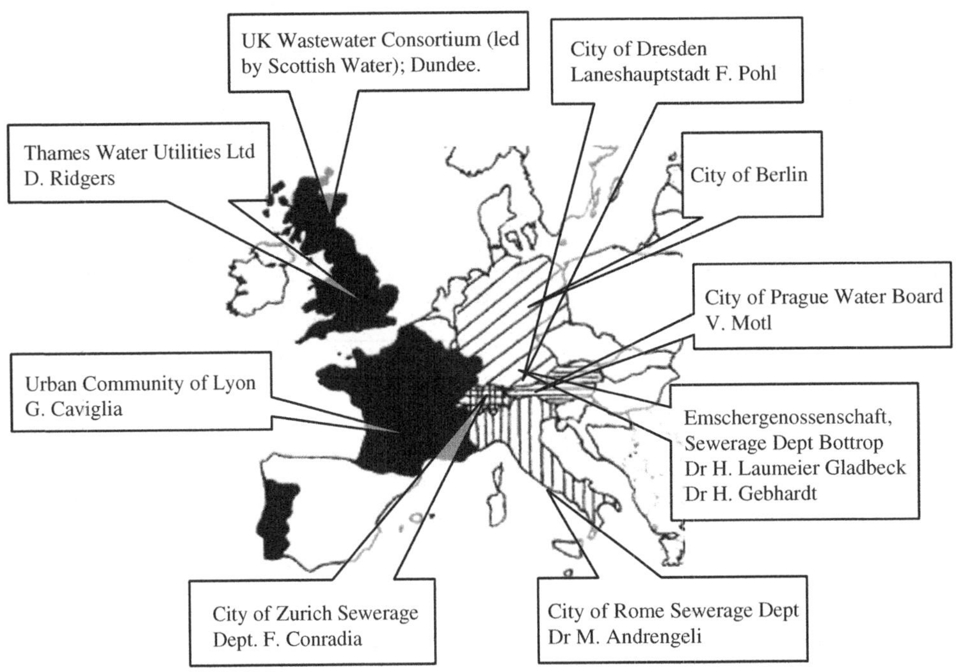

Figure 1b: Location of APUSS end-users.

The end-users represented a range of public and private wastewater utilities and water authorities, consulting environmental engineers and technical service and sewerage departments of city municipalities. Project progress was guided by annual advisory steering group meetings with cross-Work Package and individual task progress being discussed at more frequent smaller work meetings. Within Work Package WP1, postgraduate research staff working in the Italian, German, Czech and UK partner institutions were seconded for variable time periods to the Work Package leader institution in Zurich, Switzerland to familiarise themselves with the new tracer methodologies being developed by EAWAG staff. This also facilitated a harmonisation of, and common approaches to, sampling and field employment of instrumentation within the demonstration cities.

The APUSS project has also been part of the 5th R&D Framework Programme European cluster CityNet which included six individual projects dealing with integrated urban water systems; more information about CityNet is

Table 2: APUSS scientific partners, their main activities and associated end-users.

Scientific partner	Main activity	Associated end-user
INSA de Lyon (France; APUSS Project co-ordinator)	WA2, WA4	Greater Lyon
EAWAG (Switzerland)	WA1	City of Zürich
Dresden University of Technology (Germany)	WA2, WA3	Cities of Dresden and Berlin
Czech Technical University in Prague (Czech Republic)	WA1	City of Prague
DHI Hydroinform a.s. (Czech Republic)	WA3	City of Prague
Hydroprojekt a.s. (Czech Republic)	WA3	City of Prague
Middlesex University (United Kingdom)	WA1, WA2	Thames Water, London; UK Wastewater Utilities Consortium, led by Scottish Water (Dundee)
National Laboratory for Civil Engineering (Portugal)	WA4	
Emschergenossenschaft (Germany)	WA2, WA4	Cities of Bottrop and Gladbeck
IRSA-CNR (Italy)	WA2	City of Rome

available at http://citynet.unife.it. Among the five other CityNet projects, the CARE-S (Saegrov, 2006) and AISUWRS (Wolf *et al*., 2007) projects represent the most direct links and interactions with APUSS. Representatives from APUSS. CARE-S and AISUWRS attended various meetings convened by each project programme to inform themselves of complementary progress and common areas of interest being progressed within each separate EU project. In addition, partners from each project attended and presented on-going work at various international conferences in the urban drainage field during the course of the overlapping project periods. This included an international cross-project conference organised under the CityNet banner as part of the 10th International Conference on Urban Drainage (Theme L) convened in Copenhagen during August 2005 (ICUD, 2005). Table 3 identifies the themes and topics which were of complementary concern to the three closest related EU 5th Framework projects and which formed the focus for the majority of papers and discussion for the 10th ICUD meeting.

Table 3: Topic relationships between CityNet projects.

EU CityNet Project	APUSS	AISUWRS	CARE-S
APUSS		• exfiltration impacts • prediction of I/E from CCTV • case studies • modelling	• structural state of sewers • network techniques and costs
AISUWRS	• methods of quantifying I/E • modelling approaches		• groundwater level changes • rehabilitation approaches
CARE-S	• input sewer rehabilitation needs for I/E • sewer performance indicators (PIs) for I/E	• groundwater levels • sewer rehabilitation benefits	

PROJECT CONTEXT AND BACKGROUND

Modern sewer systems have been developed in European cities since the middle of the ninteenth century in order to protect public health, safety and quality of life of the urban population. This hygienist viewpoint was the starting point for modern efforts and investment in urban drainage. The collection and conveyance of wastewater was considered to provide a solution to avoid pollution of both groundwater and surface water resources. Irrespective of this early drainage infrastructure provision, negative impacts could still be observed on downstream receiving waters although such effects were usually not appropriately considered during the initial decades of sewer network construction. Nevertheless, it is undoubtedly the case that traditional sewer systems still make a substantial contribution to European public health. Dry weather flow effluents (up to 5/6 × Dry Weather Flow – DWF) are normally treated in Sewage Treatment Plants (STPs), with direct discharges to receiving waters occurring through overflow structures during storm events, and being theoretically limited to exceedance of the design flow level.

However, such historical sewer systems can frequently be considered not to have functioned in a fully satisfactory manner. This lack of functionality is largely because of a combination of factors including:

(i) ageing of pipes (some parts of sewer systems in European cities are more than 100 years old),
(ii) poor construction quality (poor quality of pipe material, bad laying conditions, ignorance or underestimation of the effects of geotechnical and road traffic conditions etc.),
(iii) lack of or insufficient maintenance,
(iv) lack of appropriate investment and rehabilitation strategies, and
(v) high costs of construction and rehabilitation.

As a result many sewer systems across Europe are not watertight and suffer from significant infiltration and/or exfiltration. Where infiltration and exfiltration are observed, repair and/or rehabilitation will normally be considered to be required. However, the investment to build or, more frequently, to renew, rehabilitate and upgrade urban sewer systems needs substantial resources which often can be beyond the short-term capacity of operating managers. Nevertheless the pressures of stricter EU regulation and directives as well as the rise of new paradigms concerning the urban water cycle and sustainable integrated urban water management are inexorably leading to fundamental re-appraisal of sewer asset condition (Le Gauffre *et al.,* 2007).

Water authorities and companies throughout Europe recognize this problem but calls for comprehensive appraisal and economic pricing encounter difficulties because of a lack of knowledge concerning the nature and processes of sewer losses and gains. It is estimated for example, that up to 40% of UK sewers have structural defects and about 1 in 12 (of the 302 000 km of public sewers) have defects with some €11.6 billion needed to alleviate the most immediate problems. Very little attention within Europe has been given to potential exfiltration losses from sewers in contrast to infiltration and only some 70 officially documented cases in the UK of sewer leakage exist over the past 60 years or so (Blackwood *et al.*, 2001). However, some 5000 sewer collapses are recorded per annum and there is widespread anecdotal evidence that suggests that sewer exfiltration may be more important than previously assumed. The UK water companies requested the UK regulator (OFWAT) for €12.3 billion in the most recent Asset Management Plan (AMP5) review of the industry in order to redress a historic neglect of sewers. It is also clear from the EU Water Framework Directive (WFD) that such diffuse, non-point sources of pollution to groundwater will be targeted for much stricter control in the future and that there will be a need to identify and quantify potential losses in Catchment Management Plans (CMPs) prior to the introduction of any Programme of Measures (PoMs) as required by the Directive.

Moreover, between 5000 to 7000 UK properties (0.1%) are flooded each year by sewage although there may be three to ten times as many external sewer flooding incidents (gardens, streets, etc.) which are unaudited (National Audit Office, 2004). In 2002–2003, the UK water companies invested over €940 million on the sewage network of which €275 million was spent on sewer network maintenance, including sewer replacement. However, little information exists on the rates of deterioration of the UK sewer network which are needed for a robust cost-benefit analysis to inform decisions on sewer flooding policy. The effects of climate change have already increased the incidence of sewer flooding and considerably raised the profile of infiltration and other pluvial contributions to such flooding (Pitt, 2008). The increased risks from future climate change and extreme events are exacerbated by new housing and legislative demands (e.g. the EU WFD) which will also impact upon future sewer performance. In addition, the WFD requires a very different holistic, integrated approach to land and water management which demands that the interfaces of groundwater, surface water and wastewater flows be considered and managed interactively. However, the cost of dealing even with the most severe internal flooding problems in the UK (associated with the 32 000 km or 10% of sewer grades 4 and 5) are estimated at €1.6 billion, with the worst external cases costing a further €0.7 billion.

In France, only a few comparative national statistics are available. A national survey made in 2002 by Berland and Juery, cited in Le Gauffre *et al.* (2007), indicated that the total length of sewers (combined and separate) was 250 000 km, with an asset value estimated from €65 to €75 billion. It was considered that approximately half of the sewers needed some rehabilitation, and that the annual amount to be spent in the near future for their rehabilitation, with a life expectancy of 60–80 years, was estimated between €800 and €1300 million. Over the last twenty years the six French Water Agencies have promoted diagnostic studies of sewer systems, one of the most important objectives being the detection and the quantification of infiltration. Following such diagnostic surveys, large investments have been devoted to sewer rehabilitation. As an example, approximately €25 billion have been spent in the EU up to now to meet the requirements of the Urban Waste Water Directive 91/271. As it is not conceivable to increase the wastewater treatment efficiency without increasing the efficiency of wastewater collection, i.e. the quality of sewer networks, considerable resources will be needed in order to reach the objectives of the European Directive.

However, the publicly-owned sections of sewer systems are only one aspect of the problem. Privately owned house connections are often the most neglected components of the whole sewer system and may contribute to as much as 50% to the total infiltration of groundwater. Exfiltration from leaky house connections can also be substantial and of dramatic importance in terms of protecting groundwater against pollution as well as preserving foundations of buildings. Rehabilitation work aimed at achieving a decrease in water infiltrating from surrounding subsoil focusses mainly on the core public sewers, thus leaving the house connections neglected in the majority of cases.

Figure 2 is a conceptual representation of the sources and routes whereby infiltrating groundwater and stormwater inflow (I/I) might enter both surface and sanitary sewers. Under rising groundwater conditions, cross-connections from the foul system can facilitate infiltration to the separate stormwater sewer with dilute groundwater entering both sewer systems and contributing to both premature overflows as well as reducing treatment plant performance. Sewer infiltration and inflow (I/I) can result in surcharging of manholes and house connections, causing raw sewage to flow into properties with potential public health impacts. The components of infiltration/inflow (I/I) include diurnal sanitary flows, lateral stormwater sub-surface inflows and groundwater infiltration. These flow components enter the sewer system via defective pipes and joints, leaking manhole walls and covers or via low disconnector traps. Such infiltration may be predominantly seasonal in nature but normally responds rapidly to storm events, decreasing as rainfall decreases. Tree roots, heavy traffic loads and pipe/joint defects can in turn generate access and exit routes for both groundwater and exfiltrating sewage effluent. As indicated by the hydrological pathways shown in Figure 3, direct contamination of the groundwater zone by downward percolation (5) via fissure and matrix flow, through the unsaturated zone is only likely to occur during prolonged dry periods when the water table falls below the invert level of the sewer pipe. The sub-lateral seepages (2 and 4) through the unsaturated zone contribute to the contamination of surface water drainage which is exacerbated by the cross-connections (3) which all too frequently occur between the foul and surface water sewer systems. The latter discharge directly and untreated to receiving watercourses and present major sources of urban runoff pollution. A principal source and pathway of sewer exfiltration may well be related to leakage from house connections (1) especially at the junction point with the main sewer line and because such connections are generally above the groundwater level for much of the time.

National approaches to the exfiltration/infiltration (I/E) issue frequently assumes that a Pareto effect exists at all levels of the wastewater system. The Pareto effect assumes that the source of the major extraneous I/E flows will be located in a relatively small part of the system, and that cost-effective, long-term solutions can be achieved by locating and rectifying these major but localised sources of losses and gains. This Pareto effect has been identified and effectively utilised by various water authorities in many countries and has provided demonstrable results. However, a well-planned approach is critical to reducing I/E problems and requires attention to more than just the identified problem areas. The root cause of I/E problems may well not be at the locations where the problems are most evident in the sewer, and remedial actions at the generating source of the offending flow may be required. Nevertheless, the risks associated with I/E do demand an accurate determination of the minimum dry weather exfiltration and infiltration rates in terms of their location, magnitude and potential. Source detection will then involve information from this measurement stage to undertake a Pareto analysis to identify the most affected sub-catchments or pipe segments and their source flows and causes.

This procedure can then lead to the development of a rehabilitation strategy to determine the most appropriate and cost-effective rehabilitation techniques to be employed to resolve the hydraulic situation. Traditionally, a combination of source detection techniques have been used including CCTV inspection, smoke and pressure testing, night flow isolation and mass balance as well as wet weather simulation studies. All of these approaches have their uncertainties and all are inherently high cost options.

Infiltration and exfiltration therefore will appear as increasingly important issues in terms of future asset performance indicators for the European water industry. However, information concerning infiltration and exfiltration from sewer systems is relatively poor and investments are frequently based on weak information and limited data sets having rather high uncertainties. The assessment of water-tightness under operational conditions is in practice normally derived by approximated assumptions of water use rather than through estimation under validated methods. Consequently, there is commonly little or no policy or operational awareness of treating this issue until the economic or social consequences have become impossible to ignore. It is sometimes argued by the water industry that as far as sewer leakage is concerned, there is little if any evidence of "victims" of general exfiltration losses. Given that this is the case, and customers are not complaining and are additionally much more concerned with keeping their bills low, then solving the problems where causal links may already be in place or in hand, they are unlikely to support spending money for no visible benefit. Neverthless, it is clear from the EU WFD as well as from current developments of individual national water regulations, that such sewer deficiencies will not be tolerated much into the future. A better knowledge of infiltration and exfiltration processes and rates is thus becoming appropriate and necessary. Such knowledge will make important contributions to the robust evaluation of performance indicators which are now increasingly frequently used as a basis to assess the structural and functional states of sewer asset systems (Abraham, 2003). Maintenance and rehabilitation strategies will also be (at least partly) based on these performance indicators.

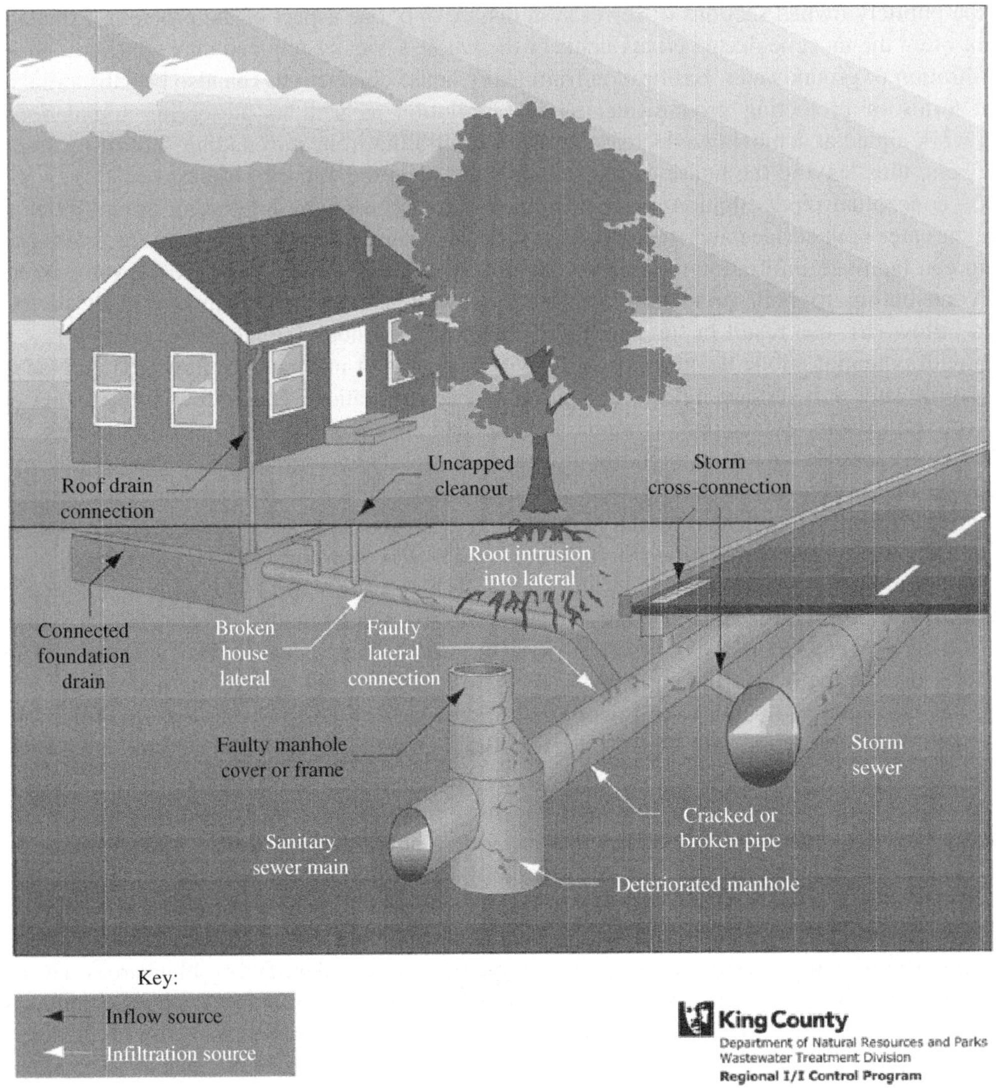

Figure 2: Sources of infiltration and inflow (I/I) to sewers. (Source: King County, Washington. US. Dept. of Natural Resources & Parks, Wastewater Treatment Division, Regional Infiltration and Inflow Control Program. Available at http://www.kingcounty.gov/environment/wastewater/II/What.aspx)

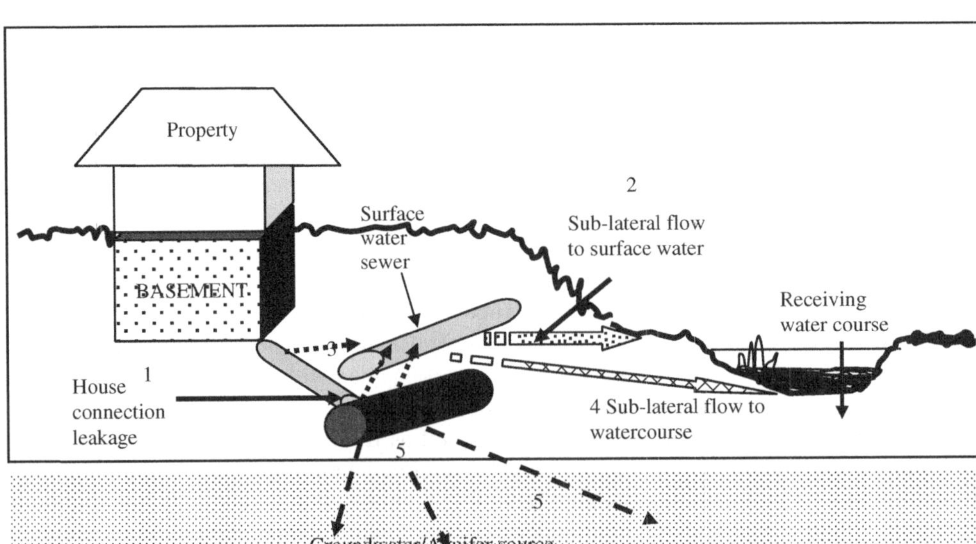

Figure 3: Hydrological pathways of sewer exfiltration.

PROJECT OBJECTIVES

The principle objectives of the APUSS project were to:

- develop and test improved methods for the evaluation of dry weather sewer I/E rates using both natural (O and N isotopes) and artificial tracers (NaCl, Rhodamine WT etc.) with volumetric methods for house connections, and to assess the effects and interactions of sewer sediment on tracer behaviour.
- calibrate and verify the developed tracer methodologies based on field experimentation within a cross-section of European cities utilising a common software database and GIS templates.
- develop conceptual and associated modelling approaches to evaluate dry weather I/E rates for both pipe and sub-catchment scales at variable time scales to allow comparison with tracer determined rates.
- develop appropriate tools for decision-making to support management strategies for I/E control based on probability analysis and performance indicators. This would incorporate economic evaluations of investment and rehabilitation strategies.

In the context of these specified objectives, the main goals of the APUSS project as developed within the various separate work packages included:

- the elaboration and testing of new measurement methods to directly estimate exfiltration and infiltration rates in sewers by means of tracers: at sub-catchment scale for infiltration and at sewer reach scale for exfiltration, with standard operating protocols (SOPs) including detailed uncertainty analysis and data processing in order to obtain accurate I/E rates.
- the development and testing of volumetric measurement methods for house connections, with a matrix approach to extrapolate measurements from specific house connections to the wider catchment scale.
- the testing and validation of the new measurement methods in various contexts with associated end-users, in order to contribute to local sewer improvement and to explore their advantages, possible difficulties and conditions of use or application.
- the development and validation of modelling approaches at various time scales, including statistical methods to i) attempt to extrapolate experimental results to wider and/or similar unmonitored sub-catchments and sewer reaches, and ii) to optimally choose the measurement locations.
- the provision of end-users with a unified software tool, including experimental data storage, automatic model calibration, graphical display of results, and links with performance indicators.
- the provision of end-users with a fully documented large scale example of the application of the methods and associated software.
- the definition of a set of specific performance indicators which take account of infiltration and exfiltration.
- proposals for a multi-criteria approach to compare and rank investment strategies to resolve I/E problems.
- the analysis of the economic value of sewer systems and an evaluation of the cost of application of the new tracer methods.
- the public dissemination of all the above elements at the end of the project.

Both this book and the accompanying APUSS website where additional information is available (http://www.insa-lyon.fr/Laboratoires/URGC-HU/apuss), put the last objective into concrete form. Whilst the APUSS WorkPackage (WP) deliverables can be currently assessed from the website and much of the work output has been made available through a variety of publications, this volume provides the added-value of a single source compendium to the work undertaken and knowledge gained in the project. Inter-relationships between the various WorkAreas (WAs) and WorkPackages (WPs) are more fully identifed in this volume and the advantages and difficulties of translating the procedures, tools and approaches into practice on the ground are also more fully developed. This will provide a valuable insight for practitioners when applying the scientific methods and software analyses as well as providing a valuable source of further academic information and insight on the scientific principles and processes governing I/E in public sewers.

BOOK STRUCTURE

The innovative tracer methodologies developed for the determination of sewer exfiltration and infiltration (I/E) rates are described in Chapter 2. Their working concepts, operational principles and experimental design are fully explained together with their practical limitations and difficulties. Standard operational procedures (SOPs) for both exfiltration and infiltration methods are freely available from the APUSS website.

It has been estimated that up to 50% of exfiltration/infiltration losses and gains in sewer lines might occur as a result of defective house connections. Chapter 3 outlines some of the difficulties and terminologies associated with the definitions of public and private sewer systems and typical lay-outs of private property drains The objectives and design of measurement surveys for house connection I/E rates are examined and background European legislation for "outside" drains and sewer systems is discussed. Chapter 4 describes the application of volumetric methods to the measurement and scale-up of I/E rates for house connections as determined in field experiments conducted in the city of Prague.

Given the documented effects of sediment solids on artificial tracer behaviour, Chapter 5 evaluates the adsorption characteristics and interactions of sewer sediment with fluorescent tracer use for exfiltration monitoring. Laboratory batch methods and both clean water and live wastewater test rig experiments have been assessed as well as the compounding effects resulting from variable defect geometries, flow heads and trench backfill conditions. Potential effects of cohesive organic sediment simulated from testing a range of synthetic materials and from live wastewater tests conducted on a field rig in Dundee, Scotland are described. The significance of scale-up and tracer rig tests are finally examined and compared to direct volumetric measurements.

The sewer rig experiments allowed detailed examination of the growth processes associated with the development of the sealing colmation layer in joint gaps within sewer pipes. Chapter 6 examines the mechanisms of colmation layer development and proposes amended definitions for the structural and growth phases of both this layer and associated layers. The concept of equilibrium breakthrough is introduced whereby sealed defects randomly exfiltrate following rupture of the colmation layer by turbulent bed flows but under varying pressure head conditions.

Chapter 7 outlines the application and outcomes of field applications and testing of the developed exfiltration and infiltration methods in cities within Germany and France, with Chapters 8 and 9 providing separate descriptions of the application and outcomes in cities within Italy and the UK respectively.

Modelling and associated software approaches for the sub-catchment scaling of inflow-outflow leakage rates are described in Chapter 10 based on uncertainty analysis and the utilisation of the AquaBase data management tool and associated GIS templates.

Chapter 11 develops a generic multi-criteria analysis for the comparison of alternative investment and rehabilitation strategies for the management of I/E. The methodology is tested by application to a semi-virtual case study for the 245 hectare Ecully catchment in Lyon, France.

Further economic and operational aspects of I/E are examined in Chapter 12 through the development and testing of performance indicators (PIs) and cost distributions for the application of the APUSS tracer methods.

Finally, Chapter 13 provides an overall perspective and concluding remarks on the outcomes and significance of the APUSS project. A full list of publications arising from the APUSS project is also given.

REFERENCES

Abraham, D.M. (2003) Sewer asset management, rehabilitation and security. *New Pipeline Technology, Security and Safety*. In: *Proc. Int. Conf. Pipeline Engineering & Construction*, (ed. M. Najafi). 13–16 July 2003. Baltimore, Maryland, US. American Soc. Civil Eng., New York, pp. 816–821. ISBN 0 7844 0690 1.

Arthur, S., Crow, H., Pedezert, L. and Karikas, H. (2008) Using serviceability to prioritise sewer maintenance. *Proc. 11th Int Conf. On Urban Drainage* (11th ICUD). Edinburgh, Scotland. CD-ROM. ISBN 978 1899796 212.

Blackwood, D.J., Cavill, S., Hendry, S. and Marr, I. (2001) *The Incidence and Impact of Exfiltration from Sewers*. Report WM-07/D. UK Water Industry Research Ltd (UKWIR). London, UK.

CEN (1996) *Drainage and Sewers Outside Buildings : Performance Requirements*. EU Standard EN 752-2. European Committee for Standardisation (CEN). Brussels, Belgium.

ICUD (2005) *Proc. 11th Int Conf. On Urban Drainage* (11th ICUD). August 2008. Edinburgh, Scotland. CD-ROM. ISBN 978 1899796 212.

Le Gauffre, P., Vasconceios, E., Breysse, D., Gibello, C. and Desmulliez, J. (2007) Performance indicators and multi-criteria decision support for sewer asset management. *J. Infrastructure Systems, Amer. Soc. Civil Eng.*, **13**(2): 105–114.

National Audit Office (2004) Out of Sight-Not Out of Mind: Ofwat and the Public Sewer Network in England & Wales. Report HC161. HMSO Stationery Office, London. UK. ISBN 0 1029 2680 8.

OFWAT (2002) *Flooding from Sewers: A Way Forward.* Consultation Paper. Office of the Water Regulator (OFWAT). Birmingham, UK.

Pitt, M. (2008) *Learning the Lessons of the 2007 Floods; The Pitt Review*. The Cabinet Office, London.

Saegrov, S. (ed.) (2006) *CARE-S: Computer Aided Rehabilitation of Sewers and Storm Water Networks*. IWA Publishing, London. UK. ISBN 9781843391159.

Wolf, L., Morris, B. and Burns, S. (2007) *Urban Water Resources Toolbox: Integrating Groundwater into Urban Water Management*. IWA Publishing, London, UK. ISBN 9781843391388.

Chapter 2

New measurement methods

Jörg Rieckermann, Oliver Kracht and Willi Gujer

INTRODUCTION

Traditionally, the assessment of infiltration and exfiltration (I/E) in sewer systems is based on rather inaccurate methods of flow measurement, simplified analysis of diurnal flow and load variations, and balancing of water input-output (e.g., Kracht *et al.*, 2008; de Bénédittis and Bertrand-Krajewski, 2005; Rutsch *et al.*, 2006; Rutsch et al., 2008). One of the key objectives of the APUSS project was to develop innovative measurement methods for I/E that use information on natural and artificial tracers in addition to discharge data. This should allow for the differentiation of sewer zones exhibiting infiltration and exfiltration, based on limited analytical effort and procedures having little environmental risk. It was a declared goal of the APUSS project that such methods might be readily applied by end-users such as municipal sewer operators and consulting engineers. Therefore, standard equipment, readily available analytical devices and open source software were used as far as possible with an exclusion of procedures that would have been either too involved or impractical (e.g., experiments using radioactive tracers).

By means of an optimised experimental design and a detailed analysis of the resulting data and of their associated uncertainties, a higher degree of accuracy and a better quality knowledge than associated with traditional methods was expected. Work Area 1 (WA1) of the project thus included the development, testing and validation of methods to accurately quantify:

- exfiltration from sewer reaches
- infiltration into sewer systems
- effects and interactions of the applied tracers with sewer solids and sediments (Chapter 4)
- infiltration and exfiltration from house connections (Chapters 2 and 3)

The developed methods are based on selected chemicals and natural isotopic tracers and have been tested and validated under various operational conditions during dry weather, at different space scales (from the pipe reach to the whole catchment) and under different flow conditions. They have been applied in different catchments chosen in the associated European cities and compared to traditional approaches based for example, only on flow rate measurements or on large scale water mass balance. Standard Operation Procedures (SOPs) and specific algorithms for data processing and detailed uncertainty analysis have been established. Detailed information on the SOPs and thc software for data analysis are freely available on the APUSS website (http://www.insa-lyon.fr/Laboratoires/URGC-HU/apuss).

EXFILTRATION METHODS

Despite several decades of research, detailed knowledge on sewer leakage is still deficient, which is partly due to the lack of reliable measurement methods (Rutsch *et al.*, 2008) . The specifically designed tracer experiments developed

Edited by Bryan Ellis and Jean-Luc Bertrand-Krajewski. ISBN: 9781843391494. Published by IWA Publishing, London, UK.

within the APUSS project are seen as providing one solution to this undesirable situation as they have several advantages over existing methods. It might be considered to be very appealing to deduce wastewater losses from tracer mass loss because tracer tests i) are relatively inexpensive, ii) are universally applicable, iii) monitor leakage under realistic flow conditions and, most of all, iv) have no disturbance or damage effects during the measuring process, such as pressure tests do. In this section, the underlying methodological concepts are explained, and then the individual methods and their subtypes are described in detail. As it is considered crucial to always report an estimate of uncertainty together with the measurement results, the uncertainty analysis associated with each method will be also outlined. Then, a brief focus on the experimental design and suggestions on how to optimally plan an exfiltration monitoring experiment (e.g., to efficiently assess the dispersion of a tracer cloud within a sewer reach), will be outlined. In a separate section, the benefits and limitations of the methods are discussed with comments on their practical relevance. The section is closed with a brief outline of the existing documentation and the information that is available on the project homepage.

Concepts

The basic working idea of the experimental tracer methods is that exfiltration is quantifiable from a mass balance of a tracer substance over the investigated sewer line (Rieckermann, 2005; Rieckermann *et al.*, 2005a). At the upstream extremity of the investigation reach, a known mass of tracer is dosed. The tracer then passes through the sewer line and, if exfiltration occurs, tracer substance is lost along with the seeping wastewater (Figure 1). From the remaining tracer mass at the end of the reach, the loss of tracer in the tested section can then be computed.

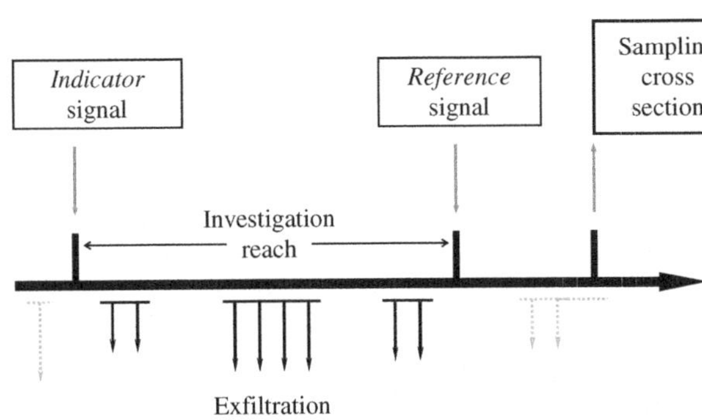

Figure 1: Conceptual sketch of the experimental set-up of the tracer method (Rieckermann *et al.*, 2005a).

Infiltration, can thus be determined as:

$$E = 1 - \frac{mass_{out}}{mass_{in}} \quad (1)$$

where E = exfiltration, expressed as a proportion of the labelled wastewater, and $mass_{in}$ and $mass_{out}$ = upstream and downstream mass of tracer, respectively.

Computation of the remaining tracer mass requires integration of the product of instantaneous tracer concentration and wastewater discharge over time. However, discharge measurement in sewers may not be very accurate, especially if portable flow meters are used, which are known to be subject to considerable bias due to insufficient calibration (Bertrand-Krajewski *et al.*, 2000; Bertrand-Krajewski *et al.*, 2003). For this reason, it is proposed to dose one or more downstream *reference* signals in addition to the upstream *indicator* signal (Figure 1). Leakage in the investigation reach can then be estimated from the relative differences in the tracer masses between the points of injection and the sampling point:

$$E = 1 - \frac{mass_{REF,in}}{mass_{IND,in}} \cdot \frac{mass_{IND,out}}{mass_{REF,out}} = 1 - \frac{mass_{REF,in}}{mass_{IND,in}} \cdot \frac{\int Q(t)C_{IND}(t)dt}{\int Q(t)C_{REF}(t)dt} \quad (2)$$

where $mass_{REF,in}$ and $mass_{IND,in}$ = introduced mass of reference and indicator tracer, $mass_{REF,out}$ and $mass_{IND,out}$ = downstream mass of reference and indicator tracer, C_{REF} and C_{IND} = downstream concentration of reference and indicator tracer, and $Q(t)$ = downstream discharge.

Note that the reference signal can be used in a practical sense to correct for systematic errors in concentration and flow measurements, as it can be regarded as an intrinsic calibration standard. In the case where the measurement devices have a (linear) bias due to insufficient calibration, this equally applies to both the indicator and reference signal and thus is cancelled out in equation 2. The computed exfiltration ratio is also not affected by leakage downstream of the addition of the reference tracer (i.e the dotted grey arrows before the sampling cross section in Figure 1), because it can be assumed that it equally affects both the indicator and reference signal. When the discharge is steady during the experiment, only downstream concentration measurements are necessary, but no discharge measurements are required.

Along with these working concepts, two generic methods have been developed which differ in dosing strategy, tracers and data analysis. The QUEST method uses instantaneous injections of a single tracer substance, whereas the QUEST-C uses a continuous dosing strategy of two different substances to be able to distinguish the indicator and reference signal at the sampling cross section (Figure 2). For each of the methods two subtypes have been developed which differ regarding their experimental effort and/or data analysis.

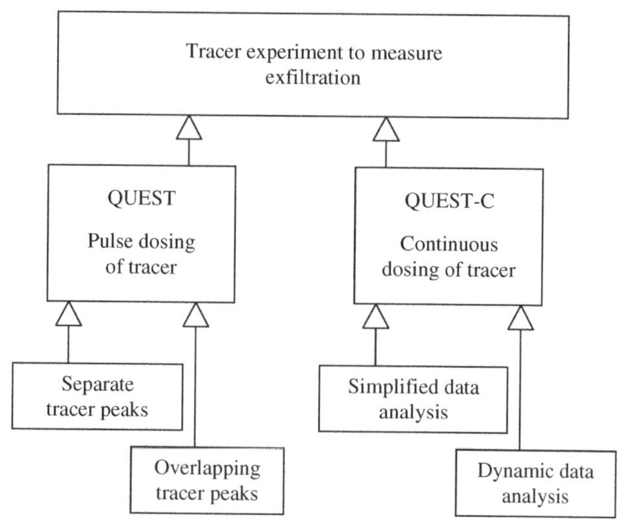

Figure 2: Overview of the different tracer methods to quantify sewer leakage.

QUEST

For the QUEST method (**Qu**antification of **E**xfiltration from **S**ewers with Artificial **T**racers) a single tracer substance is used, which is dosed in pulses both as indicator and reference signal (Rieckermann *et al.*, 2005a). It represents a novel experimental method to directly measure sewer leakage that is accompanied by a comprehensive framework to rigorously assess the accuracy of this method using reliable methods of uncertainty analysis.

Choice of tracer and measurement technique

In the course of the APUSS project, measurement campaigns have been performed mostly with Sodium Chloride (NaCl) as a tracer and, to a lesser extent, fluorescent dyes (Uranin and Rhodamine WT). This was because one explicit goal was to avoid complicated analytical techniques that are generally not available to the targeted end-users. For the QUEST method, a basic requirement is in-line measuring technology with a measurement and recording frequency of a few seconds. This is because the tracer signals resulting from the slug injections are highly dynamic and discrete sampling prohibits accurate monitoring of the tracer peaks.

Experimental Setup

A QUEST experiment only requires relatively little measuring equipment: the tracer solution, the in-line measurement device and a data logging unit. In addition, the use of a flow meter at the sampling station is highly recommended. For a detailed description of the setup and the applied devices, please consult the Standard Operation Procedures, SOP (Rieckermann and Niesel, 2005) which can be obtained from the APUSS website.

Data analysis

To calculate exfiltration from the experimental data, the tracer concentration is de-convoluted into three components: i) an indicator signal, ii) a reference signal and iii) the natural background (baseline) as shown in Figure 3. For the

QUEST method, the timing of the tracer injection can be chosen such that the measured concentration time series shows *separated tracer signals* (as seen in Figure 3) or *overlapping tracer signals* (not shown). For separated peaks, the data analysis is straight forward, as it should not be problematic to define starting and ending times of the individual tracer signals as well as an appropriate model for the baseline and the computation of exfiltration from equation 2. For the overlapping peaks, the identification of the three components is considerably more difficult, as starting and ending times cannot be assigned clearly. This can be made possible by applying models that parameterise the tracer peaks as well as the baseline. The parameters of the peak and the baseline models are then estimated from the measured data by non-linear regression techniques, as the peak models (e.g., lognormal distribution function) are generally nonlinear in the parameters. The best parameter estimates are then used to compute the remaining tracer mass m_{out}, from which exfiltration is calculated according to equation 2. Where additional flow measurements are available at the monitoring site, these should be included in the computations.

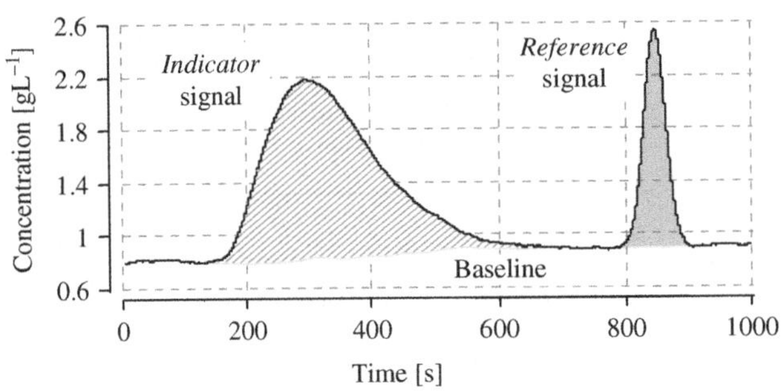

Figure 3: Exemplary data set for the QUEST tracer method.

For the analysis of uncertainty, the following error contributions have been considered explicitly: model parameter uncertainty, dosing errors, uncertainties in the calibration function of the measurement device and discharge measurements. In addition, possible errors might be introduced from fluctuations of the unknown baseline under the tracer signals. Both bootstrap and regression techniques have been developed to assess the individual error contributions. Once these are established, the uncertainty in the final leakage estimate is assessed by Monte Carlo Simulation (Rieckermann *et al.*, 2005).

The QUEST method uses pulse injections of indicator and reference tracer, which results in a dynamic and unique pattern allowing a clear distinction of the individual components. Thus, a single tracer substance can be used and the measurement errors typically cancel out to a large degree. However, a clear disadvantage is that current in-line measuring techniques for wastewater applications are still very limited regarding the availability of appropriate tracer substances. Furthermore, continuous problems were experienced with clogging and fouling of sensors, which generally lack adequate cleaning mechanisms. To overcome these difficulties, a complementary tracer experiment has been developed which uses a continuous dosing strategy of two tracer substances.

QUEST-C

The QUEST-C methodology (**Qu**antification of **E**xfiltration from **S**ewers with Artificial **T**racers- **C**ontinuous Dosing) uses two different tracer substances, basically to be able to use more selective/specific tracers and to thus overcome the field limitations experienced in the use of in-line sensors (Rieckermann *et al.*, 2007). Scientific innovations associated with this methodology concern:

(i) an improved experimental design that uses gravimetric measurements and thus reduces systematic errors,
(ii) an assessment of uncertainty for the obtained results, and
(iii) a new approach for data analysis that accounts for dynamic flow conditions.

Choice of tracers and measurement techniques

The methodology, which is in principle independent of the applied tracer substances, has been implemented mostly using Lithium (Lithium chloride) and Bromide (Sodium bromide). From working experience, these substances can be expected to be conservative for most sewer applications and to have a minimum baseline. Batch test results suggested a non-reactive behaviour of the tracers in wastewater and in the sample bottles. The analysis by ion chromatography (Metrohm, Compact IC) showed a very good reproducibility of ±1.2% of the measured value for Li^+ and ±0.5% for Br^- (single standard deviation) (Rieckermann *et al.*, 2007). However, there have been some analytical difficulties reported regarding the measurements of Lithium in wastewater from work undertaken in a complementary APUSS workpackage. Equally, during the development of the method, it was found that analysis of lithium in wastewater on ICP-OES performed unsatisfactorily, although this problem has since been resolved to a large extent. Lithium concentrations were overestimated by 45%–47%, which was essentially due to significant cross-sensitivities with other wastewater components.

Experimental Setup

The experimental effort associated with a continuous dosing experiment is more involved than for the QUEST method. As one of the methodological improvements concerns the use of gravimetric measurements to monitor the tracer injection, a balance and data acquisition unit is required at each dosing site (Figure 4). In addition, four peristaltic pumps are needed for dosing and sampling. As with a QUEST experiment, it is recommend to always install a discharge measurement device at the sampling point as it gives complementary information that can be used to check the tracer data for systematic errors. In addition, the flow data can also be used for dynamic data analysis that can account for transient flow effects. This is feasible if a second measuring device is installed at the beginning of the investigation reach and the tested section has negligible inflows. Details regarding the experimental setup are given in the corresponding SOP (Rieckermann and Bares, 2005) and Figure 5 illustrates an exemplary data set for the tracer method.

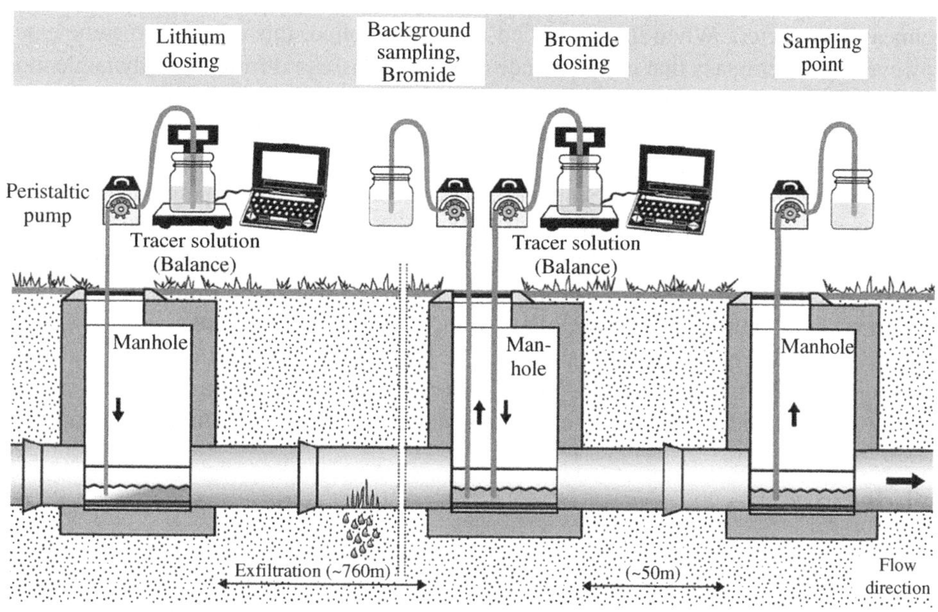

Figure 4: Setup of a QUEST-C experiment with field equipment (balances, data acquisition units and peristaltic pumps for dosing and sampling; Rieckermann *et al.*,2007).

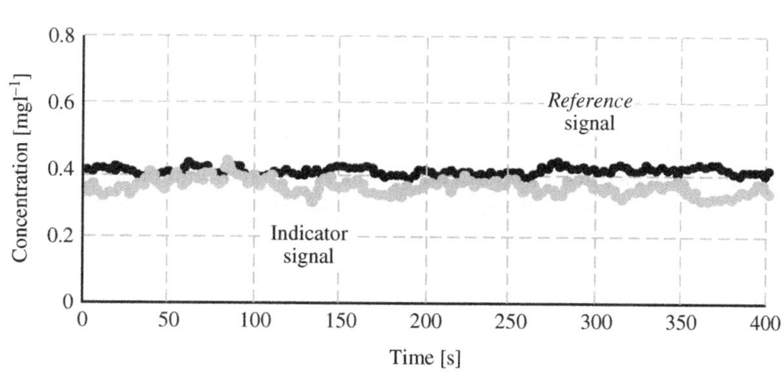

Figure 5: Exemplary data set for QUEST-C tracer methods.

Data analysis

As stated above, there are two subtypes of the QUEST-C method, which mainly differ in the data analysis procedure (Figure 2): Firstly, a *simplified data analysis* can be undertaken which assumes steady flow at the sampling point. This means that exfiltration is computed from a comparison of measured tracer concentrations. Secondly, a *dynamic data analysis* can be undertaken which explicitly uses the measured flow data to compute exfiltration. Thus, leakage is estimated from the comparison of tracer loads, and hydrodynamic modelling is used to assess the impact of dynamic flow effects of the computed exfiltration.

In the simplified approach, the discharge is assumed to be steady during the experiment. This reduces equation 2 considerably so that the data analysis can be performed in a spreadsheet calculation. Analysis of uncertainty considers seven major sources of uncertainty, the most relevant of which are: inaccuracy in the dosing rates, tracer measurement error in samples and standards, and the uncertainty of laboratory equipment. As the model is linear in its parameters, a Gaussian error propagation can be applied.

In the dynamic approach to data analysis, which requires discharge data, exfiltration is computed from the ratio of tracer loads instead of tracer concentration ratios. This approach was developed, because it was found that a certain fraction of the observed variations of the tracer ratio in the samples might have been caused by the separation of wave and fluid in the investigation reach. Discharge fluctuations over time result in waves which travel at a higher velocity (i.e., celerity) than the main water body (i.e., flow velocity). For the QUEST-C method, this means that the two tracer substances that have been dosed to the same water element are in fact diluted differently at the dosing point. The magnitude of this error basically depends on the flow conditions during the experiment and the transport characteristics of the investigation reach (i.e., length, roughness, slope, etc.). When the simplified model is applied, this effect is to some extent inherent in the captured data. However, the dilemma is that its magnitude can only be assessed from hydrodynamic modelling, for which additional discharge information is required. In the uncertainty analysis of this dynamic approach, additional errors to those mentioned for the simplified approach are thus introduced: i.e errors in discharge measurement and integration errors from different time resolution of flow monitoring and sampling. As the model for data analysis is non-linear and some of the errors are not normally distributed, Monte Carlo Simulation was selected for error propagation (Rieckermann *et al.*, 2007).

It is important to note that both the QUEST and QUEST-C tracer methods have a high degree of freedom with regard to the choice of tracer and the dosing strategy. This means that the application of any of these techniques requires the investigator to make a number of decisions regarding experimental design which influence the quality of the results (i.e., the amount of uncertainty in the final estimate of exfiltration). In the following sections, a user guide for the optimal application of methods is given. Firstly, a formal procedure is proposed for optimal experimental design. Secondly, we examine tracer mixing and dispersion in sewers. Thirdly, the applicability and limitations of the tracer methods for exfiltration monitoring is explained and finally, the available documentation for the tracer methodologies is described.

Experimental design

The tracer methods described above have a high degree of freedom with regard to the choice of tracer and the dosing strategy. To guide the end user in the optimal performance of an experiment, the use of decision analysis (DA) is suggested to select the optimal experimental design. Due to limited resources, it was only possible to develop a proof of concept for this procedure, which is for the QUEST method only (Rieckermann *et al.*, 2006). The transfer of the procedure to the QUEST-C method, which uses a continuous dosing strategy, has yet to be undertaken.

For the QUEST method, it is necessary to choose the starting time of the experimental campaign, the number of tracer additions to use, the mass of tracer used in each addition and the relative timing of additions. It was found that different experimental setups can lead to very different degrees of uncertainty in the measured exfiltration ratio. However, there is no universal answer to the question of what is the optimal choice of options that applies equally for all sewer systems (Rieckermann *et al.*, 2006).

The major finding of the study is that the best experimental design depends on how much the investigator is willing to spend on the monitoring process (in terms of effort and financial resources), and how these expenditures compare with the consequences of an incorrect determination of exfiltration. This issue can be fully addressed using the formal framework of DA. The core of the procedure consists of two important elements: a simulation model that is used to predict the outcome (or accuracy of exfiltration estimates in a reach) and a preference model that is necessary to evaluate what level of accuracy is desired. It was decided to use DA for the experimental design because it can be easily adapted to the individual requirements of the situation in which the end-user wants to measure exfiltration (e.g. a drinking water protection zone or rural area, cost limits, etc.). As such it provides an open framework that can be easily extended if necessary.

A frequent and specific question regarding a QUEST experiment is whether to choose a design with overlapping tracer peaks or to keep them separate. However, in considering that an overlap of the different tracer signals keeps the total time of passage of the pulses as short as possible, maximum information on the baseline is gained and the confounding influence of the natural tracer background is minimised. Therefore, a very general recommendation is to chose a tracer dosing strategy that leads to overlapping peaks, if i) the baseline cannot be accurately described and ii) when long distances are measured, as then passing times of the tracer peaks at the measuring point are rather large. If short distances are investigated and short passing times of the peaks are expected, and if the pattern of the baseline during the experiment is negligible or shows significant deviations from a chosen model (e.g., a linear or polynomial function), a dosing strategy with separate peaks should be selected.

Important tasks, which usually precede the experimental design, are i) the estimation of the necessary mixing length of the tracer in the reach of interest, ii) the transformation of the tracer signal on its passage and iii) the prediction of the time of passage of a tracer peak at the sampling station. Each of these requires an estimate of longitudinal dispersion of the tracer cloud.

Dispersion in sewers

To determine the transport properties of the sewer reach under investigation, which govern the mixing of the tracer in the sewer of interest, additional measurement campaigns are recommended prior to the final exfiltration experiment. However, in some cases it might be desirable to predict the development of the tracer signal from simple transport models without much experimental effort. A unique database of tracer experiments was compiled by analysing the data from 60 tracer studies for longitudinal dispersion (Rieckermann *et al.*, 2005b). The tracer tests were carried out by APUSS partners in 37 different reaches having very different structural conditions in various sewer systems and under differing climate and hydrological conditions. These datasets were completed with well-documented records drawn from the literature.

Analysis of the resulting database found that sewer dispersion coefficients are two to three orders of magnitude smaller than those measured in rivers and do not differ much from system to system. Average dispersion coefficients were in the order of 0.1 or 0.2 $m^2 s^{-1}$, with extreme values not exceeding 1.2 $m^2 s^{-1}$. (Figure 6). Suitable equations were identified to: i) deduce dispersion coefficients from tracer data and ii) to predict reasonable dispersion coefficients in sewer reaches having uniform geometry and stable flow conditions. For details regarding the methodology and dispersion models used, refer to Rieckermann *et al.*, (2005b).

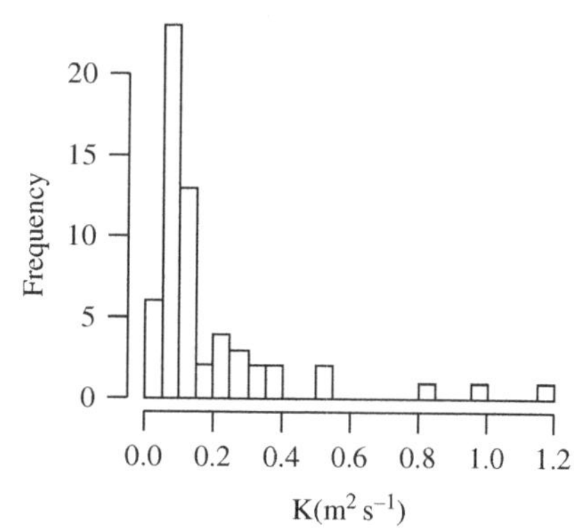

Figure 6: Empirical distribution of the dispersion coefficient K (n = 60) from experimental data analysed in the APUSS project. (Average dispersion coefficient: $0.16 m^2 s^{-1}$ [$K_{10} = 0.05$; $K_{50} = 0.10$; $K_{90} = 0.36$]).

Limitations and practical experience regarding the application of the tracer methods for exfiltration monitoring

In order to develop a sound framework of exfiltration measurements with tracers, it is crucial that limitations of the QUEST and QUEST-C methods are made clear. From the experience gathered in the APUSS project, four conceptual limitations may hinder their application or even make it impossible:

(i) insufficient flow,
(ii) incomplete mixing,
(iii) highly unsteady flow conditions and
(iv) application in certain network situations with highly non-uniform flow.

To further investigate these critical conditions, additional effort should be devoted to investigate whether it is possible to identify those sewer reaches where the transferability of tracer mass loss to meaningful exfiltration values is questionable (Rieckermann, 2005; Rieckermann *et al.*, 2009).

Uncertainty in the obtained tracer loss are expected to be, at maximum, in the order of a few percent of total flow, if the measurement campaign is prepared and performed with the required care and according to the provided documentation on error analysis. However, the question of interest is not the precision of the tracer measurements as such, but rather how this compares to the expected leakage in sewers and whether this is sufficient to detect individual leaks? Recent investigations suggest that such tracer methods are not suited for the detection of individual leaks and should be rather used for an integral screening of sewer networks. While many previous investigation were performed under laboratory conditions, partly also using drinking water as test medium, Klinger *et al.* (2005) quantified sewer leakage under operating field conditions in Rastatt, Germany. Using drop-counter measurements over several months, a mean loss of 0.0016% was observed at a typical single damage defect (an average sewer flow of 3.5 $l\ s^{-1}$ leading to a leakage of about 5 $l\ d^{-1}$). Note, that this is several orders of magnitude lower than the accuracy of the experiments conducted within the context of the APUSS project, which could be about 1%–3% (single standard deviation). In contrast to this, mean exfiltration rates for entire catchments are reported to be much larger e.g. catchment-wide sewer losses in Dresden, Germany, have been estimated by Karpf and Krebs (2005) to be 2.8%, which is within the measurement accuracy of both tracer methods described here. However, the difficulties in explaining the enormous discrepancy in the magnitude of reported leakage rates shows that our current understanding of exfiltration is still limited (Rutsch *et al.*, 2008). Possible influential factors could be leaking house connections or some single, extreme leaks but, as said, the general understanding of such phenomena is still unclear.

Given the current state of knowledge, it cannot be recommended to use tracer methods to assess single or moderate damages. However, they can be very useful to at least rule out the existence of extreme leaks. Therefore, the methods are considered to be operationally relevant and have the potential to support the development of problem-oriented sewer rehabilitation strategies. Much can be gained by rehabilitating the most leaky sewers first, especially in areas with less developed or less maintained sewer systems (e.g., in third world countries) where leakage estimates are reported to be larger than 50% (Wolf *et al.*, 2006).

Regarding their practical applicability, there are few if any obstacles as long as general precautions for working in the sewer environment are taken. Several test runs have been performed under a variety of conditions in different European cities within the APUSS project and it was found that both QUEST and QUEST-C are generally applicable in practice. However, it must be noted that within the APUSS project, practical experience with the tracer methods is limited to open channel flows and most experiments have been performed under dry weather conditions. A comparison of important features is given in Table 1. Finally, the tracer methods that have been developed are clearly not intended to replace CCTV inspections or pressure testing. This is because the measured tracer loss is an integrated measure, which provides no information about the exact location of the leaks in the investigation reach. In addition, structural damages which are not leakage-relevant (e.g., in the soffit of the pipe) can easily pass unnoticed.

Table 1: Comparison of the QUEST and QUEST-C tracer methods to measure sewer leakage.

	QUEST		**QUEST-C**	
Pro	Simple experimental setup in the field, little equipment needed (i.e., no peristaltic pumps, no balances). One single tracer substance used for both indicator and reference pulses, only one measurement error.		Advanced experimental setup in the field. Pumps and balances (incl. data acquisition) guarantee the continuous and accurate dosing of tracer solution. Off-line analysis of tracers makes it possible to use more selective substances. Natural background concentration can be sampled for one substance.	
Con	In-line monitoring devices are required. Their operation is not always trivial in the aggressive sewer environment.		Two different tracer substances have to be applied. This means two measurement errors from analytical instruments. Natural background concentration must be negligible for at least one substance	
Subtype	*Separate tracer peaks*	*Overlapping tracer peaks*	*Simplified Data Analysis*	*Dynamic Data Analysis*
Pro	Simple data analysis, no parameterization of tracer peaks needed.	Can counteract the natural background concentration of tracer to a certain degree.	Simple data analysis can be performed in a spreadsheet calculation.	Analyses ratio of tracer loads instead of ratio of tracer concentrations
Con	Requires strict separation of peaks, which leads to minimal information on the baseline during the experiment	More involved data analysis, parameterization of tracer peaks required.	Assumption of steady flow during the experiment. Analysis of concentration ratios instead of loads.	A full data analysis (including an assessment of uncertainty) is rather complicated, as it requires accurate hydrodynamic transport modeling.

Documentation

Both the QUEST and QUEST-C methods have been thoroughly documented and standard operation procedures (SOPs) for the performance of field experiments and data analysis have been provided (see APUSS website). The underlying theoretical concepts and statistical methods for data analysis and experimental design are further described in the referenced scientific publications. The algorithms for data analysis and experimental design, which have been programmed in the R language (R Development Core Team, 2007), are packed in libraries. R is available under the GNU public license; all documents, libraries and code examples are freely available on the APUSS website for public use.

INFILTRATION METHODS

The applicability of different infiltration quantification techniques clearly depends on the scope of a survey. Today, larger punctual sewer leaks can be located relatively easily by closed-circuit television or walk-through inspections. Yet, sewer infiltration typically occurs through a vast number of smaller defects, which are rather diffusively distributed over widespread areas of a sewer system. Likewise the direct metering of infiltration by physically isolating a certain sewer length (i.e. by means of plugs inserted to the sewer pipes) is only feasible for comparable small subsections and over short time periods. Most conventional infiltration/inflow studies therefore address this issue by different types of local water balances or calculate the "extraneous discharge" based on flow measurements that are causally combined with more or less complex statistical methods. However, in many cases the data analysis is simply based on the assumption that the night-time minimum of the diurnal wastewater hydrograph equals the extraneous flows. In the context of today's ever-growing metropolitan areas these practices are considered oversimplified and are likely to lead to erroneous results (Kracht *et al.*, 2008). Yet, to date little research has been devoted to the development of alternative practices.

Similar to the experimental methods developed regarding the measurement of sewer leakage, two innovative tracer methods for the quantification of extraneous discharges into sewers have been developed. Both infiltration methods are based on natural tracer signals (***Stable Isotopes*** and ***Pollutant Time Series***, respectively), but principally differ in the utilized tracer signal and the associated experimental setup and investigations. The ***Stable Isotope Method*** uses the different isotopic signatures of mains water and infiltrating water as direct natural tracers and therewith allows for a direct calculation of infiltration ratios. The method is suited to quantify infiltration, when useable differences in the oxygen or hydrogen isotope ratios of the drinking water and the infiltrating water exist. This is possible (but not necessarily granted) where drinking water originates from a distant hydrological regime, whilst infiltrating water stems from local precipitation. The ***Pollutant Time Series Method*** estimates the fraction of infiltrating water from a combined analysis of time series of continuously measured pollutant concentrations and wastewater discharge. Its application requires a certain dynamic of the wastewater hydrograph and pollutograph, which is usually suited by the diurnal variations of the wastewater discharge. The method uses automatically operating in-line devices to obtain time series of pollutant concentrations with a high temporal resolution. Measurements (concentration and discharge) are carried out at a single point in the sewer system without recourse to any additional external information.

A significant effort has been devoted to the evaluation of uncertainties and errors both in field measurements and in the data processing phase. For both methods, experimental protocols, uncertainty analysis procedures and data processing algorithms have been established and are available. The methods were tested and evaluated under field conditions and have been validated in the course of comparative experimental studies. The results confirm that the proposed tracer techniques robustly produce accurate results and are suited for routine applications on the catchment or sub-catchment scale.

CONCEPTS

The applicability of various techniques to quantify sewer infiltration from the separation of wastewater hydrographs clearly depends on the scope of the required survey. Today, most conventional infiltration/inflow studies are based on more or less complex statistical analyses that only take into account wastewater flow time series. The conclusions that can be drawn from such evaluations are often limited by the underlying assumptions, which are typically rather simplified (i.e. that the hydrograph diurnal night-time minimum is equivalent to the amount of extraneous discharge). In order to improve the accuracy of wastewater hydrograph separation and thereby obtain more reliable information on the hydraulic interaction between leaky sewers and their ambient aquifers, the use of natural tracer signals was considered to be an optimal alternative. Both novel approaches described here use intrinsic physico-chemical characteristics of the wastewater as a natural tracer for the differentiation of its constituting components (i.e. "real" foul sewage or infiltrating groundwater).

The tracer based separation of wastewater hydrographs basically makes use of a mass balance approach to differentiate discharge components from two or more sources. In a binary mixing system, the fraction of wastewater discharge that is stemming from infiltration (infiltration ratio $X_{Infiltration}$) and the corresponding amount of infiltration discharge ($Q_{Infiltration}$) are given by:

$$X_{Infiltration}(t) = \frac{C_{Foul\,Sewage}(t) - C_{Wastewater}(t)}{C_{Foul\,Sewage}(t) - C_{Infiltration}(t)} \tag{3}$$

$$Q_{Infiltration}(t) = X_{Infiltration}(t) \cdot Q_{Wastewater}(t) \tag{4}$$

where Q is the discharge and C the tracer concentration.

$C_{Wastewater}$ and $Q_{Wastewater}$ can be measured directly in the sewer. On the other hand, $C_{Infiltration}$ and $C_{Foul\,Sewage}$ are typically hardly accessible for investigations at the catchment or even sub-catchment scale. When foul sewage or infiltration enter the sewer system, they immediately mix with the existing wastewater effluent, which basically hinders the direct analysis of these individual mixing components. As a consequence, the investigation of foul sewage and infiltrating water must be performed indirectly. This leads to the idea of separately examining the properties of drinking water and the local groundwater, which constitute their intrinsic origins. Yet, infiltration cannot be quantified using artificial tracers; artificial labelling of drinking water is prohibited and homogenous distribution of tracers throughout an entire aquifer is neither feasible nor desirable for environmental reasons. Rather, specific natural characteristics of the local drinking water, groundwater and sewage have to be used as natural indicators of the mixing and dilution processes. However, since generally a vast number of dissolved species are added to the foul sewage, it is rare to find ideal natural tracers. Most potential chemical components exhibit large daily fluctuations and therewith obscure the natural tracer signals.

To overcome the above limitations, two generic quantification methods have been developed which differ in the utilized tracer signal and the associated experimental setup and investigations:

(a) the ***Stable Isotope Method*** (Kracht *et al.*, 2007) uses the stable isotope compositions of mains water and local groundwater as suitable proxies for the stable isotope compositions of the foul sewage and the infiltrating water respectively. It therewith allows for a direct calculation of infiltration ratios from eq. 3. The method is suited to quantify infiltration, where useable differences in the oxygen or hydrogen isotope ratios of the drinking water and the infiltrating water exist. The proper use of this method requires a detailed hydrologic and hydrogeological investigation to define the principal hydrologic interactions and pathways of different waters in the catchment.

(b) in contrast to the stable isotope approach, the ***Pollutant Time Series Method*** (Kracht and Gujer, 2005) does not depend on a direct investigation of drinking water or infiltration origins. The fraction of extraneous water is determined by simultaneously analysing time series of pollutant concentrations and wastewater discharge that are obtained from measurements carried out at a single point in the sewer system. A characteristic feature of this method is the use of automatically operating in-line devices to obtain time series of pollutant concentrations with a high temporal resolution. Based on the recorded time series of wastewater discharge, a modelled time series of pollutant concentrations can be calculated. By fitting this model series to the measured data, a set of parameters defining $Q_{Infiltration}$ can be estimated.

In the following paragraphs, a synthetic overview is given on the two infiltration methods and discusses important aspects regarding their practical application. In-depth information on the specific topic is provided in the cited references and is also available on the APUSS website.

The stable isotope method

Natural water (H_2O) is composed of the three stable isotopes of oxygen (^{16}O, ^{17}O and ^{18}O), the two stable isotopes of hydrogen (^{1}H and ^{2}H or D = deuterium) and the unstable isotope Tritium (^{3}H). Compared to most other natural water characteristics, the stable isotope composition of the water molecule is very robust with respect to changes in water chemistry or biological activities. Particularly, it is hardly affected by utilization in urban water infrastructures and therefore constitutes a suitable direct natural tracer.

Within the global water cycle, the stable isotope composition of water is predominantly controlled by large-scale meteorological evaporation and precipitation effects; evaporation from the oceans being the major source of atmospheric water masses. As clouds move across the continents, the heavier water molecules containing D and ^{18}O will preferentially rainout first in coastal areas and the remaining water vapour is continuously depleted in these isotopes, i.e. isotopic fractionation (Figure 7).

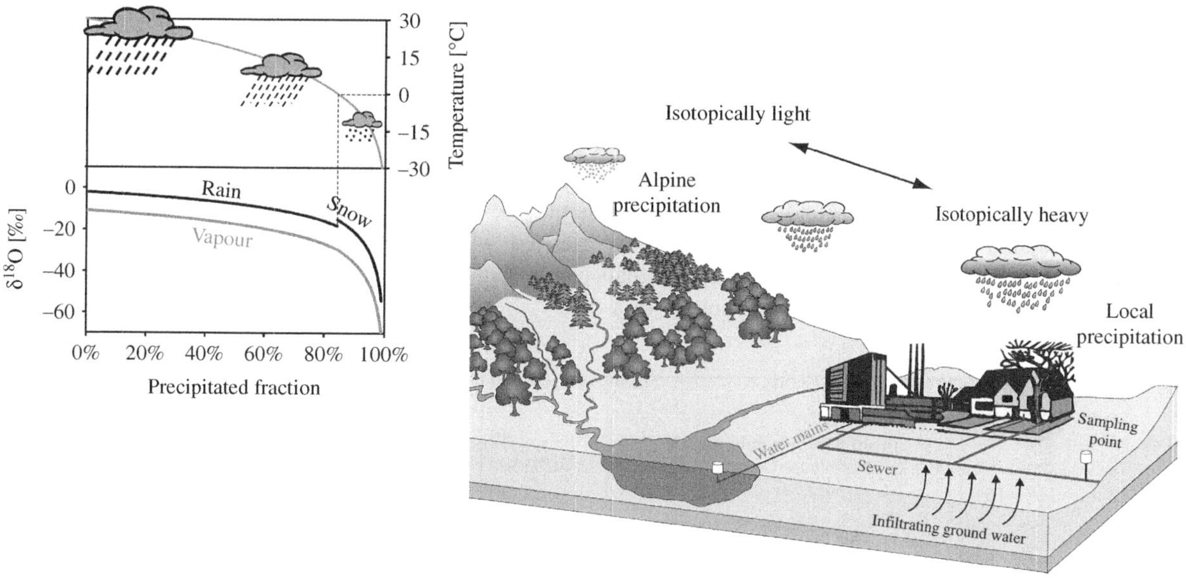

Figure 7: **Left:** Modelled evolution of the isotopic composition in precipitation from a cooling vapour mass. (The stable isotope composition of natural ground and surface waters is predominantly determined by the topographic elevation and continental setting of the area of recharge by precipitation; altitude effect, continental effect). **Right:** Isotopic separations for applications in urban hydrology (Convenient separations can exist, where drinking water originates from a distant hydrological regime, whereas extraneous flows stem from local precipitation.)

The isotopic composition of natural waters of a meteoric origin is in consequence determined by the continental setting (continental effect) and topographic elevation (altitude effect) of the region where ground water and surface water are recharged by precipitation (Craig and Gordon, 1965; Rozanski *et al.*, 1993; Rozanski *et al.*, 2001). As many urban agglomerations obtain their drinking water from remote sources, this effect is considered to be relevant for a significant number of urban catchments.

Therefore, the stable isotopes ($\boldsymbol{\delta}^{18}$O and $\boldsymbol{\delta}\,^{2}$H of water) provide a promising tool to quantify the discharge of extraneous waters in those situations where a city uses drinking water from a watershed that is situated in a different hydrological regime than the urbanized area. Ideally, the differences in the isotopic compositions of the drinking water, the wastewater and the groundwater (Figure 7) are significant enough to allow for the quantification of the various mixing ratios (Kracht *et al.*, 2003). Beyond this prerequisite of a sufficient isotopic separation between the mixing endmembers, it is critical to verify a low spatial variability of the isotopic composition within the ambient aquifer (Kracht *et al.*, 2008).

Figure 8 (left) displays measured δ^{18}O values from an experimental catchment site in Rümlang, Switzerland. The groundwater and drinking water end-member ranges are compared with δ^{18}O values of a series of 28 hourly wastewater samples taken from a connector sewer at the outlet of the urban catchment. As expected, the wastewater is enriched in ^{18}O and ^{2}H during night time, indicating a higher fraction of isotopically "heavier" infiltrating water being present in the sewer. During day time, the wastewater shows depleted isotope values, due to the larger amount of isotopically "lighter" foul sewage being discharged into the sewer system. Based on these data, the wastewater hydrograph is subdivided into its elementary components "foul sewage" and "infiltrating groundwater" by calculating the infiltration ratio eq. 3) and the corresponding infiltration discharge (Figure 8, right). In the APUSS project, the Stable Isotope method was applied more extensively in Rümlang (Switzerland), Lyon (France) and Rome (Italy) revealing some interesting results. For example, this novel method showed that sewer infiltration is not necessarily constant during the day. Based on precise measurement data, substantial intra-day fluctuations occur in the infiltration discharge. Such fluctuations can indicate irregular dispersion effects probably caused by cumulative backwater zones. A more detailed discussion of these observations is given in Kracht *et al.*, 2007.

Limitations of the stable isotope method

The underlying concept of the *Stable Isotope Method* defines parasitic infiltration as a fraction of the total wastewater that is carrying a specific isotopic signature. It is therefore generally limited to catchments or sub-catchments where both drinking water and groundwater have homogenous isotopic signatures, and where only two components (one drinking water source, one groundwater source) are interacting. The two critical system properties for the applicability of the method are the existing isotopic separation between drinking water and infiltration and the natural variability of possible infiltration sources. The question of inhomogeneities of the local groundwater or other origins of parasitic waters is often crucial for the applicability of the method. For this reason it is mandatory to always conduct a broad hydrological and hydrogeological investigation to obtain sound information on the principal hydrologic interactions and pathways of different waters in a catchment. It is inevitable that the isotopic compositions of infiltration origins are derived from direct investigations within the catchment. The principle sources of infiltration (i.e. groundwater) need to be definable, accessible for sampling and statistically describable with sufficient precision. The simple use of regional background values is not appropriate, as local groundwater compositions may be altered by processes such as drinking water losses or irrigation. Beyond this, practical difficulties are experienced which may result from regional cross-linking of individual drinking water systems. Disturbed tracer signals can be caused, when parts of a catchment area receive water from different drinking water supplies in the course of a day.

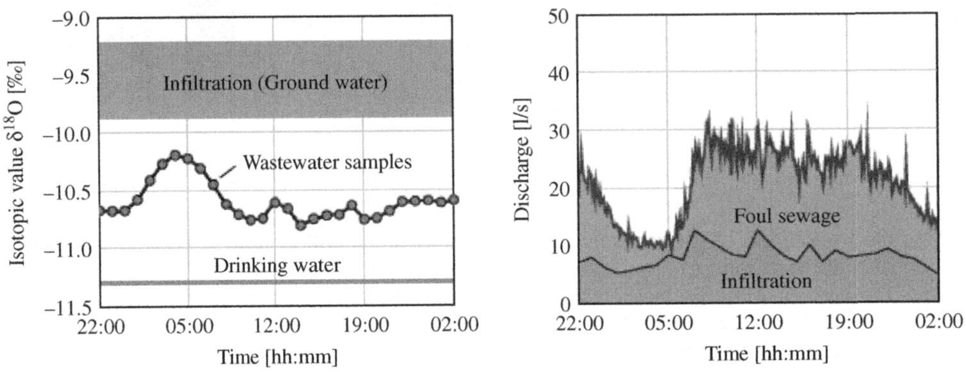

Figure 8: Isotopic characterisation and corresponding decomposition of a diurnal wastewater hydrograph into its elementary components "Foul Sewage" and "Infiltration".

The simultaneous use of $\delta^{18}O$ and $\delta^{2}H$ may also theoretically enable application where two isotopically different sources contribute to sewer infiltration. However, in relevant situations the discrimination between two different infiltration origins is likely to be impossible because of the high correlation between $\delta^{18}O$ and $\delta^{2}H$ observed in most natural waters (meteoric waterline). Secondly, the methodology would attain reduced precision and suffer the loss of redundant control with respect to the discrimination of potentially interfering evaporation effects. The stable isotopes composition of water is in principal very robust with respect to changes in water chemistry or biological activities. Nevertheless, a higher degree of evaporation can affect the isotopic values of the remaining water. The wastewater data investigated in the course of the APUSS studies have so far not shown any alteration by evaporative enrichment (no discernible deviation from the groundwater/foul sewage mixing line in the $\delta^{18}O/\delta^{2}H$-plot). Additionally, evaporative enrichment was investigated by a smaller set of exploratory experiments regarding in-household transformations due to major water consuming activities, with no relevant effects being observed (Kracht *et al.*, 2008). However, the question of possible evaporative enrichments should be carefully regarded in future studies, especially when transferring the methodology to cities in differing climate zones or where stronger industrial influences should be expected (e.g., where cooling towers are used).

The pollutant time series method

This method quantifies infiltrating non-polluted waters based on measured time series of wastewater flow and pollutant concentrations. In comparison to the stable isotope method, the time series based approach has the advantage that it does not require recourse to any additional external information: all measurements are carried out at a single point in the sewer system. Based on a transformation of eq. 3, the analysis uses a mixing model describing the pollutant concentration in terms of dependency on wastewater discharge and time (eq. 5).

In contrast to earlier tracer based methods (Hager *et al.*, 1984), the current novel approach is based on high temporally resolved measurements with in-line devices, which allows for considerable refinements of the data analysis. Specifically, it does not require the critical assumption of a stationary cyclic behaviour of the diurnal hydrograph and pollutograph. The underlying model structure can incorporate temporal fluctuations of the pollutant concentration in the foul sewage (eq. 6) as well as time dependencies of the infiltration rate (i.e. dynamics of rain-induced infiltration). For the latter, the amount of infiltrating water is conceptually divided into a constant baseflow and an exponentially receding interflow (eq. 7):

$$C_{Wastewater}, \text{model} = \frac{(Q_{Wastewater} - Q_{Infiltration}) \cdot C_{Foul\,Sewage} + Q_{Infiltration} \cdot C_{Infiltration}}{Q_{Wastewater}} \tag{5}$$

$$C_{FoulSewage} = f(t) + f(Q_{Foul\,Sewage}) \tag{6}$$

$$Q_{Infiltration} = Q_{Baseflow} + Q_{0,Interflow} \cdot e^{-k_{rec}\cdot(t-t_{0,Interflow})} \tag{7}$$

with $Q_{0,Interflow}$: initial magnitude of interflow at the time $t_{0,Interflow}$; k_{rec}: recession constant.

The application of eq. 5 requires information on the tracer concentration in the infiltrating water, which is typically not identifiable from the time series itself. A suitable parameter for this application is the chemical oxygen demand (COD), which was utilized in most of the APUSS applications. The COD concentration in the parasitic groundwater can in most cases be assumed negligible, which avoids some difficulties in characterizing the tracer mass contribution stemming from the infiltration itself. Recently available submersible spectrometer probes allow for a direct measurement of COD-equivalents in the media by the means of light absorbance in the UV-VIS range (Langergraber *et al.*, 2003). The use of automatic measuring devices considerably simplifies the recording of longer time series (several days or weeks) in a high temporal resolution, which significantly improves the documentation of the wastewater composition dynamics. $Q_{Infiltration}$ is finally identified by fitting a modelled time series of pollutant concentrations to the measured data (Kracht and Gujer, 2005). Figure 9 displays an excerpt of results from the experimental catchment site in Rümlang, Zurich. Note that although the measurements were performed during a dry weather period (last antecedent rainfall occurred three days before this measurement campaign) it was necessary to include a suitable model term to describe the continuous recession of the infiltration hydrograph (compare with Figure 8).

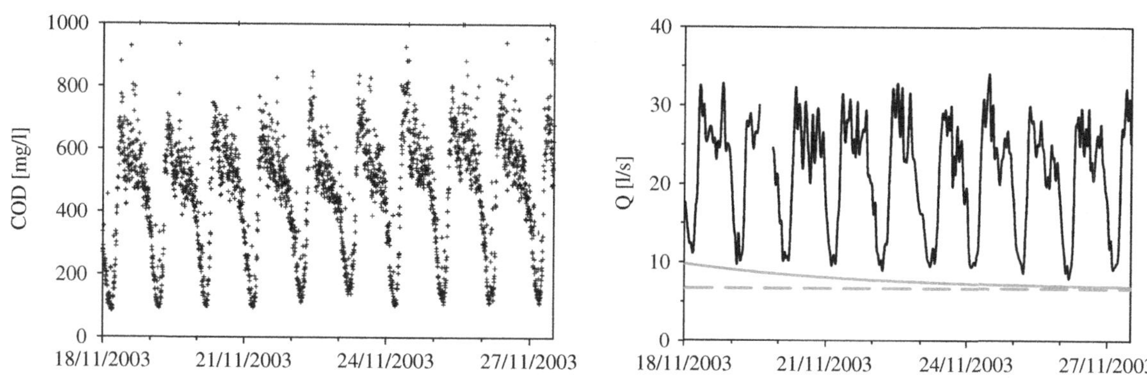

Figure 9: Decomposition of a diurnal wastewater hydrograph based on the *Pollutant Time Series Method.* **Left:** COD measurements; **Right:** Computation of the hydrograph separation (black upper series line: total wastewater discharge; lower solid line: discharge of extraneous fresh water).

Infiltration measurements with the Pollutant Time Series Method are very flexible. The capability to consider natural storage and interflow phenomena significantly broadens the practical applicability of the method and particularly simplifies the measurements in rainy seasons, when infiltration rates typically increase due to elevated groundwater tables. The experimental protocols have been tested within several measurement campaigns performed by APUSS partners in different catchments in Berlin, Bottrop, Dresden, Gladbeck, London, Lyon, and Rome.

Limitations of the pollutant time series method

From the experience gathered in the APUSS project, it can be concluded that the successful implementation of the methods crucially depends on a thorough preparation of the experimental campaign and the local boundary conditions at the investigation site (i.e flow conditions, accessibility, etc.). One of the main limitations of the method seems to be bound to the required bias and drift free operation of the submersible UV-VIS spectrometer (which was chosen by all investigators). As the experience of all case studies undertaken in the APUSS project shows, it has to be accepted that the continuous monitoring of wastewater quality parameters with optical measurements in raw sewage is obviously not a trivial task. Amongst other things, the measurement quality strongly depends on the skills and experience of the technical field team. Thus, the required investments for a successful measurement campaign should not be underestimated. The method may be used at the outlet of any sub-catchment where a continuous discharge of wastewater can be assured. However, a minimum depth of wastewater flow is required for the disturbance-free operation of the measuring devices, which may be critical during minimum night flow. Furthermore, predominant types of industrial effluents should be excluded, as this may hinder a regular data analysis.

Future prospects for the infiltration methods

The two tracer methods were specifically developed to consider information on the physico-chemical properties of the wastewater, in addition to flow measurement data. Thus, they exclude several conventional sources of error (i.e. uncertain assumptions on residual foul sewage discharge at night time, amount and pattern of drinking water consumption, and temporal behaviour of extraneous flows). They do in this way overcome critical limitations of current practice in infiltration measurements. The studies demonstrate the suitability of both approaches in practical applications under a variety of situations and locations.

The experience gathered in the APUSS project shows that the successful implementation of the tracer methods depends on the local conditions. In practice it will thus be required to assess the confidence in the obtained results on an individual case basis. Given a basic presumption that many European sewer systems suffer infiltration ingresses that approximately fall within a range of 20% to 70% of the total wastewater discharge, our experience suggest that a careful investigator can confidently estimate under suitable conditions, infiltration ratios with a 95% confidence level accuracy of better than ±4% (related to the total amount of wastewater). In this respect, the new methods do therefore fully satisfy the requirements for practical application in the performance assessment of sewer networks.

Documentation

Both the Stable Isotope and the Pollutant Time Series infiltration method have been described thoroughly in scientific publications (Kracht, 2007; Kracht *et al.*, 2007; Kracht and Gujer, 2005) and a comprehensive overview is given in Kracht (2007). A direct comparison and cross validation of the methods has also been reviewed in detail in (Kracht *et al.*, 2008). Furthermore, the performance of field experiments and data analysis has been documented in standard operation procedures. Algorithms for data analysis have been programmed in the R language (R Development Core Team, 2007) and are packed in libraries. R is available under the GNU public license; all libraries and code examples are freely available on the APUSS website for public use.

REFERENCES

Bertrand-Krajewski, J-L., Laplace, D., Joannis, C. and Chebbo, G. (2000) *Mesures en Hydrologie Urbaine et Assainissement.* Technique et Documentation, Paris, France. 794 p. ISBN 2-7430-0380-4.

Bertrand-Krajewski, J-L., Bardin, J.P., Mourad, M. and Beranger, Y. (2003) Accounting for sensor calibration, data validation, measurement and sampling uncertainties in monitoring urban drainage systems. *Water Science and Technology,* **47**(2), 95–102.

Craig, H. and Gordon, L. (1965) Deuterium and oxygen-18 variations in the ocean and the marine atmosphere. 9–130 in Tongiorgi, E., (Edit): *Stable Isotopes in Oceanographic Studies and Paleotemperatures*: Spolleto, Italy. OCLC 6454168.

de Bénédittis, J. and Bertrand-Krajewski, J-L. (2005) Measurement of exfiltration in urban sewer systems using NaCl as an artificial tracer. *Houille Blanche-Revue Internationale De L'Eau,* (5), 26–34

Hager, W.H., Bretscher, U. and Raymann, B. (1984) Methoden zur indirekten Fremdwasserermittlung in: Abwassersystemen. *Gas-Wasser-Abwasser*, v. **64**, 450–461.

Karpf, C. and Krebs, P. (2005) Assessment of extraneous water inflow in separated sewer networks. In Mikkelsen, P. (Edit): *Proc. 10th Int. Conf on Urban Drainage* (ICUD) 21–26 August 2005, Copenhagen, Denmark, IWA Publishing, London. UK.. ISBN 978 1843395 744.

Klinger, J., Wolf, L. and Hötzl, H. (2005) Leaky sewers; Measurements under operating conditions. *Proc.4th World Wide Workshop for Young Environmental Scientists* (WWW-YES2005), Urban Waters: Resource or Risk? 10–13 May 2005. Vitry-sur-Seinne, Paris, France.

Kracht, O., Gresch, M., de Bénédittis, J., Prigiobbe, V. and Gujer, W. (2003). Stable isotopes of water as a natural tracer for infiltration into urban sewer systems. *Geophysical Research Abstracts*, Vol. 5, 07852, 2 p.

Kracht, O. and Gujer, W. (2005) Innovative tracer methods for infiltration monitoring. In Mikkelsen, P. (Edit): *Proc. 10th Int. Conf on Urban Drainage* (ICUD). 21–26 August 2005. Copenhagen, Denmark, IWA Publishing, London. UK. ISBN 978 1843395 744.

Kracht, O. (2007). *Tracer-based Hydrograph Separation Methods for Sewer Systems* [PhD thesis]. Swiss Federal Institute of Technology (ETHZ). Zurich, Switzerland.

Kracht, O., Gresch, M. and Gujer, W. (2007) A stable isotope approach for the quantification of sewer infiltration. *Environ. Sci. Technol.*, v. **41**, 5839–5845.

Kracht, O., Gresch, M. and Gujer, W. (2008) Innovative tracer methods for infiltration monitoring. *Urban Water Journal*, **5**(3), 173–185.

Langergraber, G., Fleischmann, N. and Hofstaedter, F. (2003) A multivariate calibration procedure for UV/VIS spectrometric quantification of organic matter and nitrate in wastewater. *Water Science and Technology*, **47**(2), 63–71.

R Development Core Team (2007) *R: A Language and Environment for Statistical Computing.* R Foundation for Statistical Computing. Vienna, Austria, ISBN 3 900051-07-0.

Rieckermann, J. (2005). *Quantifying Sewer Exfiltration with Tracers.* PhD Thesis: Swiss Federal Institute of Technology (ETH), Zürich, Switzerland.

Rieckermann, J. and Niesel, K. (2005) *Standard Operation Procedure (SOP): Quantification of Exfiltration from Sewers with artificial Tracers (QUEST).* EAWAG, Zurich, Switzerland. p. 48.

Rieckermann, J. and Bares, V. (2005) *Standard Operation Procedure (SOP): Quantification of Exfiltration from Sewers with Artificial Tracers: Continuous Dosing (QUEST-C)*. EAWAG, Zurich, Switzerland. p. 35.

Rieckermann, J., Borsuk, M., Reichert, P. and Gujer, W. (2005) A novel tracer method for estimating sewer exfiltration. *Water Resources Research* **41**(5).

Rieckermann, J., Neumann, M., Ort, C. and Gujer, W. (2005) Dispersion coefficients of sewers from tracer experiments. *Water Science and Technology*, **52**(5), 123–133.

Rieckermann, J., Borsuk, M.E. and Guyer, W. (2006) Using decision analysis to determine optimal experimental design for monitoring Sewer exfiltration with tracers. *Water Science & Technology*. **54**(6–7): 161–168.

Rieckermann, J., Bareš, V., Kracht, O., Braun, D. and Gujer, W. (2007). Estimating sewer leakage from continuous tracer experiments. *Water Research*, **41**(9), 1960–1972.

Rieckermann, J., Sydler, D., Borsuk, M., Gujer, W and Reichert, P. (2009) "Bayesian experimental design of tracer experiments to monitor leakage from sewer networks." (in preparation).

Rozanski, K., Araguas-Araguas, L. and Gonfiantini, R. (1993) Isotopic patterns in modern global precipitation. 1–37 in Swart, P.K., Lohmann, K.L., McKenzie, J and Savin, S. (Edits): *Climate Change in Continental Isotopic Records*; Geophysical Monograph 78: American Geophysical Union, Washington, DC. US.

Rozanski, K., Froehlich, K. and Mook, W.G. (2001) In Mook, W.G (Edit): *Environmental Isotopes in the Hydrological Cycle: Principles and Applications.* Volume 3: Surface Water. UNESCO/IAEA, Paris, France. 118 p.

Rutsch, M., Rieckermann, J. and Krebs, P. (2006) Quantification of sewer leakage: A review. *Water Science and Technology* **54**(6–7), 135–144.

Rutsch, M., Rieckermann, J., Ellis, J.B., Vollertsen, J., Cullmann, J. and Krebs, P. (2008) Towards a better understanding of sewer exfiltration. *Water Research,* **42**, 2385–2394.

Wolf, L., Klinger, J. and Hötzl, H. (2006) Risk potential of leaky sewers for soil and groundwater in the example of a medium-sized town (*Gefährdungspotential von Boden und Grundwasser durch Kanalleckagen am Beispiel einer mittelgroßen Stadt)* (in german). Leaky sewers – (no) risk? ("Undichte Kanäle–(k)ein Risiko?"), Joint DWA meeting 11–12 October 2006, Frankfurt, Germany.

Chapter 3

Infiltration and exfiltration in house connections

Claude Joannis

HOUSE CONNECTIONS AND DRAINS

The EU Standard EN 13508-2 defines a "connection" as "*a general term used for the location at which one pipeline joins another pipeline or a manhole or an inspection chamber*" (European Committee for Standardisation, 2006). On the other hand, Pr EN752-1 defines a drain as a "*pipeline, usually underground, designed to carry wastewater and/or surface water from a source to a sewer*", whereas a sewer is a "*pipeline or other construction, usually underground, designed to carry wastewater and/or surface water from more than one source*" (European Committee for Standardisation, 1995). Given these difficulties, the term "drain" should be preferred to "house connection" as the latter rather denotes the downstream part of a drain, although the ambiguity remains to a certain extent.

Some further ambiguity could arise from the word "*source*" contained in the Pr EN752-1 definition of a sewer, which can be interpreted either as an individual private property (= a specific plot of land), or a collective building. But the very title of the standard "**Drains and sewers *outside buildings***" strongly suggests that sources are buildings. In this chapter sources are considered as being housing buildings (but many issues can be extended to other kinds of building as well). With this definition, a drain, (also named a "lateral" in technical literature), includes a (usually) public part and a private part[1]. Both can be concerned with infiltration and exfiltration problems, but most technical solutions are suitable only for the public part, as these solutions are extensions of techniques available for sewers. Moreover, the private part displays features, such as branches with no inspection chambers, which make them very difficult to handle and inspect. This is shown in Figure 1 from the result of a field survey of 150 drains in the city of Pornic, France (Joannis *et al.*, 2001). This survey was conducted in residential areas and direct connections, mainly present in the historic centre of the city, have been underestimated.

Thus for sewer evaluation purposes, an essential distinction must be made between public and private parts of sewer systems, in addition to the usual distinction between drains and sewers. In any case the public part of drains can be detailed as shown in Figure 2 as constituting:

(1) a connection pipe;
(2) at the downstream end of the connection pipe, a connection device, joining the drain and the sewer;
(3) in some cases, at the upstream end of the public part of the drain, a special appurtenance (inspection chamber, siphon,..) located near the limit (or cartilage) of private grounds.

[1] The word "private" is not quite accurate to make the distinction, as private collective sewers do exist (and are often considered as "laterals" by communities or operators). On the other hand, public buildings also have drains, and some of them may be used as housing. Anyway no better word has been found and it is suggested that the "public/private" distinction is retained, with a somewhat extended meaning.

 Assessing infiltration and exfiltration on the Performance of Urban Sewer Systems (APUSS). Edited by Bryan Ellis and Jean-Luc Bertrand-Krajewski. ISBN: 9781843391494. Published by IWA Publishing, London, UK.

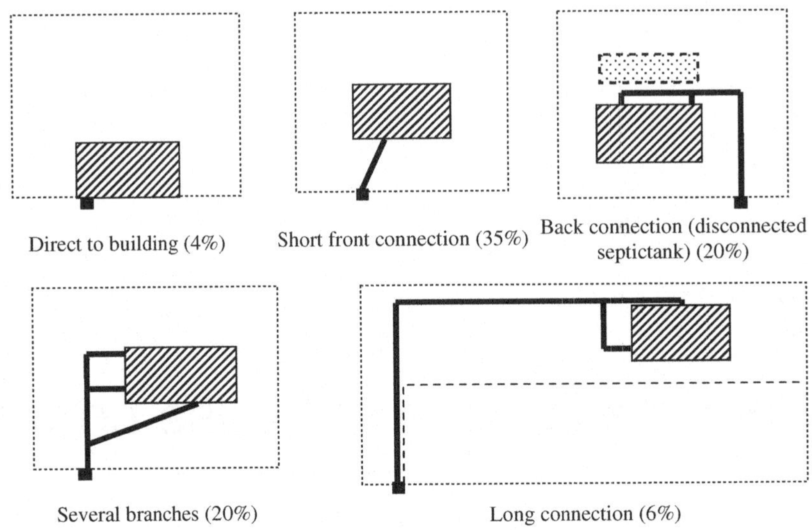

Figure 1: Typical lay-outs of drains in private grounds identified by a survey in Pornic (France).

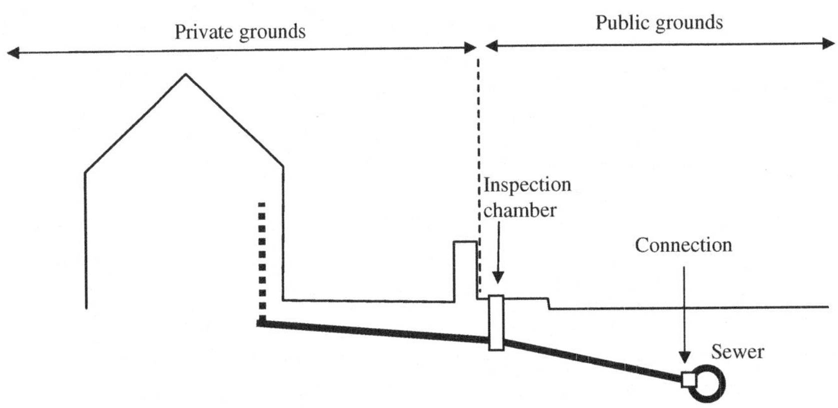

Figure 2: The main parts of a drain.

It must be emphasized that the connection between a drain and a sewer is a critical set-up, especially when the drain has not been contemporaneously built together with the sewer line. When new drains are connected to a sewer in operation, dedicated fittings such as "wyes" (Y) or "tees" (T) can be used. More often a hole is drilled or broken into the sewer, through which the drain is inserted (although a better solution is fitting the drain to a saddle which is then fitted to the sewer), and a proper sealing may or may not be applied to ensure water-tightness.

Fortunately, defects on connections can be spotted by CCTV inspection, and fixed by grouting or local "short" sleeving. Moreover, presently available techniques also enable CCTV inspection (with so-called "satellite" cameras) and subsequent sleeving of some length of the public part of a drain. Unfortunately, the practical use of these techniques is limited by various features frequently encountered in drains, especially multiple bends.

When speaking about assessing and rehabilitating drains, many communities and operators mean only the public part of drains, as this part is usually integrated into the sewerage asset that they own or operate. Yet the private part of drains should technically be considered as a part of the sewer system; some surveys have showed their cumulated length is more or less equivalent to the total length of public sewer and drains (Table 1).

According to the kind of setups occurring at both ends of the public part of a particular drain, measurements of flowrates (either infiltrated or exfiltrated) are liable to be more or less difficult to perform, and will encompass both private and public parts of the drain, as a whole or separated, or the private part alone.

Table 1: Quantitative assessment of laterals of several cities in France (Joannis *et al.*, 2001).

City	Population	% combined (/sewer length)	Drains/km of sewer Nr.	People/drain Nr.	Public grounds: Mean length of drains (m)	Public grounds: Total length of sewers/total length of drains (m/m)	Public grounds: Total length of drains (km)	Private grounds: Mean length of drains (m)	Private grounds: Total length of sewers/total length of drains (m/m)	Private grounds: Total length of drains (km)
Bordeaux	210 336	93	87	4.4	4.4	2.6	210	3.3	0.3	160
Dijon	151 000	80	60	7.6	5.7	2.9	115	12.5	1.3	250
Strasbourg	252 338	95	49	10.5	7.3	2.8	174	37	0.6	888
Amiens	131 350	0	115	4.4	3.0	2.9	91			
Nantes	244 995	58	61	7.0	6.0	2.7	211			
Caen	112 833	0	70	6.5				8.7	0.6	151
Pornic	50 000	0	68	7.9	5.7	2.6	35	22.5	0.8	120

OBJECTIVES FOR MEASURING INFILTRATION AND EXFILTRATION FLOW-RATES IN HOUSE CONNECTIONS

Methods for measuring *infiltration* usually encompass flowrates coming into public sewers and house connections as a whole, whereas *exfiltration* measurements are more specific to a particular reach of (usually public) pipe. Yet it is an important matter to be able to assess the relative contribution of drains to overall infiltration and exfiltration (I/E).

The purposes of assessing *infiltration* flow-rates may be different for house connections and sewerage as a whole. For sewerage the objectives are:

(1) establishing a relationship between water tightness defects and observed problems, such as poor performance of the treatment plant, excessive operating cost of pumping stations, separate sanitary sewers overflows, and impaired physical conditions of sewer facilities;
(2) evaluating potential effects of observed infiltration flow-rates;
(3) locating catchments or sewer segments where infiltration is high by comparison with other catchments or segments, thus being priority candidates for CCTV inspections and rehabilitation
(4) monitoring the efficiency of rehabilitation, at the scale of a catchment or a sewer segment.

Objectives 1 and 2 both contribute to assessing the need of rehabilitation programmes, whereas objective 3 aims at defining these programmes. For all these objectives, a global assessment (scale of a catchment, assessment encompassing both public sewerage and house connection) of infiltration is useful, but a local and specific assessment (scale of a sewer segment, making a difference between the inputs into sewer and the house connections) can be more suited for the objectives 3 and 4.

For house connection, the purposes of an assessment may be:

(a) assessing the overall contribution of drains to infiltration at the scale of a catchment,
(b) assessing the need for renovating a particular drain

Objective a for drains is closely related to objectives 1 and 2 for sewerage, and provides evidence about the efficiency which may be expected from an action on public facilities alone. For this purpose, both public and private parts are involved. Whereas infiltration into sewerage is easier to assess at the global scale of a catchment, and when possibly downscaled to sewer segments, infiltration into drains usually can only be monitored at the scale of individual drains, and must be up-scaled to a more global scale.

Objective b seems quite clear by itself, and it is clear indeed as far as the private part of a drain is considered. For the public part it may be easier to CCTV inspect the drain together with the sewer, rather than measuring flow-rates, which is often a difficult operation in any case.

A special feature of infiltration into house connections is that it may be very temporary, occurring in relationship with rainfall events and described as "rainfall induced infiltration" or RII (EPA, 1990). When cumulated at the scale of a catchment, RII may easily be confused with direct inflow of rainwater (misconnections of rainwater on separate systems, or regular connections on combined systems). As a matter of fact both inputs are often referred to by a single expression such as "rainfall induced infiltration/inflow (RII/I)". Although poorly documented in the literature, the issue of making a difference between infiltration and inflow during and immediately after rain events is worth considering, especially for separate sanitary sewers, as it has important operational implications. Inflow usually is fairly easy to disconnect, whereas restoring the watertightness of the private part of house drains is much more difficult, and may lead to re-building the drain. For combined systems, RII is less a concern during rain events, as it is much less important than runoff at the catchment scale.

For separate sanitary systems, identifying the respective contribution of both components to RII/I could be easier to achieve at the scale of a single connection than an entire catchment, by analysing the different dynamics of hydrographs. Thus a third objective for assessing infiltration flowrates into drains can be:

(c) Assessing the respective contribution of infiltration and inflow to rainfall induced flow-rates

This objective is especially meaningful when up-scaled to the catchment level.

In practice it is safer to combine flowrate measurements with a check of the separation of effluent (i.e; runoff water and sanitary wastewater) inside private properties and buildings, so that inflow can be better evaluated than by using hydrographs alone. Logically, this check should be followed by the disconnection of runoff sources from the sanitary sewer, so that only infiltration remains to be measured.

Removing inflow can be done systematically at the scale of a catchment or a whole city, as will be detailed below and this is a first step towards the complete control of house connections. Such a step-by-step process (check inputs - remove inflow - measure infiltration) may be a problem if it leads to removing infiltration in private grounds, as it will be very difficult to convince customers to upgrade their systems twice. But if the ultimate objective is objective a (assessing the efficiency of an action on public facilities alone), no such problem will arise.

As a conclusion, no measurement of infiltration in drains should be made, unless an inflow check has been conducted, and preferably followed by disconnecting operations when applicable (i.e. inflow of sanitary sewers).

For exfiltration the question is simpler to handle, as available measurement methods operate at a local scale, both for sewers (sewer segment) and house connections. Objectives a and b thus become:

(a) Assessing the overall contribution of drains to exfiltration at the scale of a sewer segment;

(b) Assessing the need for renovating a particular drain.

Usually no up-scaling is needed to achieve objective a' other than a mere summation of results obtained for every individual drain on the segment.

DESIGNING MEASURING SURVEYS

Once the objectives for a survey have been defined, the precise modalities of the survey must be design. A first guideline has been proposed above, which is checking effluent separation in private properties, and preferably removing misconnections, before infiltration measurements are performed. This principle might be extended to removing infiltration in the public part of the system (including drains) before trying to assess the inputs for the private part. In some cases this strategy may cancel the need for surveying drains, if residual infiltration appears as being negligible but this is liable to happen only for very small catchments. Moreover this *post-eriori* evaluation is hardly consistent with objective a presented above, for which an *a-priori* evaluation is preferred.

Another important matter is timing, because continuous measurement cannot be performed on drains over a period of time exceeding a few days. So this period must be carefully chosen, especially for infiltration measurement. As infiltration into drains is liable to be an important contribution to peak flows, measurement should be performed during high water table conditions (usually in winter), immediately after a rainy period. If several drains are to be surveyed, it is better to survey them at the same time. It is important for interpreting the results that the context of measurement must be precisely documented, so a continuous monitoring of flowrates at the outlet of a catchment including the surveyed drains is essential. This monitoring may also be used for triggering the survey when observed conditions display (very) high infiltration rates.

Table 2 clearly shows that the practical ability to perform flow-rates measurements, either for the private or the public part of drains strongly depends on the lay out of each particular drain. So, if such measurements have to be conducted on a somewhat systematic basis, a preliminary knowledge of the asset is needed to design the survey and evaluate the necessary means, especially for manpower time. This knowledge is needed for both infiltration and exfiltration measurements, and is also valuable for providing some reliability to an up-scaling process.

Table 2: Possibilities for measuring flow-rates according to the setup of a drain.

Setup of the drain		**Possibilities for measuring flow-rates**		
Upstream	Connection drain/ sewer	Flow-rates from private part	Flow-rates from public part	Flowrates from public + private parts measured as a whole
Inspection chamber	Man-entry sewer or manhole, with sufficient drop height	Direct or difference with/without ballooning	Easy (i.e. volumetric + Upstream ballooning)	Easy (i.e. volumetric)
Inspection chamber	"clean" connection to non man-entry sewer pipe	Preferably direct or difference with/without ballooning	Not easy (in pipe volumetric – double ballooning of drain)	Not easy (in pipe volumetric – downstream ballooning of drain)
Inspection chamber	"crude" connection to non man-entry sewer pipe	Preferably direct or difference with/without ballooning	Not easy (in pipe volumetric – ballooning of drain + double ballooning of sewer)	Not easy (in pipe volumetric – double ballooning of sewer)
Not sufficient access	Man-entry sewer or manhole, with sufficient drop height	Impossible, unless an inspection chamber is created	Impossible, unless an inspection chamber is created	Easy (i.e. volumetric)
Not sufficient access	"clean" connection to non man-entry sewer pipe	Impossible, unless an inspection chamber is created	Impossible, unless an inspection chamber is created	Not easy (in pipe volumetric – downstream ballooning of drain)
Not sufficient access	"crude" connection to non man-entry sewer pipe	Impossible, unless an inspection chamber is created	Impossible, unless an inspection chamber is created	Not easy (in pipe volumetric – double ballooning of sewer)

This knowledge can be obtained by different means:

(1) processing available databases;
(2) sampling paper files in the archives;
(3) field survey.

The first one of course relies on the availability of databases containing information about drains. Such databases do exist in many urban communities, either dedicated to drains for filing systematic checks of effluent separation in private properties (Joannis and Calvar, 2006), or as a part of a more general GIS, encompassing sewerage, buildings and so on. Some of them include a fairly complete description of private collection systems, but most are not designed specifically for that purpose. The important thing is that a process of data collection and management is already organised, and can be extended to collect other data. In the meanwhile, some information can be "guessed" from available data. Interesting clues can thus be derived from the following information:

- material and size of the sewer pipe
- connections built together with the sewer, or afterwards
- approximate year of setting up the drain, and contractor (the latter may be derived from the former, and both provide some clues on techniques which have been used).

What is true for databases also holds for paper files, with the difference that processing such data needs much more manpower. So a complete processing of all available data is not feasible. Random sampling of a few hundred drains can provide a fairly good view of the asset of drains in a city or a catchment, especially for the public part of drains (Joannis *et al.*, 2001). Several sampling processes can be used:

- direct random sampling in the list of customers;
- two step sampling: first a random sampling of streets, followed by a random sampling of drains in streets;
- stratified sampling, with suitable criteria (which are more or less the same as listed above: features of the sewer, year of construction, etc.). These criteria must be documented for the whole asset, so databases are again needed.

When paper files do not include sufficient data, especially for the private part of drains, the same kind of sampling process can be applied for field surveying a few hundreds of private properties. A suitable sampling process is sampling streets, and then sampling every drain in the streets, because it may be difficult to explain to customers why only a few (or even only one) of them are checked in the same street. Anyway no suitable criteria are readily available for stratification of the private part, so other possible sampling strategies imply more random sampling, provided a proper public information programme is scheduled.

According to the features which appear to be present in the asset, a decision must be made about coping with no suitable access (i.e. inspection chamber at the limit of private property) or creating accesses when needed.

METHODS FOR QUANTIFYING INFILTRATION AND EXFILTRATION FOR ONE PARTICULAR HC

Measuring infiltration in drains is not an easy task for two reasons:

- Infiltration flow-rates are low: a mean flowrate of only one litre/minute in individual drains would lead to a total of 50 or 100 m^3/day/km when cumulated over a catchment, which is quite significant compared with infiltration into sewer lines;
- Foul water is quite disturbing, as it displays sudden and strong variations of flow-rate, and conveys "big or gross solids".

Therefore, infiltration measurements should preferably be performed in conditions when no foul water is present. This can be achieved by operating during night, and/or by stopping drinking water supply during an hour or so. Both strategies of course imply a proper and early information of customers.

The so-called "volumetric" method, combining volume and time measurements, is quite efficient and flexible for measuring low flow rates. It comes in several versions:

- Basic vessel filling method;
- Tipping buckets;
- Pumping;
- In-pipe storage.

For the two first methods a drop in water level is needed, which can be created by a temporary weir. A drop in level of a few centimetres is sufficient for the basic method, provided a soft measuring vessel is used. Tipping buckets with a capacity of a several litres are available (or can be built on demand), and extend the range of use of the technology commonly used for raingauges. The capacity of buckets can be adapted to the flow rate and the available level drop.

Pumping is a means of conveying the flow to a place where it is easier to measure by a volumetric or other method. In some cases the same result can be obtained by gravity: a rubber stopper, or "balloon" is inserted at the downstream end of the drain, and a hose going through the balloon conveys the flow towards a more suitable place for measurement, for instance a manhole (see Chapter 4). Much time may be needed for the water level inside the drain to attain a sufficient level and for flow-rate to stabilize, but the process can be accelerated by adding a volume of water into the drain. Besides this application, ballooning is a general method for isolating a facility on which measurements are carried out, and may be used for separating contributions from the public and private parts, from the drain and the connection, or for handling rough connections (see Figures 3a to e).

In pipe-storage uses the house connection pipe itself as the measuring vessel. The downstream end is blocked by a balloon (see Figure 3d), and the level of water accumulating in the drain is monitored along with time. As infiltration is supposed to occur first in the lower part of the drain, no loss by exfiltration should be encountered. The accumulation of water in the drain modifies the infiltration flow-rate, and caution is required when interpreting the results.

METHODS FOR QUANTIFYING EXFILTRATION FROM ONE PARTICULAR HC

For exfiltration, very few methods may be suitable, because of the very short travel time between a house to a sewer, and the very low exfiltration flowrates which can be expected for a single drain, unless a huge defect is present. The only practical method which has been tested for HCs in the APUSS project is an in-pipe volumetric method (Kohout *et al.*, 2004), which implies a modification of hydraulic conditions, so it is rather a test for water-tightness than a measurement of actual exfiltration. For this method the drain is ballooned at its down stream end (or the connection is double-ballooned), then is filled up with water, and the water level is monitored along with time. This water-tightness testing method is also available in an air-pressure version, with an upstream and downstream ballooning (Miami Dade WSD, 2006).

SINGING TOMORROWS: SUGGESTIONS FOR OPERATING THE EXISTING ASSET AND DESIGNING NEW HC

Many problems encountered when surveying drains, either for performing the survey, or for dealing with the defects which have been highlighted by the survey, are a consequence of a poor control of drains, either new ones, or ones that have been in operation for years. Control can be obtained by two means:

- specification and check of new drains;
- checking and fixing drains in operation.

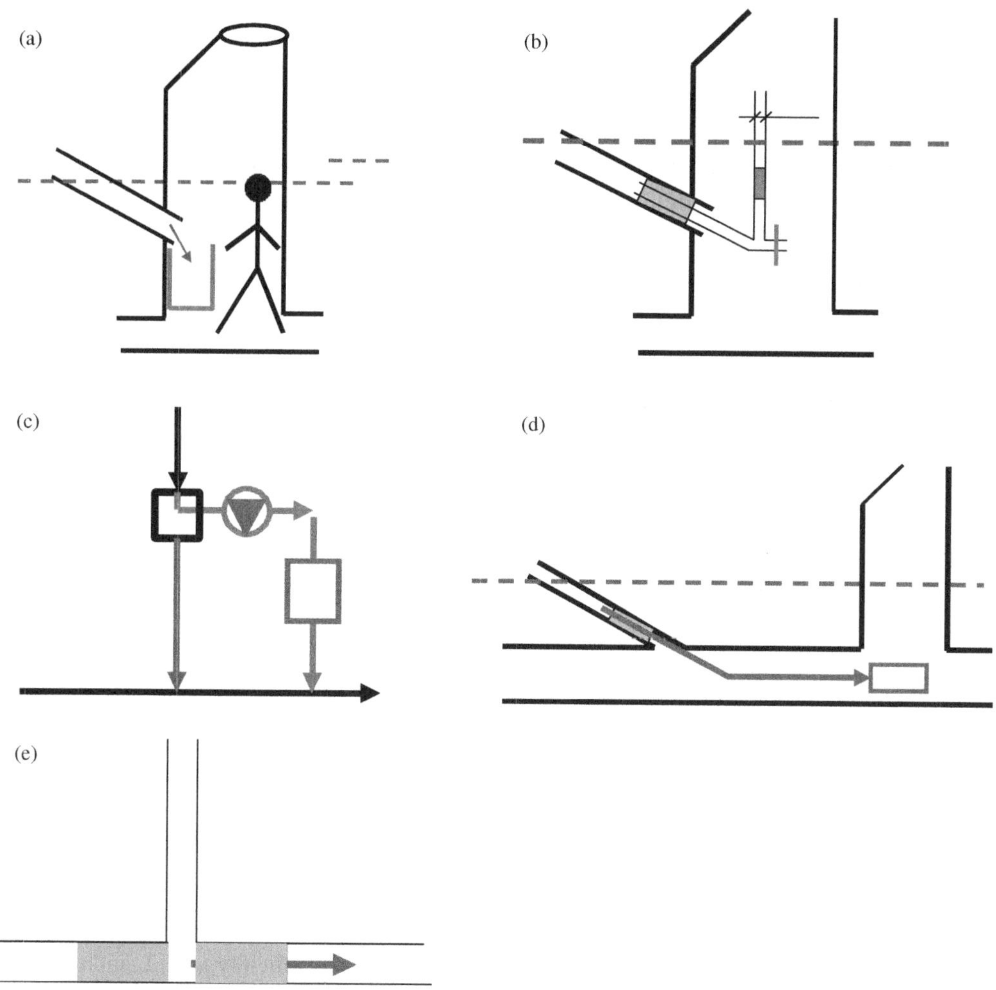

Figure 3: Methods for Infiltration Measurement. (a) Basic vessel filling volumetric method. (b) Volumetric method with in-pipe storing. (c) Using a pump for displacing measuring point (d) Ballooning for displacing measuring point. (e) Double ballooning at connection.

Specifying new drains

Relevant European regulations are EN752-1 to 752-8 "**Drain and sewer systems outside buildings**" (a new version of which is being currently discussed), and EN1610 "**Construction and testing of drains and sewers**" (European Committee for Standardisation, 1997). They both specify that drains should be water-tight, and set-up with standard fittings, so that infiltration and exfiltration problems should no longer be relevant for drains specified according to these standards.

However, no specification is useful unless its application is properly checked. Moreover new drains may lose their qualities as years go by. So additional specifications are needed for enabling an easy check of drains. These are not included in the standards quoted above, but national (MELT-MEF, 2004) or local regulations often address this issue. Such specifications include:

- straight lay-out of the drain, at least between the limit of a property and the sewer;
- inspection chambers at each bend and connection, with a maximum interval of 35 m;
- inspection chamber near the limit of a property, preferably on the public side of the limit, with sufficient dimensions for inserting a balloon or a camera (for instance Φ250 to Φ600 according to depth);
- separation of effluent inside private properties, even if they are served by a combined public sewer system.

Other features worth considering are:

- no drop or a limited drop height at the connection with a non-man entry sewer pipe;
- connection on manholes (with a sufficient drop, for instance 15–20 cm) rather than on non-man sewer pipes whenever possible;
- inspection chambers or similar fittings at every output of a building.

The specification for "straightness" should include the vicinity of the connection, which means that no bend should be allowed for adjusting the direction. This implies that tees are often preferred to wyes (that is connection with a 90° angle rather than 67°30). Together with the "no drop" condition, this may adversely affect the hydraulic efficiency of the connection, and trade-offs will be necessary.

In most cases there should be one drain and one connection for each building (or staircase in the case of collective housing buildings). Anyway, some buildings may have several separate drains, and some collectivities accept that one connection may be "shared" by several drains and buildings.

Checking and fixing drains

Technical means available for controlling drains include:

- direct observation of new facilities, before burying;
- direct observation of specific facilities, like inspection chambers, siphons, vents, septic tanks etc..
- dye or smoke testing, to check the correct routing of wastewater and rainwater into separate systems;
- CCTV inspection, either from an inspection chamber, or from the sewer pipe (by using specially equipped cameras);
- water or air tightness proofing, of sections isolated by the means of rubber stoppers.

The control of drains can be exerted in two different contexts. The first one is the control of new drains, and implies an immediate fixing. The second one relies on checking drains which have been set-up in the past, and fixing may be more difficult to obtain (Joannis *et al.*, 2000)

New drains

New drains are controlled in many urban communities throughout Europe including France; a rapid survey could easily identify more than a hundred of them. Controlling new laterals outside the properties and/or the installations inside properties can be performed in three different situations:

- extending public sewers;
- building new private sewerage;
- connecting new buildings to existing public sewers.

The building of public sewers provides an opportunity to build at the same time, laterals in public grounds, at least for each plot with existing buildings. This solution is both technically and economically sound, and enables a check of the water-tightness of new facilities including laterals. It can be extended by servicing plots with no existing building. In that case a consultation with the owners of each plot is needed, to implement the lateral at the right place and of course to obtain a payment for it.

The situation should be even better when a private sewer is being built, together with housing, because this may enable a control over the whole system, including drains inside properties. But the private contracting authority may have different priorities, and the public authority must have the means (staff and regulations) to check what is being done. Nowadays, city sewerage rules often include regulations for this situation.

The case of new connections on existing sewers is harder to control, because the check of the quality of facilities cannot be as thorough as for connections built together with the sewer, especially for water tightness. Moreover these operations are often very short, and not planned a long time ahead, so that an efficient scheduling of checks is difficult to achieve. Irrespective of this, they give an opportunity for checking the installations inside properties. Sewerage rules may indicate that the connection to public sewer will be carried out unless the private drain is in good condition, but once again this cannot be put into practice unless precise procedures are applied.

Drains in operation

Checking drains in operation is much more difficult than checking new drains. It is often limited to checking the connection of wastewater and rainwater to the proper pipes (in separate systems), and in some cases to checking the presence of some particular facilities (venting, inspection chamber...). Anyway a variety of situations can be turned into opportunities for such checks:

- renovation of roads and/or sewers;
- change of ownership;
- modification of buildings (either with a building permit, or a mere declaration);
- problems for operating the sewer (blockage, odour...);
- sewer evaluation study (they usually include checks of laterals on sample areas).

Many urban communities do use these opportunities, especially the first three of them. But an efficient control of laterals usually needs a more systematic action, aiming at controlling every lateral in a street, and every street in a city. This can be done only if a strong political will is present, and maintained over years. A recent survey (IFEN, 2006) shows that among a total number of 8,300 cities with a population over 1,000 people, 2,600 had checked their drains over their entire territory, and 2,100 more had performed such checks on a part of their territory. Some examples show that a complete check of every drain in a city can be achieved within a period of ten years or so, provided sufficient staff is allocated to this task; according to the size of the community, about 1 to 3 field teams, with a full time secretary may be needed.

Rather surprisingly, checking is usually sufficient to trigger fixing actions, notwithstanding their cost, which is charged to the owners of the premises. In some cases more stringent actions may be needed, such as doubling the sewerage fee. But the best incentive is completing checking with advice, for designing suitable solutions, choosing a competent craftsman, and in some case applying for financial support. This advice can be provided by the operator of the sewer system or by the staff of the community.

A key point in lateral control is the follow-up. For instance, "sample" checks performed in sewer evaluation studies often lack efficiency, because no proper organisation is provided to file the result or for harassing reluctant owners. On the other hand, a systematic control, with follow-ups, advice, counter checks etc. is really efficient, but also needs procedures for checking the evolution of compliant laterals. Compliancy is not guaranteed forever, and must be checked from time to time. The opportunities listed above may be a means to update the available information on laterals.

CONCLUSION

House connections, including their private extension, are a full part of sewer systems and should not be overlooked by sewer evaluation studies and rehabilitation programmes. Technical means and methods are presently available for checking individual house connections, and include effluent separation checks by smoke and dye testing, CCTV inspection, air or water testing, at least of the private part, and in some cases of the private part of house connections. Consistently applied over years, in connection with other sources of information and a proper data management system, they provide not only a precise assessment of each individual installation, but also an overview of the whole asset. Many communities and operators have put these techniques into practice nowadays.

Flow-rate measurements are presently not included in this toolbox. The APUSS project has showed that such measurements are feasible, but need much man-power, and must be completed by other investigations (CCTV) if the objective is the precise assessment of one particular drain. Then the question is about the possible integration of such measurements into regular management and operation processes.

It seems that the most interesting scale for analysing infiltration flowrates measurements in drains is the catchment. As the systematic assessment of each and every drain is not feasible, the key point is the sampling strategy and the uncertainties affecting the extrapolation process. Some suggestions have been made to address these issues, and partially tested, but more investigations are needed to validate these guidelines.

Flowrate measurement on one (or a few) drains would also provide very interesting data for research purposes, by giving experimental clues about rainfall induced infiltration but for operational purposes, the possible use of flowrate measurement at the scale of a particular drain is not quite clear. It may be interesting for the private parts when it shows that no significant infiltration is present, but exfiltration should also be checked. For the public part of house connections, flowrate measurement may be an interesting investigation means when CCTV inspection is not feasible.

Anyway, for both parts of the drain, exfiltration measurement or air testing may be considered as a sufficient means for investigating the condition of a house connection, and assessing its potential for actual infiltration or exfiltration. It can then be considered as a possible preliminary step before CCTV inspection (Miami Dade WSD, 2006). For exfiltration, the only really feasible method can be achieved by water testing, which is already a regular investigation means for individual drains. Its up-scaling to a whole catchment could be considered on the same basis as infiltration.

REFERENCES

EPA (1990) *Rainfall Induced Infiltration into Sewer Systems*. Report to Congress, EPA Report No. 430990005. Environmental Protection Agency, Washington DC, US.

European Committee for Standardisation (1995) *Drain and Sewer Systems Outside Buildings*. EN752-1. EU, Brussels, Belgium.

European Committee for Standardisation (1997) *Construction and Testing of Drains and Sewers*. EN1610. EU, Brussels, Belgium.

European Committee for Standardisation (2006) *Conditions of Drain and Sewer Systems Outside Buildings*. Visual Inspection Coding System. EN13508-2. EU, Brussels, Belgium.

Institut Francais de l'Environnement (IFEN) (2006) *L'assainissement en France en 1998 et 2001*. Dossier de l'Ifen No.03. ISSN 1776–8411. (*http://www.ifen.fr/publications/dossiers/dossier03.zip*).

Joannis, C., Poychicot, B., Le Gat, Y., Labbe, P. and Chantre, P. (2001) Une enquete sur les caracterisques des branchements domestique a l'assainissement de quelques villes francaises. *Techniques Sciences et Methodes*, **96**(11), 39–45.

Joannis, C., Blin, D., Guerin, J.-F., Magnier, P. and Salic, C. (2000) Vers la Maitrise des Branchements Domestiques a l'assainissement. *Techniques Sciences et Methodes*, **96**(11),

Joannis, C. and Calvar, J. (2006) Controlling Drains of Housing in France. *Proc. 2nd Conference on Sewer Operation & Maintenance*. 26–27 October 2006. Vienna, Austria. IWA London (To be published in *Water Practice and Technology* on-line journal).

Kohout, D., Princ, I. and Pollert, J. (2004) *Methodology for the Assessment of Infiltration/Exfiltration From House Connections in a Large Catchment.* APUSS Deliverable No. 3.2. (*http://www.insa-lyon.fr/Laboratoires/URGC-HU/apuss/APUSS%20deliverable%2003-2%20version%20January%202004.pdf*)

Miami-Dade Water & Sewer Department (2006) *Comprehensive Lateral Investigation Program*. Preliminary Draft Report. (*http://www.miamidade.gov/wasd/SSO/library/Lateral_Testing.pdf*)

MELT-MEF, Ministere de l'Equipement du Logement et des Transports; Ministere de l'Ecomomie et des Finances. (2004) *Marches Publics de Travaux; Cahier des Clauses Techniques Generales*. Fascicule No. 70; Ouvrages d'assainissement.

Chapter 4

Measuring and scale-up of infiltration and exfiltration in house connections

David Kohout, Ivan Princ and Jaraslov Pollert

INTRODUCTION

Infiltration and exfiltration (I/E) from leaky house connections (HCs) can be substantial and estimates of up to 40% loss rates have been quoted in the literature (Hayward, 2002), although they are largely unsubstantiated by direct field measurements. Measurements undertaken in the context of the APUSS project in the city of Prague for a 0.25 ha residential housing development in the Troja district, indicated an average exfiltration loss of 0.23 l/s which expressed in terms of unit pipe length is equivalent to 3.5×10^{-3} l/s per metre length (Kahout *et al.*, 2003). This would represent some 75% of the connecting pipe hydraulic capacity and would suggest a considerable potential for leakage loss from such HC sources. Such high exfiltration rates might not be typical of the large majority of house connections in western Europe, although anecdotal evidence from the UK water industry might suggest that such sources can provide significant leakages to the surrounding soil and groundwater environment, particularly via defects at the junction between the house connection pipe and the main public sewer pipe. Figure 1a shows a new and properly fitted small diameter HC joining the public sewer line outside the property curtilage, whilst Figure 1b shows a severely damaged connection (ATV, 1999; grading 2, serious defect) which is allowing the majority of the household wastewater flow to escape directly to ground.

Such HCs or "lateral drains" as they are often termed, include the pipe from the house to the main sewage collection pipe as illustrated in Figure 1a, and legally fall within the responsibility of the house owner (see Chapter 3). General practice is to use a Y connector pipe (Figure 1a) rather than a T form, as they offer least resistance to flow and possible clogging. The best practice tends to adopt a one-eigth bend with the Y branch (Princ and Kohout, 2003). The basic methodological approach for I/E HC volumetric measurement is illustrated in Figure 2 which shows an external measuring tank with the monitored pipe section being filled with hydrant tank water to measure exfiltration losses or ballooned to monitor infiltration. An air compressor is used to inflate the balloons in order to achieve an effective and water-tight sealing, with fixed steel hawser wire used to remove them after the test run. In the Prague tests a combination of MULLER and IPONOR balloons were used to provide the pipe sealing.

HCs comprise a highly sensitive component of the wastewater sewer system, particularly given that they frequently receive less attention during construction as well as under operational conditions. Under normal operating conditions, house service laterals having gradients typically less than 2%–3% possess low hydraulic heads which should theoretically minimise exfiltration losses except where severe damage occurs as seen in Figure 1b. However, studies in the United States have suggested that the common parallel and adjacent alignment of surface stormwater and wastewater pipes within property boundaries can lead to cross-inflow from the stormwater pipe, thus increasing pressure and backup in the sanitary pipe (Amick and Burgess, 2000). This situation not only can generate exfiltration loss from the

 Assessing infiltration and exfiltration on the Performance of Urban Sewer Systems (APUSS). Edited by Bryan Ellis and Jean-Luc Bertrand-Krajewski. ISBN: 9781843391494. Published by IWA Publishing, London, UK.

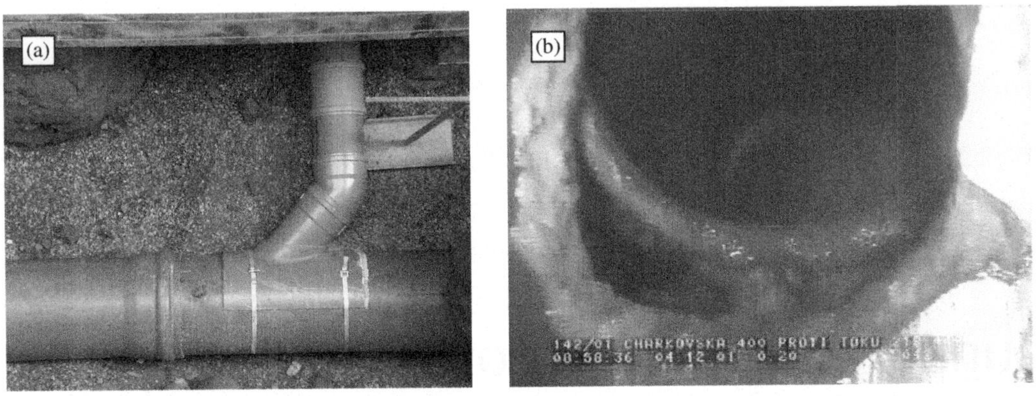

Figure 1: House Connections to the Public Sewer Line.

wastewater pipe, but can also lead to basement flooding from backup in the pipe. Typically, 55%–60% of sewage flooding of properties in the UK is attributable to "other causes" which include basement flooding from house connection backups (Ellis *et al.*, 2004). A substantial proportion of such flooding is associated with older pre-1936 properties and can largely be related to structural defects in the HC pipe as determined from dye tracer studies and surcharging tests.

Leakage into and out of small diameter HCs is likely to occur where:

- the pipe is damaged for example, at joints, at connection to the main sewer line or at other fracture points in the pipe,
- the defect/damage has a transition zone and underlying soil material which is permeable in character,
- the level of the surrounding water table (or inflowing lateral rainwater) lies above the defect/damage leading to groundwater infiltration and infiltration inflow (I/I) potential,
- the groundwater level is below the level of the piping providing potential for exfiltration loss.

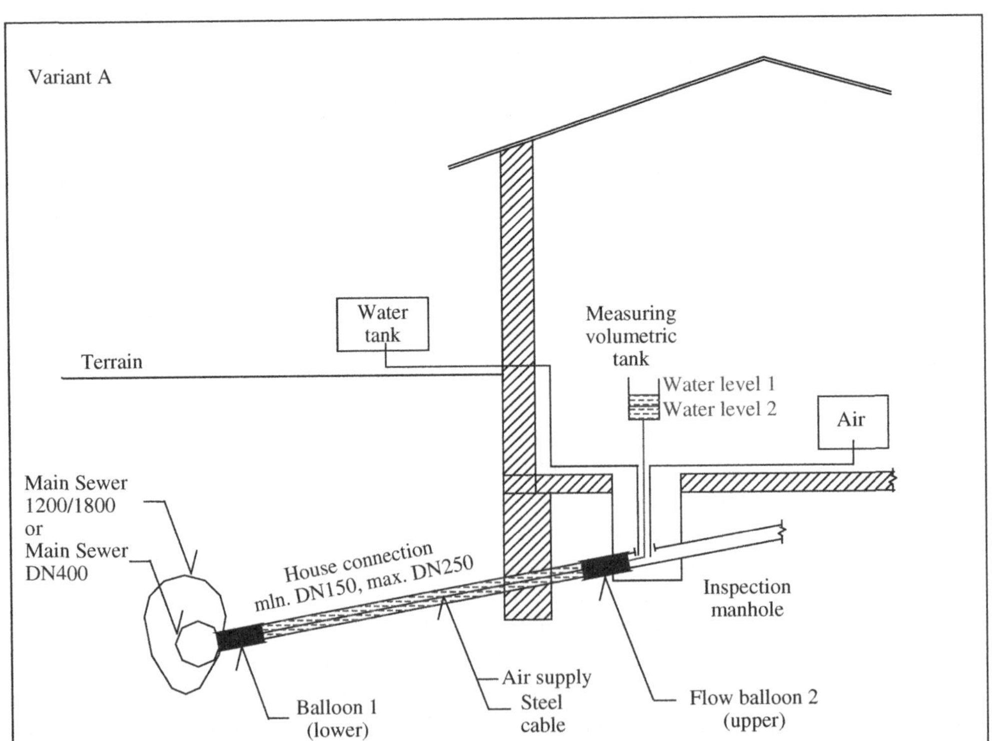

Figure 2: Layout of the HC pipe for I/E monitoring.

The following sections describe the measurement approaches that have been adopted and implemented to determine both infiltration gains and exfiltration losses in HCs that have been undertaken within the context of the APUSS project and carried out within the city of Prague.

MEASUREMENT APPROACHES

The APUSS tracer measurement methods described in Chapter 2 are applicable to infiltration and exfiltration in public sewers pipes, but it is well known that house connections (HC) are also key components in I/E phenomena and are largely in private ownership (see Chapter 3). Given difficulties of access as well as problems related to the small size (<160 mm) typical of HC pipes, the application of tracer methods was not considered to be appropriate and standard volumetric methods have been tested and applied on experimental sites to measure I/E losses and gains. Measurements have been performed on house connections within selected catchments in the city of Prague, in order to obtain a sufficiently large data set from different types of built-up area, sewer pipe ages, geology and hydrology conditions. It is estimated that there are nearly 51,000 HCs in the city of Prague totalling 518 km in length, constituting 21% of the total sewer system, and having an average age of 58 years. Given this multiplicity of HCs it was clearly not possible in practice to measure I/E in all of them. Thus the results obtained from selected sampled sub-catchments of differing urban habitat type were used as a basis to develop a matrix procedure for extrapolating individual HC results to a catchment-wide scale based on similarity characteristics.

INFILTRATION

Infiltration into HCs occurs principally as a result of one or more of the following conditions:

- direct pipe damage,
- the level of surrounding water (groundwater or inflow percolation of rainwater) lies above the damaged pipe section,
- a damaged pipe section is surrounded by permeable soil or a soil opening.

The evaluation of the amount of water infiltrated into a HC pipe includes i) the identification of the defect and ii) the measurement itself and the application of a particular measurement method depends on the arrangement of the HC to the sewer system. Typical connection arrangements of HCs to main sewers are as follows:

- a HC connection to a manhole,
- a HC connection to a man-entry sewer,
- a HC connection to a non man-entry sewer where the sewer diameter is smaller than 0.8 m.

In the first two cases, a proper volumetric measurement is possible but in the last case, it is impossible to measure individually each house connection although it is possible to measure a group of HCs, which can be essentially considered to make up a small sub-catchment. Two infiltration measurement methods have been tested:

- one is based on accurate flow monitoring over a few days at the outlet of a HC or in a manhole connecting a few houses to the public sewer. This allows estimation of the infiltration rate as equivalent to the minimum night flow. As the discharges to be measured are very small, measuring weir and water level sensors have to be installed on site as well as appropriate collector tanks.
- an alternative method is based on chemical analyses of samples collected during the night for comparison with the minimal discharges, with the chemical substances indicating the dilution of wastewater by infiltrating clean groundwater. This result is compared with values of typical polluted wastewater effluent obtained during daylight hours. Appropriate indicators of such "pollution" are typical wastewater quality indicators such as COD ($mg \cdot L^{-1}$), N-NH4^{+} ($mg \cdot L^{-1}$), temperature (°C) and conductivity ($S \cdot m^{-1}$).

EXFILTRATION

The methodology of exfiltration measurement is based on standard water tightness tests of sewers and is thus not a strict methodology for the determination of volumetric exfiltration loss. The value of leakage must be compared to the permissible values as stated in European or national technical standards (e.g. the German ATV Standard M149; ATV, 1999).

The leakage of sewage water from a HC suffering from a particular damage occurs only upon the fulfilment of the following conditions as previously stated:

- the pipe is damaged (cracks, broken joints, holes, ...),
- the ground water level (GWL) is lower than the level of the pipe or when the hydrostatic pressure in the pipe exceeds the groundwater pressure,
- the pipe is placed in a permeable environment.

The exfiltration assessment includes i) the evaluation of pipe condition as well as catchment conditions (i.e. relation between type of damage and main controlling factors, soil environment, age, …), and ii) the measurement itself. Two exfiltration measurement methods have been tested: the pressure method and the free surface method. The pressure method consists of blocking the HC pipe with sealing balloons, filling the isolated pipe section with water, and measuring the decrease of the water level as a function of time, from which the exfiltration rate is calculated (Figure 2). The free surface method consists of discharging a given volume of water into the HC pipe, collecting the conveyed volume at a fixed distance downstream of the HC pipe and calculating the exfiltration rate from the comparison of both volumes. The first method overestimates exfiltration because of the pressure, while the second one requires ease of access to the public sewer downstream of the HC, which is often not possible. A CCTV inspection of the connector pipes was made prior to the test runs which were conducted over a period of 30 to 60 minutes and repeated at least once within 2 to 12 hours in order to evaluate reproducibility of the result.

The urban catchment can be characterised by the relative exfiltration rates obtained from the field study and by values obtained from the interactive matrix (see Table 1) which is completed from CCTV inspection records to evaluate the structural state of the HC and its context. The exfiltration from a small reference catchment (i.e. a few house connections) is generally a function of the discharge Q and of the factor w which reflects the catchment evaluation given from the interactive matrix:

$$\mathrm{Exf}_{\mathrm{catchment}} = \mathrm{f(Q, w)} \tag{1}$$

EXPERIMENTAL SITES

Lazne Bohdanee

This experimental site is located in the Lazne spa city which lies in the Polabska lowlands near to the city of Pardubice. Given the close proximity (<1.0 m) of the water table to the ground surface, the adjacency of the River Labe and a permeable sub-soil whose hydraulic conductivity is 1×10^{-4} m/s, a high infiltration potential was expected. The site consists of a small reference catchment made up of 22 earthen 300 mm diameter HC pipes comprising a total length of 232.5 m and a larger sub-catchment of 54 HCs consisting of similar pipes but with a total 593 m length. During drainage improvements in 2000–2001, parasitic water entering the drainage systems via shafts and land drains was eliminated which resulted in a local rise in the groundwater level. Two measurement campaigns were undertaken during 2004 using a Parshall flume and ultrasonic level indicators installed in the outlet manhole to the smaller reference catchment. Figure 3 show a view looking down into the inspection manhole with the Parshall flume installed together with the water level meter. Average rates of exfiltration were determined as varying between 5%–6% during the winter increasing to nearly 11% in the summer period.

Prague-Petriny

This 0.5 ha experimental site monitored a 6.1 m length of ceramic, 200 mm diameter HC for a block of flats (having 30 inhabitants) constructed in 1960. Pipe damage established from CCTV inspection showed that 46% of the joints to fall into ATV class 4 or 5 (small/minor damage). The plan layout of the experimental site is shown in Figure 3 with a view of the drainage system shown in Figure 5. Exfiltration was determined by blocking a section of the pipe between two manholes, filling the section with hydrant water to an overpressure of 0.01MPa, and then monitoring the decrease in water level over time (Figure 4). The average exfiltration rate was determined as being 0.0038 l/s per metre length, equivalent to 0.004 l/s per joint. The exfiltration rate relative per 10 m^2 wetted surface was 2 l/minute as compared to the permitted Czech standard of 0.16 l/minute.

Figure 6 shows the temporal trends of the exfiltration rates based on variation in the water level in the pipe for differing gap sizes and in comparison to the base night-time level which averages 0.004 l/s. For the 1 cm joint, the average exfiltration rate over the test run period was 0.067 l/s, and was 0.187 l/s for the 2 cm joint and 0.285 l/s for the 5 cm joint. A laboratory test rig on a PVC 250 mm pipe simulating the field conditions, but without trench backfill, produced very similar exfiltration outcomes.

METHOD IMPROVEMENT

The assessment of exfiltration from house connections could be improved with tracers following the approach of Bechteler and Gunthert (2001). The current state-of-the-art is to check watertightness through pressure testing, which is frequently criticised for producing biased results. Leakage might be overestimated because:

(i) pressurised conditions, which do not normally occur in house connections, activate defects in the upper part of the pipe and at "weak" connection points.

Figure 3: Measuring equipment in HC inspection manhole.

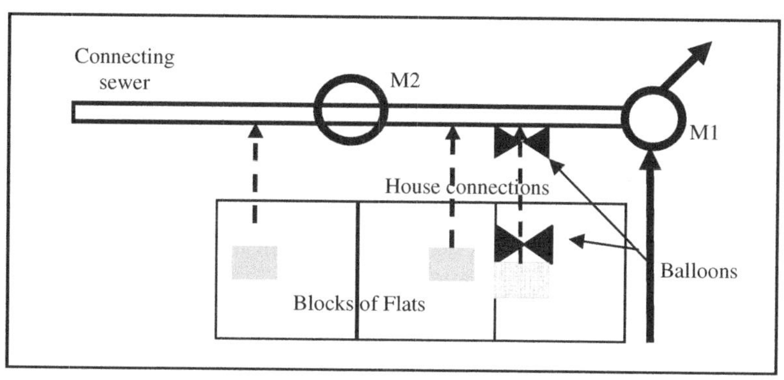

Figure 4: Plan layout of HC monitoring approach.

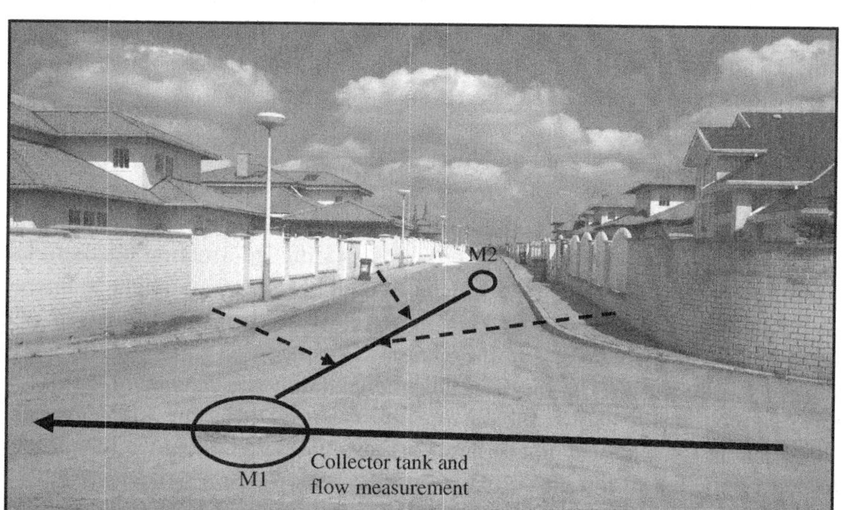

Figure 5: HC monitoring in Petriny.

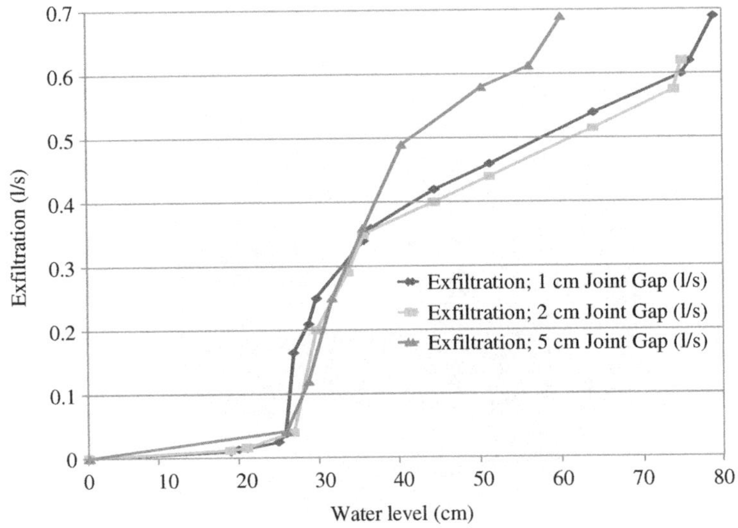

Figure 6: HC Exfiltration in Petriny for varying joint damage.

(ii) the applied pressure as well as the injection of the testing devices themselves can cause damage to weak pipes and

(iii) eventual self-sealing layers might be destroyed.

A more reliable estimate of leakage might be obtained if the house connection is blocked and a pump installed at the downstream end, which is then used to circulate an exact volume of water in the house connection. Adding tracers to the circulating water, which passes through the connection in free surface flow before it is pumped back, would allow any exfiltration losses to be identified and quantified given that such losses (or gains) occur in the same pipe reach.

A further alternative would be to copy a technique from pipeline monitoring. In oil pipeline investigations, leak detection is often performed by first pumping radioactive fluids through the pipe and later passing a detector through the line when oil is being transported again. At locations where fluid has left the pipe, radioactivity is detected by the sensor, which allows for a location of the leaks, which could be done on an automatic basis given full GIS support. If an environmental tracer was available for such application e.g. such as magnetically coated nano-particles, a similar procedure could be adopted for leakage detection in sewers. Firstly, the quantitative flow information is gained from application of the QUEST tracer methods and on completion of these tests, a detector is passed down the pipe reach to identify the exact locations of the leakage.

SCALING-UP METHOD

For obvious practical reasons, it is not possible to measure all HCs in a given sub-catchment where large numbers of properties are involved (>10–12). However, it has been suggested (Princ and Kohout, 2003) that the I/E rate for all HCs within a larger catchment can be assessed by extrapolation of measurement date obtained within a smaller reference catchment i.e. by using specifically selected reference HCs which can be regarded as being fully representative of the larger residential lot types within the larger catchment.

The procedure used in the scaling-up methodology is as follows:

(1) detailed evaluation of the pipe state for the reference catchment: including type and classification of damages and full evaluation of the catchment conditions (e.g. groundwater level, backfill type and condition, soil permeability, traffic type/condition etc.). This provides specific information on the relationship between pipe damage and controlling urban habitat factors which can then be used to complete an interactive matrix (Table 1). This approach is similar to that adopted by Davies *et al.* (2001) in their matrix analysis of the factors influencing the development of sewer pipe damages in the Thames Water region of Greater London.

(2) detailed I/E measurements within HC pipes in the smaller reference catchment.

(3) detailed pipe evaluation in the larger sub-catchment covering the same range of control factors as in the reference sub-catchment i.e. classification of damages, evaluation of catchment conditions (ground water level, backfill, . . .), and completion of the interactive matrix.

(4) Extrapolation of results from the reference catchment to the larger sub-catchment, by means of the following equations:

Table 1: Structure of the HC interactive matrix.

Exfiltration		Damages: Leakage joints			Blocking				Corrosion		Cracks			Others		Factors	
		Bad joints	Connections m. sewer	Connections manhole	Sediments	Incrustation	Roots	Overlap pipe to m sew.	Outside	Inside	Lengthwise	Crosswise	Radial	Abrasion	Deformation	Sum	Rank
Factors	High permeability	5	5	5	2	1	2	1	3	1	5	5	5	1	3	3.1	9
	Med permeability	3	3	3	1	1	2	1	2	1	3	3	3	1	1	2.0	1
	Material	4	4	4	1	1	2	1	1	1	3	3	1	1	4	2.2	2
	Area	3	3	3	3	2	5	1	2	2	2	2	2	2	3	2.5	5
	Rain	5	5	5	5	5	3	5	1	1	3	3	2	3	1	3.4	11
	Life time	5	5	5	3	5	5	2	3	3	5	5	5	3	5	4.2	13
	Maintainance	3	3	3	3	3	2	3	3	3	5	5	5	3	1	3.2	10
	Implemantation	5	5	5	1	1	–	3	1	–	2	4	1	1	5	2.8	8
	Aggressive background	3	3	3	1	1	1	1	5	–	3	3	3	1	1	2.2	3
	Aggressive water	2	2	2	2	5	1	1	1	5	3	3	3	3	1	2.4	4
	Heavy traffic	5	5	5	5	5	5	2	1	1	5	5	1	1	5	2.8	7
	Pulsating GWL	3	3	3	3	3	3	3	3	1	2	5	1	1	5	2.5	5
	Ege	4	4	4	2	4	3	1	4	4	4	4	4	4	4	3.6	12
Damages	Sum	3.8	3.8	3.8	2.0	2.4	2.5	1.9	2.3	2.1	3.5	3.8	2.8	1.9	3.0		
	Rank	11	11	11	3	6	7	1	5	4	10	11	8	1	9		

Significant Values

Influencing Factor on Damages:

0% 25% 50% 75% 100%

$$I_{LC} = I_{RC} \cdot \frac{\sum L_{LCi}}{\sum L_{RCi}} \cdot \frac{(A - w_{LC})}{(A - w_{RC})} \tag{2}$$

$$E_{LC} = E_{RC} \cdot \frac{\sum L_{LCi}}{\sum L_{RCi}} \cdot \frac{(A - w_{LC})}{(A - w_{RC})} \tag{3}$$

where:

I_{LC}, resp. ELC	infiltration/exfiltration in larger catchment
I_{RC}, resp. ELC	infiltration/exfiltration in reference sub-catchment
$\sum L_{RCi}$	total length of HC pipe in reference catchment
$\sum L_{LCi}$	total length of HC pipe in large catchment

wRC reference sub-catchment evaluation–based on evaluation using the interactive matrix
wLC larger catchment evaluation – based on evaluation using the interactive matrix.
A value expressing the best pipe condition or pipe without any damage according to standards used for classification of damages under optic CCTV inspection.

For the Lazne Bohdanee catchment, the measurement campaign determined an average infiltration rate of 0.308 l/s and an exfiltration rate of 0.0044 l/s for a single 150 mm HC within the residential villa estate. For the purposes of scaling-up this single HC value to the larger estate sub-catchment of 9 HCs having similar 150 mm pipes with an estimated ATV damage level of the same grading (i.e. A = 5), the following values of the equation parameters were:

- reference sub-catchment evaluation from the matrix of damage, $W_{RC} = 3.8$
- larger catchment damage evaluation from the matrix, $W_{LC} = 3.6$
- total length of HC pipe in the reference sub-catchment, $L_{RC} = 4.1$ m
- total length of HC pipes in the larger catchment, $L_{LC} = 40.0$ m
- exfiltration rate in the reference sub-catchment, $E_{RC} = 0.0044$ l/s

Total exfiltration rate from the larger catchment based on eq. (3) is therefore:

$E_{LC} = 0.0044.\ [\{40/4.1\}.\ (5–3.6)/(5–3.8)] = 0.05$ l/s which reflects a 10% and 16% loss rate from the reference catchment and larger sub-catchment respectively.

CONCLUSIONS

When estimating I/E in HCs, the most important thing is to correctly classify pipe defects and to evaluate the catchment conditions and qualities, particularily if the intention is to use a reference pipe (or pipes) as a basis for extrapolating up to a larger sub-catchment scale.

The pressure testing method may be limited by the water volumes available for the tested HC section and access to a public hydrant supply is strongly recommended. However, it is nearly always impossible to accurately evaluate the amount of exfiltration in HCs having large-scale defects, e.g. a missing pipe segment or a very large gap between two pipes, etc. However, these cases are extreme and pipes having such large defects should be excluded from testing.

Separate measurement of both infiltration and exfiltration from an individual HC is possible only in especially favourable cases. For this reason, it is suggested that several HCs of similar character, age and housing type should be sought for I/E testing to facilitate differing urban habitat groupings. In this case, connecting sewer groups having similar characteristics to an individually tested HC can be reasonably considered to be equivalent to the studied house connection. The layout of the measurement procedure thus depends strongly on local conditions mainly related to house type and age. For scaling-up, it is important to correctly evaluate both the pipe condition in the reference catchment and all pipe conditions within the larger sub-catchment.

Transferability of results between different sub-catchments is possible only under the following conditions:

- where there is a close reference between individual house types within larger sub-catchments in the same city.
- where the reference and larger catchments are structurally related or show similar characteristics as demonstrated from an inspection of the interactive matrix.

REFERENCES

Amick, R.S. and Burgess, E.H. (2000) *Exfiltration in Sewer Systems.* Report EPA/600/R-01/034. US Environment Protection Agency, National Risk Management Research Laboratory, Cincinatti, Ohio, US.

ATV (1999) *Recording, Classification and Assessment of the Condition of Drainage Systems Outside Buildings.* Advisory Leaflet M149. DWA, Hennef, Germany.

Bechteler, W. and Gunthert, F.W. (2001) Zerstorungsfreie dichthetsprufurg von grundstick-sentwasserungleitungen. Vol. 7. Mitteilungen des Inst. fur Wasserwesen. University of Munich, Oldenbourg Verlag Munchen, Munich. Germany. (In German).

Davies, J.P., Clarke, B.A., Whiter, J.T. and Cunningham, R.J. (2001) Factors influencing the structural deterioration and collapse of rigid sewer pipes. *Urban Water J.*, **3**, 73–89.

Ellis, J.B., Revitt, D.M., Blackwood, D.J. and Gilmour, D. (2004) Leaky sewers: Assessing the hydrology and impact of exfiltration in urban sewers. In *Proc. Hydrology: Science & Practice for the 21st Century.* Imperial College, London. 12–16 July 2004. CD-ROM. Vol. II. British Hydrological Society, Inst. Civil Engineers, London, UK, pp. 266–271.

Hayward, P. (2002) Sewerage: Fairytale spending. *Water & Waste Treatment*, **35**(8), 30–31.

Kahout, D., Princ, I. and Pollert, J. (2003) *Measurement of House Connection (HC) Exfiltration.* Report, May 2003. CTU Lermo, Prague. EU 5th Framework APUSS Project.

Princ, I. and Kohout, D. (2003) Assessing infiltration to sewer systems in the locality of Lazne Bohdanee. In: *Proc. 3rd Int. Conf. Hydrosphere 2004.* Brno, Czech Republic, pp. 275–278.

Chapter 5

Sediment impacts and exfiltration

J. Bryan Ellis and D. Michael Revitt

INTRODUCTION

The behaviour of chemical tracers may be strongly affected by the presence of both inorganic and organic solids through adsorption, flow dispersion and turbulence as well as "tailing" effects. Thc accurate quantification of such tracer adsorption effects is clearly difficult under operational sewer conditions. Baseline trends and processes need to be initially established under controlled laboratory conditions prior to scale-up to the field situation. A principal objective of the work described here was to determine how the physico-biochemical characteristics and properties of sewer sediments influence the efficiency of tracer operation and to define experimental interferences. The application of controlled batch reactor and experimental rig approaches enables a better definition of the baseline effect of sewer sediments on the robustness and reproducibility of the tracer methodology under varying flow, sediment and sewer defect conditions.

The application of an experimental sewer rig also enables a closer examination of the self-sealing effects of contained sewage solids on sewer defects as well as achieving better definitions of the mechanisms of the exfiltration process. As the APUSS project commenced, there was a generally increasing interest in the development and application of experimental rigs for the testing of live wastewater exfiltration rates (Dohmann *et al.*, 1999; Forschergruppe-Kanalleckagen, 2002; Vollertsen *et al.*, 2002). This interest reflected the national concerns for sewer integrity contained in Article 4 on diffuse pollution and in EU Standard EN 752-2 within the EU Water Framework Directive (WFD) as well as a growing need in the European and US water industry (Selvakumar *et al.*, 2001) for confirmatory methods to evaluate sewer losses and monitor sewer performance.

The following sections describe the results of batch reactor testing of the adsorption characteristics of sewage solids and their interaction with a variety of tracer substances under controlled laboratory conditions as well as the outcomes from preliminary testing in a sewer test rig and from field testing. The patterns and temporal trends of exfiltration losses for varying sewer defects and geometries have also been tested under controlled clear water conditions in an experimental test rig for differing types of artificial and natural sediment types. A series of experiments on live wastewater flows in an experimental sewer rig was also undertaken to enable scale-up to field conditions in order to examine the self-sealing mechanisms of conveyed sewage solids and their interactions with potential tracer substances.

TRACER ADSORPTION AND INTERACTION

A series of laboratory batch tests were carried out to investigate the interactions of fluorescent and non-fluorescent tracers with a range of in-pipe sewer sediments and fresh raw sewage samples in order to assess the impact of sewage solids and adsorption effects under sewer flow conditions. Methodological protocols for laboratory batch rector

Edited by Bryan Ellis and Jean-Luc Bertrand-Krajewski. ISBN: 9781843391494. Published by IWA Publishing, London, UK.

determinations of tracer adsorption to solids were developed (Deliverable D2.2, *Methodological Protocols for Batch Reactor Testing of Tracer Adsorption. www.insa-lyon.fr/Laboratoires/URGC-HU/apuss*) for a range of standard sewer sediment types as sampled from a major trunk sewer and representing the principal solid groups found in sewers (Types A – E as categorised by Crabtree *et al.*, 1991). The in-pipe sewer samples included non-cohesive granular solids from a sewer invert (Type A), organic pipe-wall biofilm slimes (Type D) and fine-grained organic/mineral mixtures typical of surface slurry in storage tanks and sediment chambers (Type E). In addition, suspended solids from raw sewage effluent and highway surface sediments were also sampled and tested in the batch reactor experiments. Both lithium and Rhodamine WT adsorption tests were conducted over extended time periods to achieve equilibrium and to determine adsorption characteristics as well as any effects of turbidity on the measurement signal employed.

Little evidence of lithium adsorption (<2% over 24 hours) was observed for any of the tested sediments, including sewer slimes and algal mats. However, the overall impact over the same time period for Rhodamine WT was an adsorption of up to 10% as illustrated in Figure 1 for the sewer slime adsorption curves. Tests on raw and settled sewage samples showed no significant impact on lithium tracer concentrations according to whether the sewage was fully mixed due to stirring (to simulate turbulent sewer flow) or was allowed to settle under quiescent conditions (simulating sewer backwater effects). Similar results were observed when bromide was used as a tracer but Rhodamine WT demonstrated more complex interactions in which the aqueous tracer concentration continued to decrease during the stirring process resulting in a measured reduction in the fluorescent signal approaching 10%.

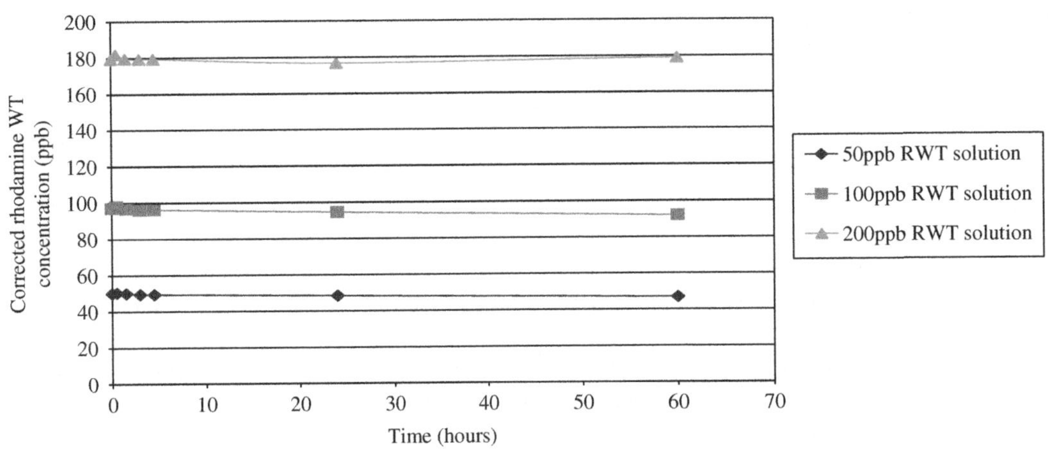

Figure 1: Temporal variation in rhodamine WT concentrations for sewer slimes.

A clear linear relationship between turbidity and an increase in the fluorescent signal was detected by the fluorescent probe (a Turner design SCUFA probe) as can be seen from Figure 2 and so turbidity corrections were applied to compensate for the effect of turbulence. The use of other fluorescent tracers (Sulphorhodamine B and Sodium Fluorescein) with raw sewage suggests that the former is more susceptible to adsorption (25%–35%) but that Sodium Fluorescein is less effected by adsorption with some sewage samples showing no impacts. The adsorption characteristics of Rhodamine WT were accurately modelled by Freundlich isotherms suggesting that removal occurs by heterogeneous surface binding rather than as a result of a surface monolayer formation.

In-sewer tests using the fluorescent probe, and following the QUEST procedure outlined in Chapter 2, at two locations in N London demonstrated the practicality of the instrumental procedure for real-time measurement over periods of up to 30 minutes. The on-line sewer tests indicated that the Rhodamine WT tracer provided a direct, high sensitivity measurement method having a low detection limit and high measurement frequency (at 1–30 second intervals). Both turbidity and temperature compensation could be satisfactorily achieved and few mixing difficulties were encountered in the turbulent sewage flows. Given the near-zero natural background of the tracer in sewage effluent, low tracer concentrations can be employed and adsorption characteristics can be pre-determined.

In addition, the indicator and reference signals can be clearly deciphered and provide a clear baseline for time series analysis.

However, the experimental work would suggest limitations to the use of fluorescent tracers as quantitative indicators of exfiltration in long sewer runs having extended travel times within the sewer network. Thus the technique will be most reliably applied for investigations over individual short reaches of sewer length. In contrast, lithium and bromide demonstrated negligible adsorption effects although, in both cases, considerable laboratory effort was required

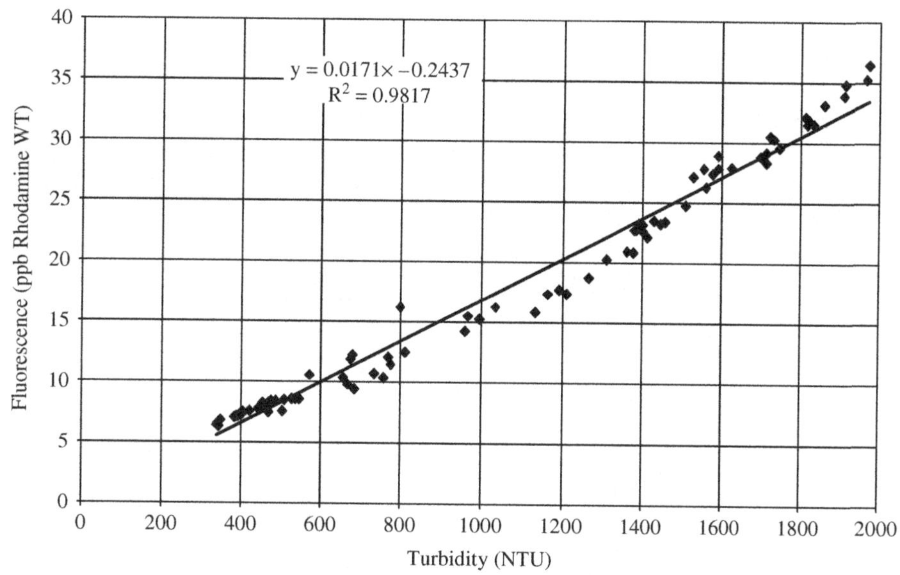

Figure 2: Calibration graph for rhodamine WT fluorescence with turbidity for raw sewage.

to overcome the significant analytical interference. In the case of lithium this was due to the high level of sodium in sewage. Further detail of these sewer tracer tests are provided in a later section of this chapter.

THE EFFECT OF SEWER SOLIDS AND SEDIMENTS ON EXFILTRATION

Clean water laboratory experiments

Effects of joint size and flow head variation

A re-circulating laboratory experimental sewer rig using clean water passing through a 150 mm diameter perspex pipe was established to evaluate exfiltration losses under varying joint openings (1.0 to 3.5 mm with varying geometries) and flow conditions (0.01–7.8 l/s). Defects were simulated by placing inserts between the joint flanges with the inserts cut to different thickness and geometries. Three geometries were considered: a 10 mm hole at invert level, a radial opening extending halfway round the pipe from invert to soffit in a vertical configuration in the pipe and a radial opening extending halfway around the pipe from the invert to the middle of the pipe wall on each side (half-horizontal mode configuration).

Figure 3a shows an overall view of the test rig with additional detail illustrating the simulated gravel trench infill (Figure 3b) and the outflow box (Figure 3c). Exfiltration gap geometries varied between a basal hole (1 mm gap width ×10 mm length) and half-horizontal and half-vertical slots (to simulate cracks in the sewer wall). In addition to free draining tests, runs were also undertake with a 1 m × 0.65 m × 0.85 m trench backfill surrounding the joint/gap opening with experiments conducted under both dry gravel and fully saturated gravel conditions. In the latter case, the "groundwater" level in the gravel backfill was maintained at surface level in the simulation trench.

The test outcomes for free draining conditions confirm a general initial linear increase in exfiltration rate against both static and pressure head for all 1 mm gap geometry experiments as well as for all 10 mm hole experiments at the 2 mm and 2.5 mm gap sizes. Figure 4 shows typical curves obtained under varying head and flow levels for clean-water runs in respect of exfiltration rates determined for 1.44 mm and 3.65 mm gaps in the half-horizontal configuration. Whilst the lower joint sizes tend to exhibit linear exfiltration rates with increasing head, there is a tendency for an exponential increase in exfiltration rates for both the larger gap sizes, with serious failure out of the joint occurring at high head as half-full pipe flow is exceeded. This exponential trend of increasing exfiltration rates with head and flow rate appear to be confirmed from independent studies carried out on the testing of storm sewers in Aachen (Dohmann *et al.*, 1999). Similar results have been experimentally recorded by Vollertsen and Hvitved-Jacobsen (2003) using sewer rig studies on wastewater effluent. At a 40 mm head flow, the leakage rate for the 3.65 mm joint was recorded in the test rig as being some three times greater than for the 1.44 mm joint. Under dry weather flow (DWF) conditions (assumed to be at a level of 10%–15% pipe-full flow; Q_{10-15}), the percentage flow loss due to exfiltration increased from 2.8% for the 1.44 mm gap to 4.0% for the 3.65 mm gap.

Figure 3: Views of the experimental sewer rig, trench backfill and outflow box.

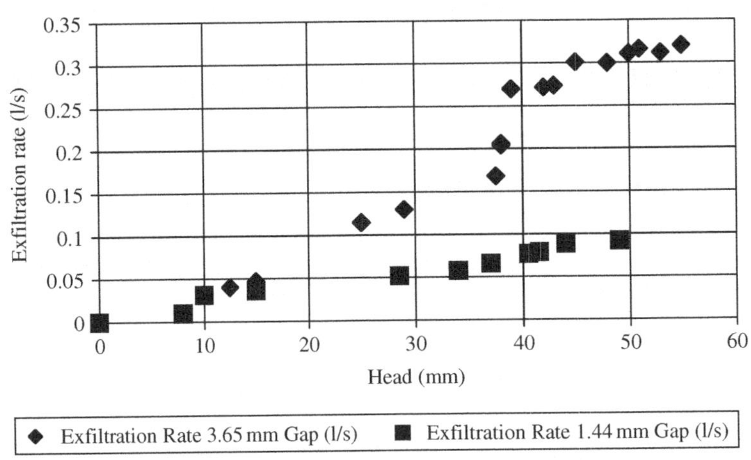

Figure 4: Exfiltration rates for the free-draining 1.44 mm and 3.65 mm joints with half-horizontal gap configuration.

Experiments were also conducted to assess the effects of surface area on the exfiltration rate using 10 mm "holes" in the pipe invert and half horizontal and vertical gaps in the sewer pipe wall. Figure 5 compares the loss rates for these varying surface area configurations based on the 2 mm gap width sizes. There is a general initial linear loss rate up to 30–40 mm head after which there is evidence of throttling of the exfiltration rates for higher flow heads, and even a decrease with increasing head as illustrated by the plot for the 2 mm × 10 mm "hole". Figure 5 also illustrates the enhanced exfiltration rates associated with a half-horizontal gap compared to the corresponding vertical gap with a 50% increase in loss rate at a 30 mm head (Q_{20}). At discharge rates above Q_{25} (Q_n corresponding to a n% pipe filling ratio expressed as the fraction of the sewer height), the forward velocity is sufficient to convey the flow across the 20 mm^2 gap without incurring additional exfiltration loss. The maximum loss rates are restricted to head levels at or below the DWF equivalent of 10%–15% pipe-full flow. At such DWF levels, the overall losses appear to be in the order of 1%–2%.

Effects of trench backfill

A 1 m section either side of the pipe joint openings was backfilled to 0.75 m depth with gravel/sand and sealed tight to simulate natural trench conditions with test runs carried out until a time-based equilibrium rate was achieved. 6 mm plastic tubes were used to underdrain the backfilled section enabling leakage rates over time to be determined. Rapid

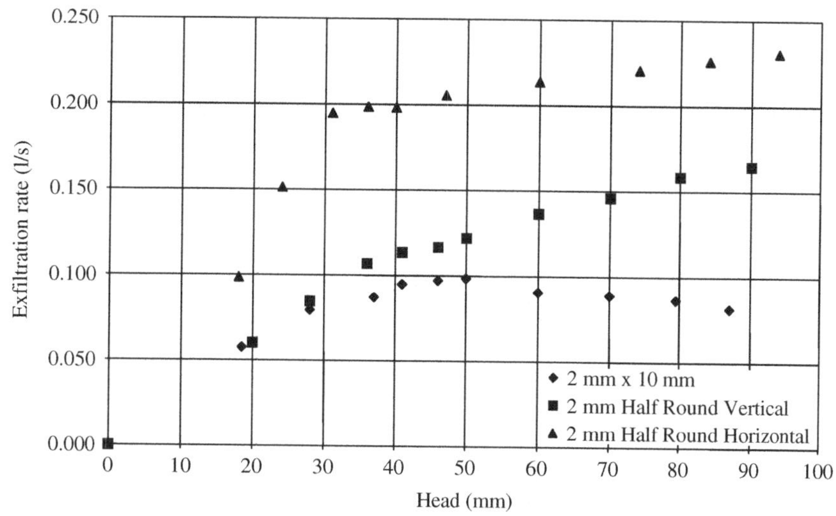

Figure 5: Free draining exfiltration rates for varying gap surface areas, based on a gap width of 2 mm.

exfiltration occurred over the first 20 minutes of the tests due to the pressure differential between the hydraulic head in the pipe and the backfill hydraulic head which forces water out of the pipe into the surrounding sand/gravel jacket. However, following this initial exfiltration, rates from the 1 m length test section very rapidly gained an equilibrium level over time (Figure 6). Although not shown in Figure 6, following the initial leakage, exfiltration rates also decreased rapidly with increasing water level. Very similar observations were made by Rauch and Stegner (1994) who noted that steady-state conditions were reached within one hour even when the joint surface area was quite large. They attributed the equilibrium to be the result of settleable solids in the wastewater effluent reducing the permeability of the bedding material and effectively preventing exfiltration. In the experimental tests being described here, clean water was used and the equilibrium is explained by the clogging of the macropores by fine particles. The test outcomes would suggest that sewer invert depth relative to local groundwater depth may not always be (at least on its own), a useful measure of exfiltration susceptibility.

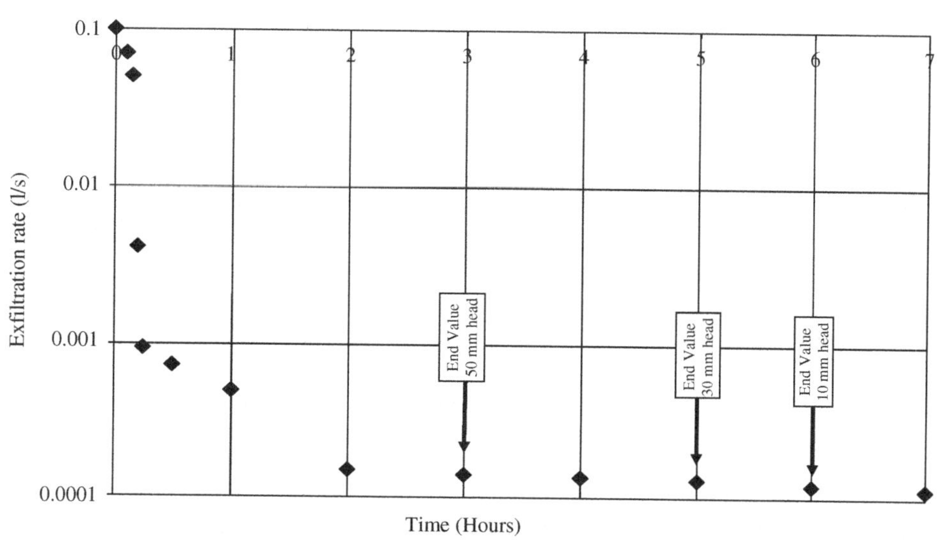

Figure 6: Equilibrium exfiltration rates over time for a Gravel-Jacketed 1.44 mm half-horizontal joint.

The ultimate equilibrium exfiltration rates for each flow depth compared to the previous open joint calibration tests are reduced by a factor of 10^2 to 10^3 to total amounts which are less than 2%–3% of DWF levels at Q_{20} pipe flow. Thus exfiltration from joints, cracks, gaps and other structural wall defects in real sewer systems is likely to be substantially suppressed in the presence of a "tight" trench backfill and particularly if groundwater in the backfill is at or near the sewer invert level. Figure 7 illustrates this for a short time experimental run for a 25 mm^2 surface area "hole" (2.5 mm × 10 mm gap) during which the equilibrium conditions described in Figure 6 were not achieved. The suppression in exfiltration loss is particularly noticeable for dry backfill gravel conditions at flow heads above the DWF level. However, there is very rapid initial exfiltration with increasing head when the gap is surrounded by a wet gravel jacket, which may be the result of strong suction and capillary effects in the immediate vicinity of the outflow gap.

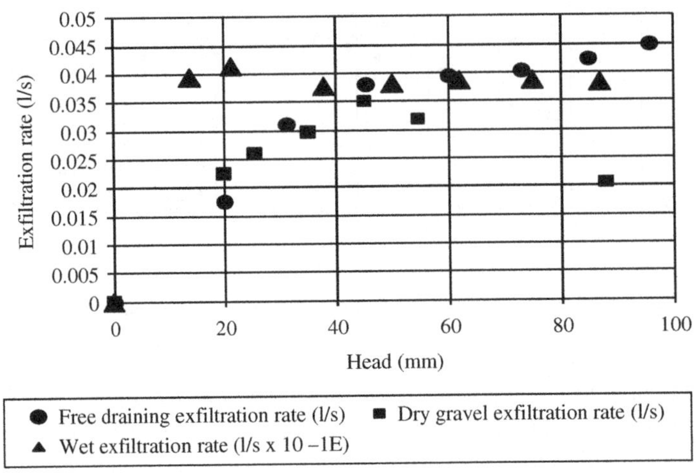

Figure 7: Comparison of exfiltration rates through a 2.5 mm × 10 mm gap for free draining, dry and wet gravel backfill conditions.

Under saturated backfill conditions, where the groundwater level is at the surface, high initial exfiltration loss occurs but rapidly reaches an equilibrium level which is similar to that obtained under free-draining conditions. The hydrostatic pressure exerted in the fully saturated trench backfill, combined with the gravel blocking, clearly prevent major exfiltration from the pipe even at high flow and pressure heads in the sewer pipe. There was a tendency for exponential increasing exfiltration rates to occur for all dry and wet trench backfill test runs with increasing inflow and pressure heads apart from the 10 mm hole rates where very low linear losses were recorded.

In-pipe sediment effects

Experimental runs under free-draining conditions using toilet paper and a variety of sediment types (sand/peat mixes, synthetic non-cohesive particles and mixes) showed that plugging and bridging of the joint gaps occurs such that leakage rates exponentially decline to $10^{-4/-5}$ l/s at DWF, with the joints becoming effectively sealed over the long term (Ellis *et al.*, 2003). Figure 8 shows the effect of differing synthetic materials on the exfiltration rate for 1 mm × half-horizontal gap with the inorganic olivestone providing little if any sealing effect in comparison to the freely-discharging situation without sediment. However, when the mono-size synthetic sediment was replaced by a mixture of crushed olivestone (<70 μm) and highway sediment (50–1500 μm), there is a considerable reduction in the exfiltration rate particularly at low flow rates up to the Q_{25} level as shown in Figure 8 (by the sediment mix curve, Δ). When the sediment mix was allowed to settle out and accumulate on the pipe bed it tends to become lodged in the gap and become compacted. Subsequent increases in flow level fail to dislodge this bridging material until Q_{50-60} levels were achieved. At these flow levels, with enhanced boundary shear and bed turbulence, the sediment matrix is "lifted" out of the gap.

The settling column tests of Karpf *et al.* (2008) confirm the two stage colmation mechanism of initial particle settlement followed by compaction due to a combination of physical and biochemical processes reducing the open pore radius and causing an over-proportional reduction in the hydraulic conductivity (see Chapter 6).

Typical leakage rates for non-cohesive road sediments are shown in Figure 9 which indicates sediment entrainment and removal of particles from a 1.44 mm joint occurring at 6.8 l/s (Q_{60}) and which was followed by rapid increases in the exfiltration rate. However, following flow reduction back down to a Q_{10} DWF level condition, leakage rates became re-established at the 10^{-4} l/s level as the mobilised particulate was re-deposited and blocked the joint gap.

Based on the initial clean-water exfiltration calibration curve for the open 1.4 mm joint, the leakage rates at DWF were reduced from 5×10^{-2} l/s to a minimum of 5×10^{-5} l/s when the joint became effectively sealed. The initial increase in exfiltration rate with increasing flow depths is clearly not as pronounced as in clear water, being "dampened" by the clogging effect of the particulate. A 50:50 sand/peat mix provided a more cohesive infill to the joint, which

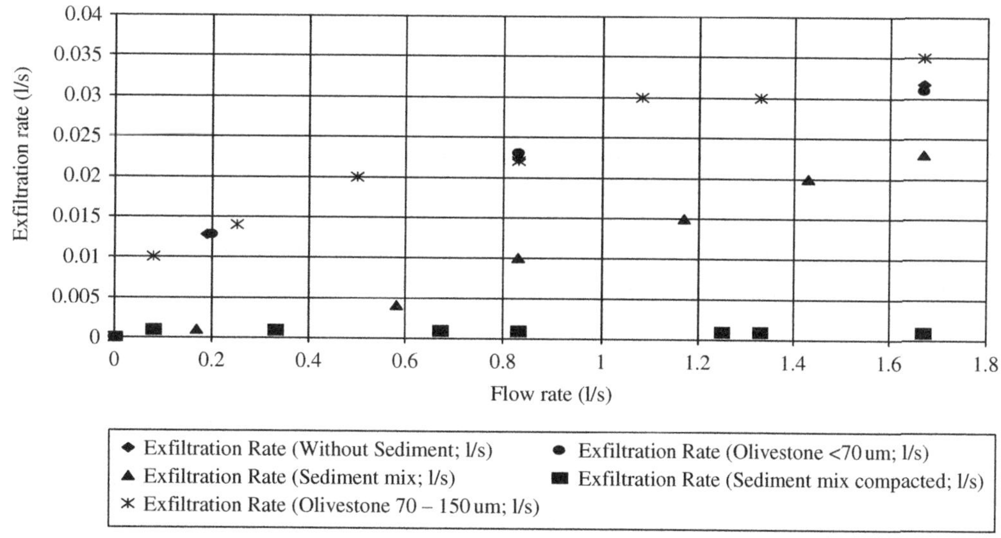

Figure 8: Exfiltration rates for synthetic sediment.

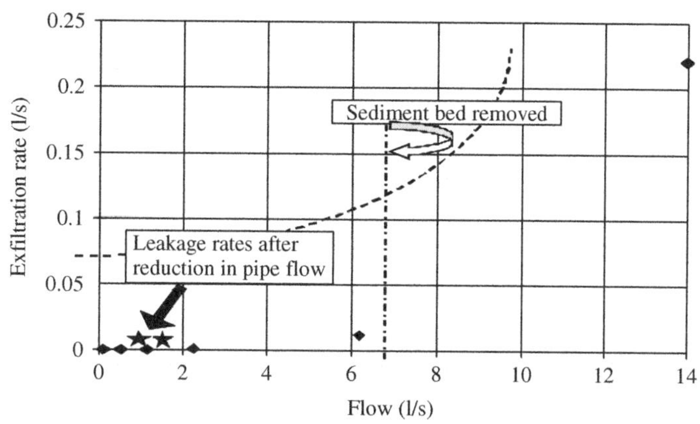

Figure 9: Exfiltration rates through a 1.44 mm half-horizontal joint for pipe discharges containing road surface sediment.

effectively sealed tight at flow rates up to nearly 6 l/s. Once shear turbulence had been reached, exfiltration rates followed the same pattern as that exhibited by the road sediment, with leakage rates falling back to near zero once Q_{10} DWF had been re-established. Clearly, expansion of the organic peat in the sand/peat mix following swelling and saturation provides a considerable cohesion and effectively "plugs" the open joint.

Wetted and shredded double-ply toilet tissue draped across a 1.44 mm joint gap tended to become "plugged" into the gap at flows up to the Q_{50} level with leakage rates declining linearly with increasing flow rate (up to 1.2 l/s). Following reduction back to DWF Q_{10} condition, it was noted that there was an almost complete sealing effect across the joint presumably as a result of enhanced surface adhesion of the tissue brought about by the increased hydrostatic pressure. These results tend to confirm those observations in London sewers where increased sewer discharges appear to block infiltrating water due to the bridging and sealing of gaps/cracks by shredded toilet paper held within a sediment matrix. Exfiltration is re-imposed when boundary shear and bed turbulence "lifts" the bridging paper from the gap. In the clean-water test rig runs, this occurred at flow levels equivalent to $Q_{50/60}$. "Cleaning" of the joint gap may well be more effectively achieved by increased hydrostatic pressure exerted from outside the pipe under rising groundwater conditions and with DWF conditions within the sewer.

An organic "cream" layer up to 15 mm thickness has been observed from endoscopic studies in Paris and Marseille sewers at the sediment-water interface and this Type A near-bed sediment is believed to constitute shredded toilet paper embedded in glutinous organic slime (Oms *et al.*, 2002). These studies in French sewers note the principal occurrence of the "cream" layer to be in hollows, slack zones and under low flow conditions which provides confirmatory evidence for the experimental rig observations.

LIVE WASETWATER EXPERIMENTS

Wastewater field experiments

A pilot field experimental sewer rig was also established at Dundee, Scotland to reproduce the laboratory test conditions on live wastewater. The test rig was constructed on the concrete cover slab of an overflow chamber at a Scottish Water/ United Utilities combined sewer site at Riverside, Dundee (Figure 10). The rig consisted of a 10 m length of 150 mm clear plastic pipes running through three, 2 m high boxes, with flanged joints between the plastic pipes inside each of the boxes (see Figures 11 and 12). Defects were simulated in the same way as described for the clear water rig by placing inserts between the flanges of the joints with the inserts cut to different thickness and geometries. Gap sizes of between 2 mm and 10 mm were tested and the same three basic geometries were considered; a 10 mm invert hole at invert level, a vertical radial opening and a radial half-horizontal opening.

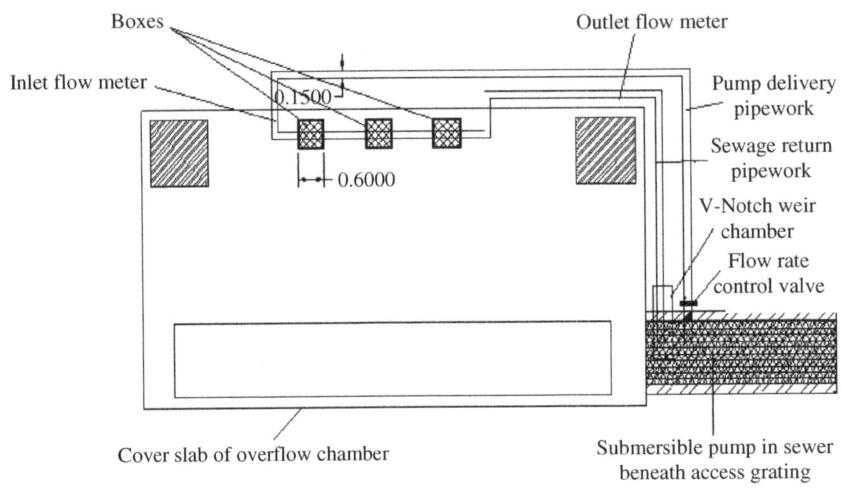

Figure 10: Plan view of wastewater test rig layout.

Figure 11: Overall view of the sewer test rig showing location of test pipe relative to the test boxes (see Figure 12).

The exfiltration rate for a particular gap size, geometry and flow head was determined by collecting and measuring the volume of exfiltrate over time, as it discharged from a pipe at the bottom of the sealed box. Sewage was drawn by a submersible pump, which was placed in a live sewer 3.5 m below the rig and which was frequently withdrawn to remove ragging and blockage from the pump. Two flow monitors measured the inflow and outflow of the rig and a V-notch weir was constructed at the outlet from the rig to enable accurate measurement of low flows. The V-notch assembly then discharged back into the sewer.

Figure 12: Interior of each test box showing variable sized pipe joint.

Complete sealing was eventually observed in 50% of the experiments for the 10 mm hole and the other runs produced lowest recorded exfiltration rates of between 10^{-2} and 10^{-4} l/s. More effective sealing was associated with higher heads, except in the case of the widest gap (10 mm). There was a considerable variation in the time taken to reach the lowest recorded exfiltration rates and this is probably due to randomness in the time of arrival of suitable sealing material in the sewage flow. The lowest recorded levels in each experiment for radial defects with 3 mm and 6 mm gap sizes were also in the order of 10^{-2} to 10^{-4} l/s, with more effective sealing occurring at higher heads. However, much higher exfiltration rates, of the order of 10^{-1} l/s, were recorded for the 10 mm gap width. Generally, the exfiltration rates were tending towards equilibrium values after between 60 and 120 minutes, although it was noticeable that occasional and random short-term pulse losses occurred even after apparent ultimate equilibrium had been achieved.

The experimental runs on the pumped live wastewater flows at Dundee exhibit similar trends to those noted in the clean-water rig. Figure 13 shows test runs for varying 3 mm gap geometries (hole, half horizontal and vertical joint openings) at near constant head under free-draining conditions; the high initial exfiltration rates and rapid decreases noted by other workers are clearly replicated with ultimate equilibrium rates for all gap geometries being just under 2×10^{-3} l/s and for the 40 mm^2 hole being 1×10^{-3} l/s. The reduction in final equilibrium level from 0.23 l/s in the clearwater test runs to 2×10^{-3} l/s in the equivalent wastewater rig tests illustrates the effect of sediment on the loss rates. A similar but lower reduction in exfiltration rate from 0.045 l/s to 0.001 l/s was observed for the 10 mm long hole. Figure 14 shows the exfiltration rates for varying 10 mm gap geometries and illustrates the effects obtained when the pipe is wrapped in a compacted dry gravel backfill. A similar pattern of rapid reduction, particularly for the half-round gap configurations, was observed tending towards a final equilibrium rate of between 0.001 and 0.006 l/s.

The rates of exfiltration as a percentage of flow in the pipe were also calculated and, in most cases, were between 0.1% and 1%. However, values of around 5% for lower heads and between 15 and 30% for the largest gaps were found. Components of the data sets have been analysed by a grouping process to provide a prediction of exfiltration rates after 4, 8 and 16 hours. Whilst the predicted range was wide, the extrapolation suggests that exfiltration rates would reduce from those observed above by an order of magnitude to 10^{-4} to 10^{-5} l/s over an 8 hour period and approach near-zero after around 16 hours. This finding reinforces the previous clean-water laboratory rig work but it should be noted that the limited nature of the data set and the degree of fit of the selected function, prevents any clear conclusions from

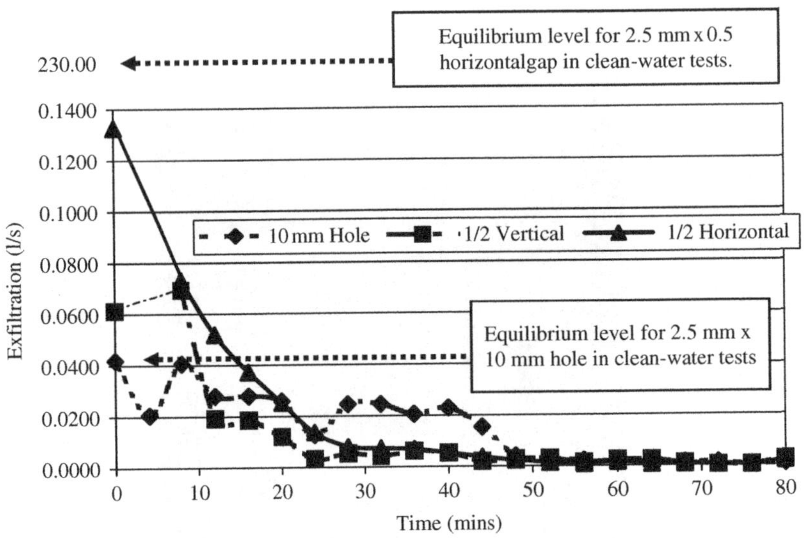

Figure 13: Temporal trends in exfiltration rates for varying 3 mm gap geometries and constant head (42–44 mm) in a live wastewater test rig.

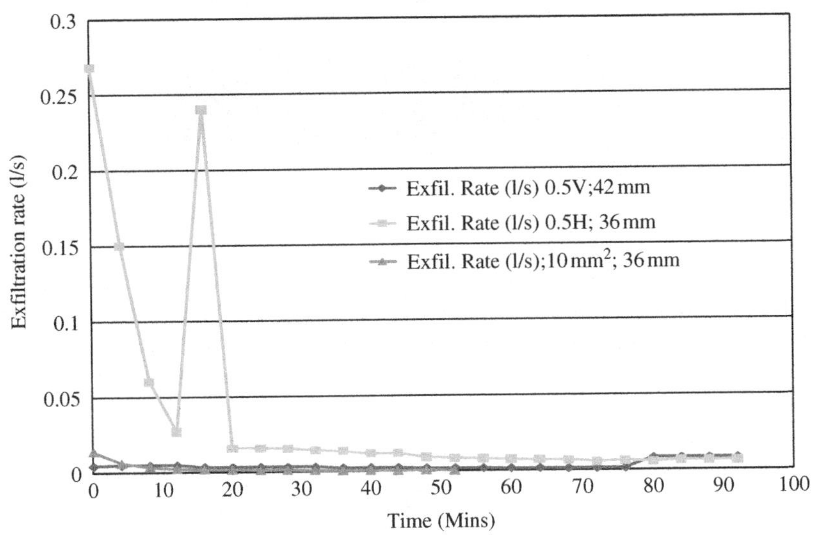

Figure 14: A comparison of exfiltration rates for different 10 mm gap geometries for dry gravel trench backfill conditions in a live wastewater rig.

being drawn. Overall, the presence of a gravel surround leads to more effective sealing of the full range of defects compared to those observed for the free-draining runs. Full detail of the experimental rig, test run data and the study outcomes are given in Blackwood *et al.* (2005a, 2005b) and complete reports are available on the APUSS website.

SCALE-UP OF SEWER RIG RESULTS

Table 1 provides a summary of published estimates of exfiltration rates reported from various European studies. The German and UK sewer networks have been noted to be similar in size, age and structural conditions, with an estimated 22% of the length of German sewers being quoted as sufficiently defective to allow exfiltration (Decker and Risse, 1993). This figure compares well with the 23% of UK sewers believed to have defects which might similarly allow leakage (Reynolds and Barrett, 2003). It is clear that the exfiltration estimates quoted in Table 1 cover a wide range of values with the highest losses coming from laboratory sewer rig tests.

Based on the upper limit of 2 l/s/km extrapolation derived from the Dundee wastewater rig, a recharge from sewers alone of 1400 mm/year would be expected for the 756 km of sewers in the 34 km^2 Nottingham area. This is much higher than the annual rainfall (650 mm/year) for the city and the total recharge of 211 mm/year for the aquifer

Table 1: Estimates of sewer exfiltration rates from various european studies.

Study	Exfiltration Rate	Method
Direct Test Rig Studies		
Rauch & Stegner (1994)	0.87–52 l/d/cm^2	Flowing wastewater through a leaky pipe
Vollertsen & Hvitved-Jacobsen (2003)	0.02–0.06 l/d/cm^2	Recirculated wastewater through a leaky pipe
Blackwood *et al.* (2005a)	9–86 l/d/cm^2 (0.1–2 l/s/km)	Flowing live wastewater through a leaky pipe
Rutsch *et al.* (2008)	2–167 l/d/cm^2	Mass balance of artificial tracer loads
Dohmann *et al.* (1999)	0.07–109 l/d/cm^2	Pressure testing of field sewer pipe
Klinger *et al.* (2005)	7–56 l/d/cm^2 (0.0002 l/s/km)	Collector tank below live sewer pipe
Indirect Modelling Studies		
Karpf & Krebs (2004)	0.0014 l/s/km (2.8% DWF)	Balancing time series on catchment scale
Eisworth & Hotzl (1997)	0.000014 l/s/km	Groundwater monitoring/water balance studies
Harig (1991)	~1.7 × 10^{-9} l/s/km	Groundwater flow modelling and solute balance
CIRIA (1993)	0.062–0.179 l/s/km	Leakage estimates (441–904 Ml/year)
Yang *et al.* (1999)	0.014 l/s/km	Groundwater flow modelling and solute balance
Morris *et al.* (2006)	0.03–0.075 l/s/km	Water balance estimate
Trauth *et al.* (1995)	<7.5 × 10^{-9} l/s/km	Balancing time series

(Yang *et al.*, 1999). The average estimated sewer recharge to the unconfined aquifer of urban Nottingham is quoted as 6–13 mm/year by Yang *et al.* (1999), corresponding to an average value of 10 mm/year. This would yield an exfiltration rate of 0.014 l/s/km for the 756 km length of sewers in the urban area. It is therefore reasonable to discount the extreme exfiltration rates recorded in the laboratory and field test rig studies and based on examination of the data in Table 1, the remainder lie in the range of 0.008–0.7 l/s/km. Even the higher end of this spectrum might be discounted as giving improbably high recharge rates as serious groundwater rise problems might be expected if such rates were active. With these considerations in mind, it is likely that real sewer exfiltration rates might reasonably be expected to be in the region of 0.01–0.2 l/s/km.

The results of the Dundee wastewater test rig experiments have been scaled up to estimate leakage rates in sewer lengths and in associated subcatchments (Ellis *et al.*, 2004). As an initial estimation, assuming that sewer joints occur at intervals of 2.5 m and that 10% of joints are defective in a 1 km sewer length, the wastewater rig outcomes for 10 mm wide defects in pipes having dry gravel trench surrounds, would imply a worst-case ultimate overall loss rate of between 130–1300 m^3/km.year. If this were scaled-up to the 25% of grade 5 and 4 sewers of the total 64,374 km network of Thames Water, this would yield losses due to exfiltration of between 2.1M and 21M m^3/year. Whilst this figure is much less than previously reported (e.g. a CIRIA (1996) 3% estimate of 109M m^3/year), it is still a concern for areas having abstractive groundwater beneath cities in the light of the requirements of the Water Framework Directive.

The pilot-scale experimental work also strongly suggests that sewers carrying a high sediment load are likely to seal gaps and cracks in the sewer walls and invert over relatively short periods of time, and thus even the reduced exfiltration values are likely to be over-estimates. The work would also suggest that sewer cleaning may substantially reduce the integrity of the sewer by re-opening the joints and this might also occur during periods of sewer surcharging under flood conditions.

TRACER STUDIES

The laboratory adsorption beaker tests previously described showed lithium to be potentially the most appropriate tracer for sewer exfiltration work but a major disadvantage of this tracer is the lack of a sensitive probe capable of continuous monitoring and the relatively high concentrations necessary for reliable detection (Deliverable D2.3, *Protocols for Assesing Sediment Effects on Tracer Behaviour Under Laboratory Simulated Flow and Quiescent Conditions. www.insa-lyon.fr/Laboratoires/URGC-HU/apuss*). Rhodamine WT (RWT) does not suffer from these disadvantages and field tests using a portable fluorimeter indicate that reliable temporal monitoring of both fluorescence and turbidity can be achieved. Figure 15 shows the relationship between fluorescence and turbidity variations within a foul sewer during a 4 hour period and suggests that providing turbidity is taken into account, RWT can be confidently used as a tracer even in the presence of high concentrations of organic-rich solids.

A maximum fluorescent signal reduction of 15%, due to sediment adsorption, was recorded for RWT in contact with raw sewage from a large Thames Water sewage works inlet at Reading in S E England and this observation is consistent with the findings reported for the use of this tracer in aquifer studies (Vasudevan *et al.*, 2001). This phenomenon needs

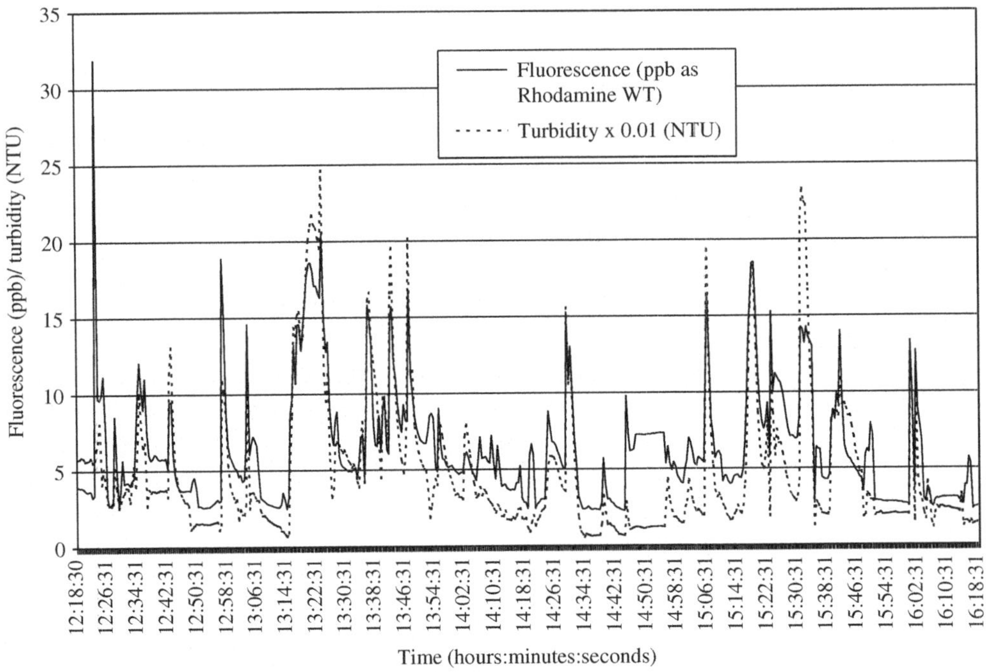

Figure 15: Temporal variation in fluorescence and turbidity within a foul sewer.

to be addressed when using tracers in association with mass balance studies, where only short term (<30 minute) experiments are likely to be feasible, but does not detract from the advantages of using fluorescent dyes with respect to sensitivity and ease of monitoring. The required dosing conditions (flow rates and concentrations) have also been optimised for the use of Rhodamine WT and lithium tracers to determine exfiltration rates within the experimental sewer rig. Tracer solutions (500 ml) were introduced into the pipe a constant dosing rate of 20 ml/minute (0.3333 ml/s) using a Masterflex peristaltic pump.

The fluorescent signal in the exfiltration samples collected at 4 minute intervals was measured directly using a Turner Designs SCUFA probe, which had previously been calibrated between 'zero' (using raw sewage) and 200 ppb Rhodamine WT (using a solution prepared in raw sewage). Hence, simultaneous measurements were obtained for the exfiltration volume and the Rhodamine WT concentration in the collected exfiltrate. In the case of the lithium tracer, it was not possible to measure the concentrations directly in the field. Therefore, samples of the collected exfiltrate were returned to the laboratory for lithium analysis by flame emission spectrophotometry. The instrument was calibrated using standards made up in raw sewage. This was necessary to compensate for the interference which occurs to the lithium signal due to the high background concentrations of sodium within sewage.

The initial tracer experiments were conducted with the 6mm spacer in the half-vertical configuration fitted into the pipe at the exfiltration point and with a sewage flow rate of 1.221 to 1.242 l/s (head within the pipe = 39 mm). This experiment was repeated using lithium as the tracer injected into a sewage flow of 1.071 to 1.092 l/s (head within the pipe = 36 mm). As identical flow conditions were not achieved in both experiments, the results are compared in terms of percentage exfiltration relative to the sewage inflow (Figure 16). The overall trends in both experiments clearly demonstrate the previously identified decreasing exfiltration over time as effective sealing of the gap occurs.

For both tracers, the estimated percentage exfiltration predicted from tracer loadings is consistently higher than that determined by measurement of exfiltration volume. This is the opposite to what would have been expected if adsorption of the tracer was occurring. Lithium has been shown to be a conservative tracer in the presence of sewage and therefore the results of these experiments indicate a similar behaviour for Rhodamine WT under the conditions existing within the sewer rig. The consistent deviation between the tracer and flow predictions for exfiltration may be due to the initial experimental design which involves collecting the exfiltrate in a holding reservoir (capacity ~10 l) before it reaches the measuring apparatus. The effect of this 'mixing volume' will be to average out the changes in exfiltrate tracer concentration which occur relative to the pipe flow rates and the variable exfiltration rates thus preventing a rapid and completely realistic response in tracer concentration in relation to other changes within the system.

A more practical design for monitoring the tracer loading in the exfiltrate would be to keep the overall exfiltration flow rate constant by topping up with sewage as required, and allowing the tracer concentration to be the only variable. Such an approach would be particularly appropriate when using Rhodamine WT because continuous monitoring of the tracer concentration can be efficiently recorded using the fluorimeter probe and data logger. The determination of such

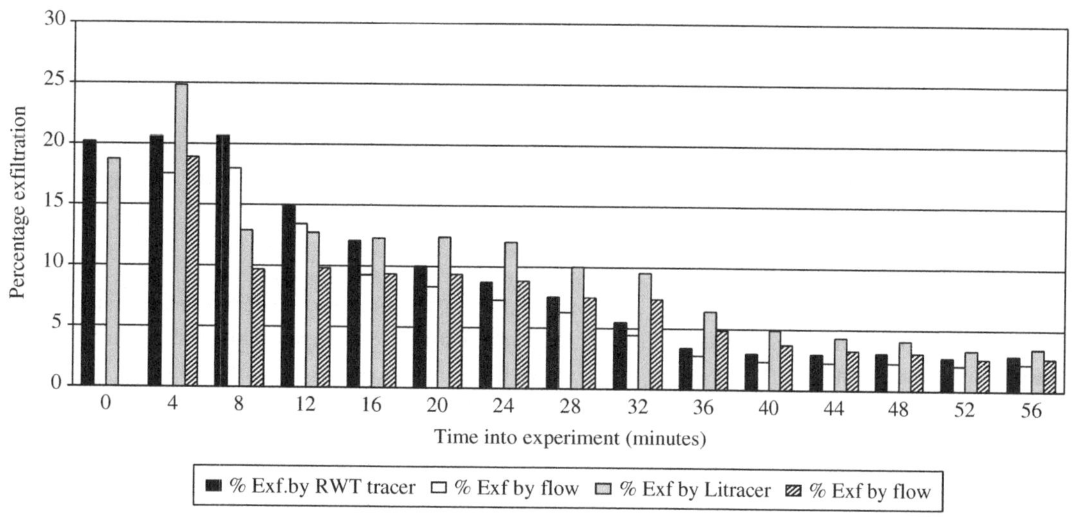

Figure 16: Comparison of percentage exfiltration by tracer and flow measurements through a 6 mm × half-vertical gap configuration.

a continuous exfiltration response avoids the requirement for any physical flow measurements and eliminates the need to average results over a set time period (e.g. 4 minutes).

A second series of tracer experiments was conducted with the exfiltration gap set at 3 mm in the ½ horizontal configuration but using only Rhodamine WT as the tracer,. Three sets of experiments were completed at sewage heads within the pipe of 60 mm, 45 mm and 41 mm, respectively. The results are shown in Figure 17. The percentage exfiltrations measured for the smaller gap size are generally lower than those observed for the 6 mm gap and, as a consequence, there appears to be much better agreement between the results based on flow and those based on tracer measurements. This may be because tracer concentrations in the exfiltrate are less influenced by the 'mixing volume', prior to collection of the exfiltration sample, than was the case in the previous series of experiments. The lower percentage exfiltration values observed for the 60 mm head experiment are partly a consequence of the higher average flow rates (3.21 l/s) compared to the 45 mm head (1.269 l/s) and 41 mm head (1.167 l/s) runs. However, the initial exfiltration flow rates were slightly lower for the higher flow rate experiment supporting the results discussed previously in this report. All three experiments demonstrated a temporal decrease in percentage exfiltration as progressive sealing of the 3 mm gap occurred.

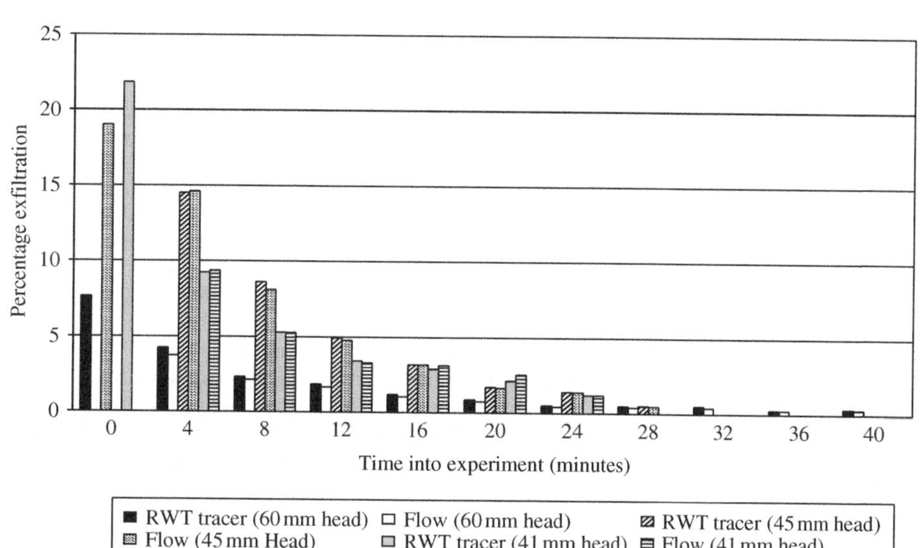

Figure 17: Comparison of percentage exfiltration determined by tracer and flow measurements for a 3 mm × half-horizontal gap configuration.

The performances of both tracers have been successfully compared with direct volumetric measurements for the determination of percentile exfiltration rates. The good agreement between the two approaches appears to be improved for smaller gap sizes but procedures have been identified to overcome these discrepancies through modifications to the experimental design. No evidence was found for adsorption of Rhodamine WT on to sewage solids during the pathways existing within the sewer rig and the investigations suggest that the ease of detection of this tracer makes it ideal for further pilot-scale experiments and for use in real sewer tracer experiments over short time scales.

ACKNOWLEDGEMENTS

The authors wish to acknowledge the full and efficient technical assistance provided by Peter Lister, Chris Willgress, Andrew Buckley and Nick Paterakis (Urban Pollution Research Centre, Middlesex University) and David Blackwood, David Gilmour and Alan Staines (Urban Wastewater Technology Centre, Abertay University). In addition the financial support of the UK Water Industry Consortium for the work undertaken on the Dundee wastewater test rig is gratefully acknowledged.

REFERENCES

Blackwood, D.J., Ellis, J.B., Revitt, D.M. and Gilmour, D.J. (2005a) Factors influencing exfiltration processes in sewers. *Water Sci. & Tech.*, **51**(2), 147–154.

Blackwood, D.J., Gilmour, D.J., Ellis, J.B., Revitt, D.M. and Stainer, A. (2005b) Exfiltration from sewers: Is it a serious problem? In Mikkelsen, P. (Ed.): *Proc. 10th Int.Conf on Urban Drainage (ICUD).* 21–26 August 2005. Copenhagen, Denmark. IWA Publishing, London, UK. ISBN 978 1843395 744.

CIRIA. (1993) *A Study of the Impact of Urbanisation on the Thames Gravels Aquifer.* Report R129M. Construction Industry Research & Information Association, London.

CIRIA. (1996) *Reliability of Sewers in Environmentally Vulnerable Areas.* Project Report 44. Construction Industry Research & Information Association, London. UK.

Crabtree, R.W., Ashley, R.M. and Saul, A.J. (1991) *Review of Research into Sediments in Sewers and Ancillary Structures.* Report FR0205. Water Research Centre (WRc), Swindon, Wilts, UK.

Decker, J. and Risse, A. (1993). Investigations about quantitative and qualitative pollution load of subsoil, ground and surface water by leaking sewers. 1591–1596 in Marsalek, J. and Torno, H. (eds.): *Proc. 6th Int.Conf. on Urban Storm Drainage.* Seapoint Publishing, Victoria, British Columbia, Canada. ISBN 1 55056 254 1.

Dohmann, M., Decker, J. and Menzenbach, B. (1999). Untersuchungen zur quantitativen und qualitativen belastung von boden, grun und oberflachenwasser undichte kanale. *1–82 in Dohmann, M (ed.)*; *Weassergefahrdung Durch Undichte Kanale, Erfassung und Bewertung.* Springer Verlag, Berlin. Germany.

Eisworth, M. and Hotzl, H. (1997). The impact of leaky sewers on urban groundwater. 399–404 in Chilton, J (ed.): *Groundwater in the Urban Environment: Problems, Processes and Management.* A A Balkema, Rotterdam, The Netherlands. ISBN 90 54109238.

Ellis, J.B., Revitt, D.M., Lister, P., Willgress, C. and Buckley, A. (2003). Experimental studies of sewer exfiltration. *Water Sci. & Tech.*, **47**(4), 61–67.

Ellis, J.B., Revitt, D.M., Blackwood, D.J. and Gilmour, D.J. (2004) Leaky sewers: Assessing the hydrology and impact of exfiltration in urban sewers. 266–271 in *Proc. Hydrology: Science & Practice for the 21st Century.* Imperial College, London. 12–16 July 2004. CD-ROM. Vol. II. British Hydrological Society, Inst.Civil Engineers, London, UK.

Forschergruppe-Kanalleckagen. (2002) *Gefahrdungspotential von Abwasser aus Undichten Kanalen fur Boden und Grundwasser.* Zwischenbericht an die DFG, University of Karlsruhe, Karlsruhe, Germany.

Harig, F. (1991) *Effects on Groundwater Due to Water Exchange Between Leaky Sewers and Aquifer.* PhD Thesis. Mitteilungen des Institutes fur Wasserwirtschaft, Hydrologie und Wasserbau, University of Hannover, Germany.

Karpf, C. and Krebs, P. (2004) Application of the leakage model to assess exfiltration. 217–224 in *Proc. Int. Conf. on Urban Drainage Modelling* (UDM). 15–17 September 2004. Dresden, Germany. Also in Koegst, J. and Krebs, P. (eds): *Urban Drainage Modelling.* IWA Publishing, London, UK. ISBN 978 1843394976.

Karpf, C., Traenckner, J. and Krebs, P. (2008) Hydraulic modelling of sewage exfiltration. *Proc. 11th Int. Conf on Urban Drainage.* August 2008 Edinburgh, Scotland. CD-ROM. IWA London, UK. ISBN 978 1899796 212.

Klinger, J., Wolf, L and Hotzxl, H. (2005) Leaky sewers: Measurements under operating conditions. *Proc 4th Worlwide Workshop for Young Environmental Scientists (WWW-YES).* Domaine de Cherioux, Vitry-sur-Seine, France.

Morris, B.L., Darling, W.G., Cronin, A.A., Rueedi, J., Whiytehead, E.J. and Gooddy, D.C. (2006) Assessing the impact of modern recharge on a sandstone aquifer beneath a suburb of Doncaster. *Hydrogeology Journal*, **14**(6), 979–997.

Oms, C., Gromaire-Mertz, M.C. and Chebbo, G. (2002) In-situ observations of the water-sediment interface in combined sewers using endoscopy. 5–12 in: *Proc. 3rd Int. Conf. Sewer Processes and Networks.* GRAIE, Villeurbanne Cedex, Lyon, France. Also in Bertrand-Krajewski, J.-L. and Chebbo, G. (eds): *Proc. 3rd Int. Conf. Sewer Processes and Networks.* IWA Publishing. London, UK. ISBN 978 1843394365.

Rauch, W. and Stegner, T. (1994) The colmation of leaks in sewer systems during dry weather flow. *Water Sci. & Tech.*, **30**(1), 205–210.

Reynolds, J.H. and Barrett, H.M. (2003) A review of the effects of sewer leakage on groundwater quality. *Journ. Chart. Inst. Water & Environ. Mangt.*, **17**(1), 34–39.

Rutsch, M., Rieckermann, J., Ellis, J.B., Vollertsen, J., Cullmann, J and Krebs, P. (2008) Towards a better understanding of sewer exfiltration. *Water Research*, **42**, 2385–2394.

Selvakumar, A., Amick, R.S. and Burgess, E.H. (2001) Exfiltration in sewer systems: Is it a national problem? In: Phelps, D. and Sehlke, G. (eds); *Bridging the Gap: Meeting the World's Water & Environmental Resources Challenges*. Proc. Word Water Congress, 20–24 May 2001. Orlando, Florida. American Society of Civil Engineers, Reston, Virginia, US.

Trauth, R., Hahn, H.H. and Xanthopoulos, C. (1995) In-situ method to determine water exchange between sewerage and groundwater. *Abwassertechnik*, **4**, 55–57.

Vasudevan, D., Fimmen, R.L. and Fransisco, A.B. (2001) Tracer-grade Rhodamine WT: Structure of constituent isomers and their sorption behaviour. *Environment Science & Tech.*, **35**(20), 4089–4096.

Vollertsen, J. and Hvitved-Jacobsen, T. (2003) Exfiltration from gravity sewers; A pilot scale study. *Water Sci & Tech.*, **47**(4), 69–76.

Vollertsen, J., Vorkamp, K., Hvitved-Jacobsen, T and Jensen, N.A. (2002) *Udsiving af Spildevand fra Aflobssystemer*. (Exfiltration of Wastewater from Sewers) Project Nr. 685, Danish Environmental Protection Agency, Ministry of the Environment, Copenhagen, Denmark.

Yang, Y., Lerner, D.N., Barrett, M.N. and Tellam, J.H. (1999) Quantification of groundwater recharge in the city of Nottingham. *Environment. Geol.*, **38**(3), 183–198.

Chapter 6

The exfiltration process and the colmation layer

J. Bryan Ellis, D. Michael Revitt, Christian Karpf and Peter Krebs

INTRODUCTION

Sewer exfiltration is a frequent, widespread and expected phenomenon which has been consistently observed on both experimental and field studies throughout the world (Selvakumar *et al.*, 2004; Rutsch *et al.*, 2008) although estimates of leakage rates cover a wide range of values with the highest losses coming from laboratory sewer test rigs as noted in Table 1 of Chapter 5. It is argued in Chapter 5 that real sewer exfiltration rates might be expected to lie within the region of 0.01–0,2 l/s/km which in most cases is considerably lower than determined from direct test rig studies and normally higher than rates indicated by indirect modelling approaches (Ellis *et al.*, 2008). It is therefore open to question as to how robust the loss rates are as quantified from experimental sewer studies. In addition, it can be questioned as to whether equilibrium exfiltration loss rates are able to be maintained over extended time periods. Both direct and indirect approaches are costly in terms of effort and time, and being largely focussed on individual pipes, require a scaling-up procedure to derive estimations of catchment-wide exfiltration rates. Thus it is inevitable that recourse is frequently made to the application of numerical modelling approaches to the determination of leakage loss, largely based on the basic Darcy equation and its various refinements (Rauch and Stegner, 1994; De Silva *et al.*, 2004; Karpf *et al.*, 2008).

The principal cause of the rapid decline in exfiltration to limiting loss levels in the region of 10^{-1}–10^{-3} l/s per km sewer length has been explained by sediment sealing of sewer pipe gaps and joints; an assertion which has been widely accepted. In most direct rig studies, periods of total sealing have also been recorded with zero ultimate exfiltration losses reported (Ab-Wahab *et al.*, 2004; Blackwood *et al.*, 2005a), although the sampling and reporting times are frequently less than 30 to 60 minutes and thus insufficient time may have elapsed before the true equilibrium condition had become established. With very short sampling times (sometimes less than 2–4 minutes), detectable exfiltration flows cannot really be expected under ultimate steady-state conditions. In addition, the throttling of exfiltration could at least be partly due to increased saturation and impermeability of the underlying bedding layers over the test run periods. However, whilst the results allow an understanding of rates and patterns of initial sealing, the short term testing ($<$24 to 36 hours) and sampling times cannot be representative of exfiltration losses that will occur over extended time periods of up to 10 years and beyond. Nevertheless, whilst it is admittedly difficult to deduce parameters from such experimentational design and accurately predict real exfiltration rates, such laboratory and field sewer rigs do provide basic working knowledge on process details. This knowledge can then be used to develop more robust exfiltration models which support exfiltration assessment within certain given boundary conditions and levels of uncertainty.

Despite this, there is a belief that even in deteriorating sewer systems having frequent joint gaps and cracks, the sealing or colmation process can be sufficiently developed to prevent further exfiltration losses of any real magnitude over extended time periods once the ultimate equilibrium level is established. However, it is highly unlikely that such sealing will be permanent and the prevailing hypothesis would argue that the onset of elevated hydraulic pressure

 Assessing infiltration and exfiltration on the Performance of Urban Sewer Systems (APUSS). Edited by Bryan Ellis and Jean-Luc Bertrand-Krajewski. ISBN: 9781843391494. Published by IWA Publishing, London, UK.

conditions provides the appropriate criteria and threshold conditions for the "rupture" of sealed joints and cracks. Such failure could then result in the catastrophic and sudden leakage of any "pooled" pollutants, e.g. in sewer invert slack sections, resulting in the release of contaminants to the sub-surface system and onward transmission to groundwater. The laboratory batch reactor, settling column and experimental rig tests described in Chapter 5 have enabled a further investigation of the exfiltration process and this chapter examines the hypothesis of colmation formation and re-examines the nature and definitions of the sealing layer(s) associated with the blockage of sewer pipe gaps and joints.

MECHANISMS OF COLMATION LAYER DEVELOPMENT

The conventional view of joint and gap sealing in sewers considers the infill of the sewer defect as a function of the settling of particles carried in the overlying sewage effluent, primarily under steady dry weather flow (DWF) conditions. Figure 1 shows a typical colmation layer developed within a half-circular joint opening in an experimental sewer and through which exfiltration has occurred. The depth of the colmation layer is roughly equal to the thickness of the pipe wall, plus the depth of any cavity that may lie below the joint opening (Figure 2). Figures 3a and b illustrate the dark-coloured exfiltration imprint left by leaking wastewater in the joint opening and immediate backfill material which is clearly seen in Figure 3b following the lifting out of the pipe from the enveloping soil material. This underlying semi-saturated area of sub-soil clogging which lies immediately below the joint opening has been termed the transition zone by Rauch and Stegner (1994).

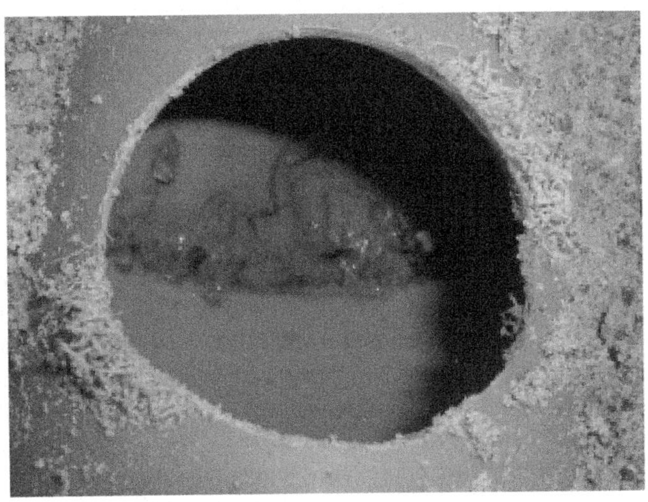

Figure 1: View of a colmation layer in a half-circular joint gap.

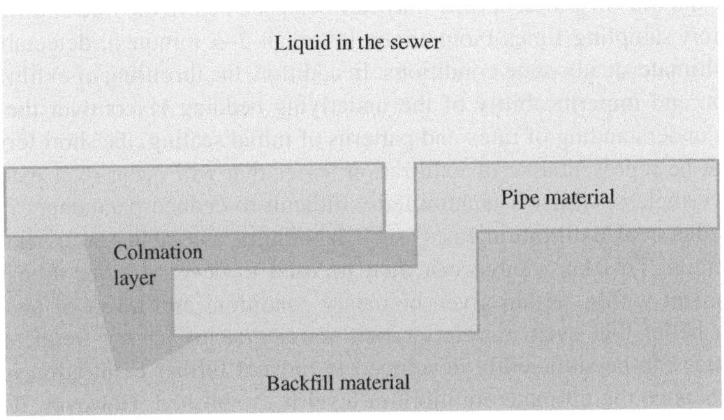

Figure 2: Diagrammatic representation of a leaking joint in a concrete sewer pipe.

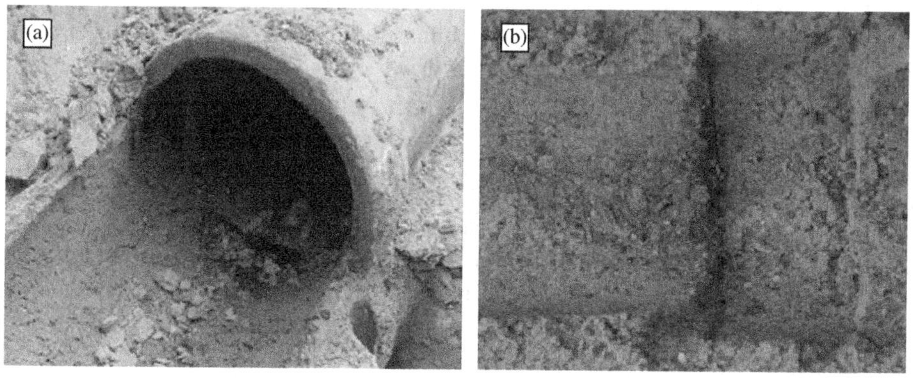

Figure 3: The imprint of exfiltrating wastewater from a leaking joint.

Current state-of-the-art approaches for modelling leakage (Q_{ex}) are based on Darcy's Law following the baseline studies of Rauch and Stegner (1994) and are illustrated in Figure 4, wherein a linear dependency between hydraulic gradient, cross-sectional leakage area (A_{leak}) and discharge is assumed to be:

$$Q_{ex} = k_{colmation} \cdot (\Lambda/h_{colmation}) \cdot A_{leak} \tag{1}$$

where $k_{colmation}$ is the permeability coefficient of the soil column below the leak, $h_{colmation}$ is the thickness of the colmation layer and ΔH is the hydraulic potential. The leakage factor (L) is therefore defined as:

$$L = \frac{k_{colmation}}{h_{colmation}} = \frac{Q_{ex}}{A_{Leak} \cdot h_{fill}} \tag{2}$$

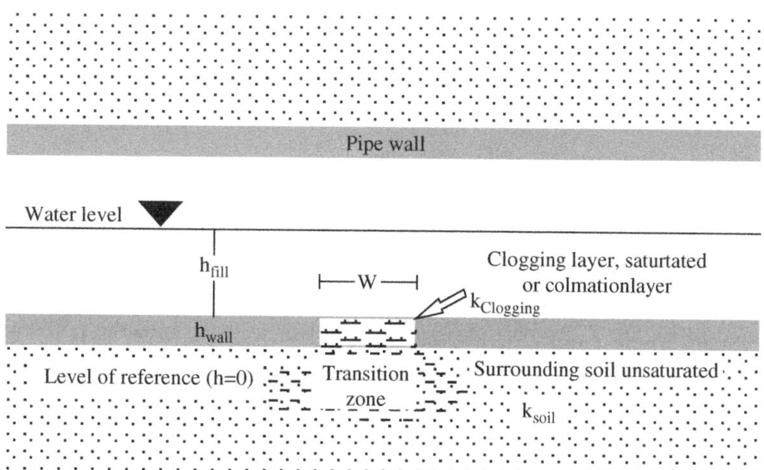

Figure 4: Schematic representation of a blocked leak in a sewer pipe.

The exfiltration flux not only increases with the hydraulic presuure ($h_{fill}+h_{colmation}$; where h_{fill} is the water level in the pipe), but also with the matrix potential, Ψ (Beal *et al.*, 2004). The matrix potential depends on the soil type and the water content (where under saturated conditions $\Psi = 0$). The latter depends on the hydraulic pressure i.e steady-state conditions of the matrix potential are approached depending on the exfiltration rate and subsequently on the colmation characteristics.

Initially, exfiltration will be governed by the permeability and saturation of the surrounding bedding material (the backfill material). Within a time span of minutes to hours, particles contained in the exfiltrating liquid will cover the surface of this coarser backfill material and initates the first phase of the colmation process (Vollertsen, 2007). This initial settling phase shown in Figure 5a leads to a covering of the coarser backfill material, gradually clogging the pores of the upper sub-soil layers and causing a reduction in hydraulic conductivity (K_c) and in the leakage factor (L). The initial colmation formation of phase 1 ($h_{colmation1}$) is dominated by physico-chemical processes and by the size and mass characteristics of the settling suspended solids (M_{TSS}) which can be defined as:

$$h_{colmation1} = M_{TSS}/[\rho_s\{r^2 - r^2_{colmation}\}\pi \cdot n_p] \tag{3}$$

where ρ_s is the density of the accumulating sludge in the initial colmation layer, r^2 is the pore radius in the underlying sub-soil (or in the colmation layer) and n_p is the number of pore openings. Karpf *et al.* (2008) give equations to solve the pore radii for the non-clogged soil layer and estimates for the number of pores. The second compaction phase (Figure 5b) leads to a gradual penetration of solid particles in to the pores of the sub-soil and to bio-clogging, both of which lower the conductivity of the colmation layer. Further, sedimentation and erosion of particles in the sewer defect, respectively increase or decrease the thickness of the colmation layer. An increase of the colmation layer by sedimentation is defined by Karpf *et al.*, (2008) as the ratio of mass solids (M_{TSS}) to the sludge volume (ϕ_s) and density (ρ_s) accumulating in the developing colmation layer:

$$h_{colmation2} = [M_{TSS}/\{1 - \phi_s\}\rho_s] \tag{4}$$

Figure 5b illustrates the penetration and solidification of settling particles in the upper layer of the sub-soil and which matures to form the more fully developed colmation layer; this compaction process typically takes between days to weeks to fully form.

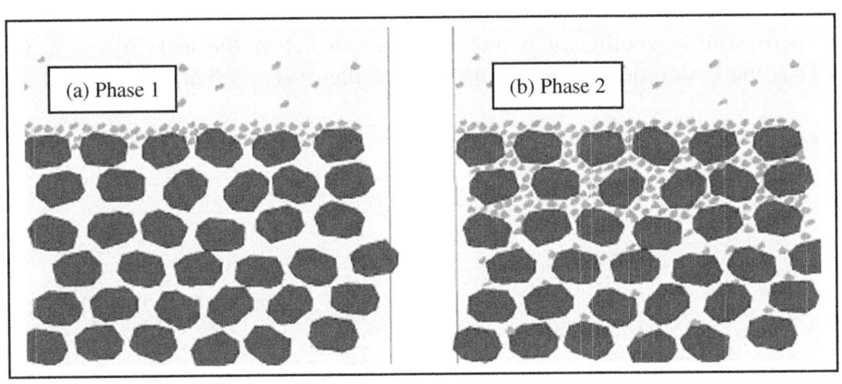

Figure 5: Representation of the phase growth of the colmation layer.

Figure 6 shows the matrix potential increasing with decreasing sewer fill depth and with increasing soil cohesiveness but also as a prime function of the degree of sub-soil saturation. Thus changes in the rainfall-runoff inflow rates or infiltration depth and capacity of the enveloping soil layer can strongly influence the matrix potential and thus the exfiltration flux as indicated in Figure 7. As particles from the sewer migrate into the pores and compact in the second colmation phase, they form the permanent colmation layer which Karpf *et al* (2008) estimate will average 7 mm in thickness with a range of 0.002–0.015 m depending on whether or not the underlying sub-soil is saturated. The upper figure in this range is consonant with that quoted (0.02 m) by Vollertsen and Hvitved-Jacobsen (2003) and as noted in field studies by Wolf *et al.* (2007) and Blackwood *et al.* (2008). As indicated by the sewer rig studies described in Chapter 5, this two stage physico-chemical process of colmation growth occurs very rapidly and leads to the de-coupling of leakage rates from the influence of matrix potential and overlying hydraulic gradient. This colmation mechanism explains the predominant exponential tendency for declining exfiltration rates noted in the majority of sewer rig tests. In addition, the rapid throttling of leakage rates tends to also establish an early apparent ultimate equilibrium level.

Figures 6 and 7 represent the impact of the hydraulic potential and the matrix potential respectively, under constant conditions in the colmation zone. In experiments it was found that the clogging layer is influenced by the hydraulic potential (Rice, 1974; Karpf *et al.*, 2008). More pressure leads to a deeper particle penetration into the underlying transition soil layer. Thus the matrix potential and the water level in the sewer pipe are both important for the development of the colmation layer.

Growth of bacteria and algae in the underlying soil pores of the transition zone will further reinforce this colmation layer. There can also also be a variable biofilm and/or basal sediment layer overlying the colmation zone. However, it has been argued by Velickovic (2005) that the most important part of the colmation structure is the external colmation which comprises the transition zone within the underlying bedding zone to the joint or gap opening, rather than the infill

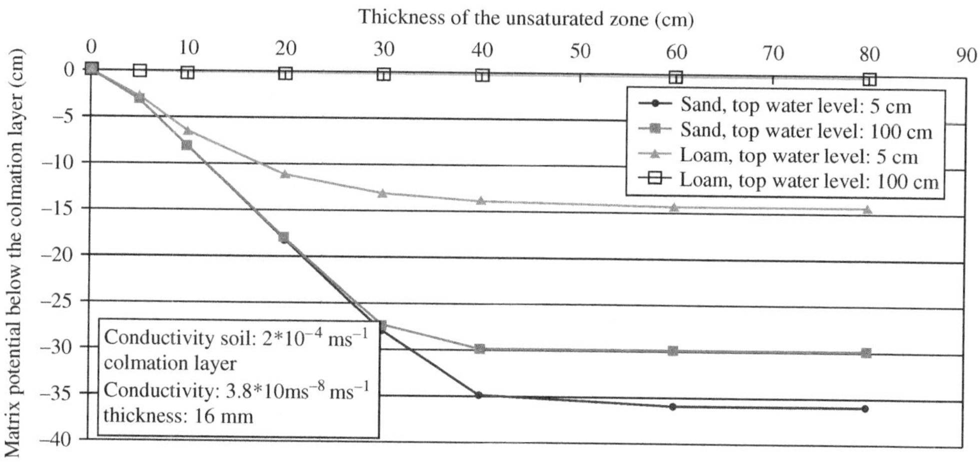

Figure 6: Variation in matrix potential with unsaturated zone depth, sub-soil type and sewage depth.

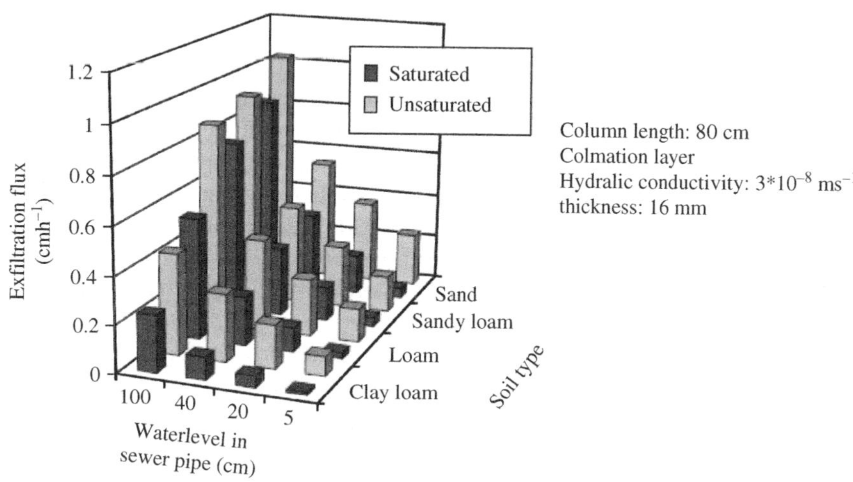

Figure 7: Influence of sub-soil type and saturation on exfiltration rate.

in the joint/gap opening itself. This assertion is to some extent confirmed by hydraulic modelling studies which show that under saturated conditions (and decreasing hydraulic conductivity), in the enveloping sub-soil, initial exfiltration rates are increased compared to saturated conditions. This is essentially because the matrix potential around the transition zone increases the tightness of the overlying colmation layer.

The application of Darcy's Law assumes that exfiltration and leakage rate will be a function of the sewer fill depth i.e pressure head and matrix potential, as well as the leakage area and colmation properties. The above equations provide a deterministic basis for the assumption that the flow depth in the sewer will have a direct influence on the exfiltration rate, since the pressure differential between the hydraulic head in the sewer and that of the underlying groundwater will force effluent out of the sewer openings into the surrounding bedding/soil material. The strength and direction of the hydraulic gradient has been stated as being the "*key factor*" influencing the occurrence of exfiltration (Amick and Burgess, 2000), and this assumption forms the basis for the US national "depth-to-groundwater" distributional mapping of exfiltration rates. Whilst Darcy's Law, as a mechanistic approach, may be considered to be appropriate to model leakage of clear water, dynamic changes in wastewater composition, processes in the underlying soil and changing soil/groundwater properties such as the degree of saturation might require additional input factors. Nevertheless, Darcy's Law has been widely used in various studies of sewage exfiltration and in the prediction of indicative amounts of sewage exfiltration. Irrespective of this, and for practical reasons, it can be questioned whether the necessary information on leakage area, soil properties, and water level are available in sufficient accuracy to reliably and robustly apply the Darcian approach for a real sewer situation (Ellis *et al.*, 2009).

DEFINITION OF THE COLMATION LAYER

Leaks from sewers typically form a colmation layer that is relatively enriched (by up to 25%) in organic matter. This colmation layer (or "schmutzdecker" as it would be termed by conventional wastewater treatment operators) serves to limit the volume of leakage. The layer is assumed to be thin and varying between 0.01 to 0.05 m according to various workers and possessing a relatively low permeability. The Darcian conditions described above, would restrict the thickness of the colmation layer to the pipe wall thickness (d_{wall}), plus the thickness of any biofilm layer forming over the leakage area on the pipe invert. Such an assumption would derive a colmation layer in the order of 0.01 to 0.02 m. However some workers also include within the colmation layer the transition zone occurring immediately below the leak (Figures 4 and 8). Inclusion of this transition zone within the definition of the colmation layer would increase its total depth to as much as 0.05 m and is also likely to increase its permeability value.

Biofilm can be formed on almost any surface exposed to sewage effluent and frequently in sewer situations, a fibrous organic matrix several millimetres thick can develop between the bacterial cell layer and the pipe wall or colmation surface (Rocher *et al.*, 2002). Norsker et al. (1994) determined biofilm growth to be of the order of 80 μm per day with little if any steady state thickness being established over a period of 3 weeks although sloughing occurred within 15 days. The biofilm remained no thicker than 1000–2000 μm during the entire period of their experiments and thus it is liable to be highly vulnerable to disruption and removal by elevated bed shear stresses, especially in its immature growth stages.

This biofilm and associated bacterial growth is therefore developed on top of the colmation layer, being embedded or rooted in the interstices and pores of the colmation matrix. The nature and structure of the biofilm is such that it will not significantly affect in any way the rate of exfiltration through a leak and thus could be regarded as forming an "integrated" component of the layer itself. If sediments are in continuous movement along the sewer invert (as a moving or tractive bed), biofilm will not develop. In addition to these organic compounds, the joint may also be covered by sewage solids and shredded toilet tissue temporarily deposited from the wastewater. Such materials can form a bridging layer lying over and straddling the colmation layer. This bridging layer can be further cemented by fat, oil and grease (FOGs) to form a cohesive congealed matrix (Ellis *et al.*, 2008). These coatings become frequently adultered through contact with detergents and cleaning solutions. In such situations, the bridging layer not only protects the underlying colmation layer but also increases the impermeability and yield strength of the combined material. The bridging layer effect caused by deposition of materials from the sewage effluent has been remarked upon by other workers (Ellis *et al.*, 2003; Blackwood *et al.*, 2005b) and has been termed a "cream" layer in French exfiltration studies (Oms *et al.*, 2002). This terminology reflects the inclusion of degraded shredded toilet paper embedded in glutinous organic and fatty slimes. The inclusion of fat globules in the wastewater flow will also lead to sealing of vertical defects in the pipe wall which clearly requires floating solids in comparison to the heavier solids which serve to seal the invert.

There is therefore considerable laxity in the literature on the definition of the colmation layer which strictly should be confined to that part of the infill material which fills the joint or gap opening and the immediate transition zone down to the saturated basal level as illustrated in Figure 8 (for sewer pipes lying above the groundwater level). Any overlying biofilm and/or sediment layer should be referred to as a bridging layer and not included in the definition (or quantification) of the colmation layer *senso stricto*. Where all of these various layers are being included in the definition, it is best to refer to a general clogging layer rather than a colmation layer. The separate terms of colmation layer, bridging layer and clogging layer are frequently used in an interchangeable manner in the literature but this misuse ignores the independent exfiltration processes which underlie their formation and growth. It is the confusion of colmation infill/transition zone, bridging layer (biofilm/sediment) and clogging layer which can help to explain the differences in hydraulic conductivities of the various clogging layers reported by test studies which have used medium sand as bedding material (Figure 9). The differences may also of course, be partly related to differing "ages" of the undisturbed clogging colmation layer as well as to whether or not the underlying transition zone was saturated. It is also the case that application of the Rauch and Stegner model gives rise to a broad band of uncertainty in defining the hydraulic conductivity (or permeability) for the overall clogging layer comprising the colmation/transition layer and any other overlying bridging layer.

EQUILIBRIUM BREAKTHROUGH

There is a general implicit assumption in the defining of an equilibrium steady-state exfiltration condition that such lower limiting losses will be maintained over extended periods of time. Such an assumption is often accompanied by an acceptance that when equilibrium steady-state is represented by zero detectable losses, that this provides evidence of total and extended sealing. However, such assumptions are normally based on evidence of short term testing and sampling times and there is little evidence of semi-permanent sealing for DWF and low flow conditions. On the contrary, both laboratory and field tests provide considerable evidence for persistent variability and breakthrough of exfiltration following the establishment of steady-state condition. Figure 10 shows the exfiltration losses recorded for a sewer in Rastatt, German

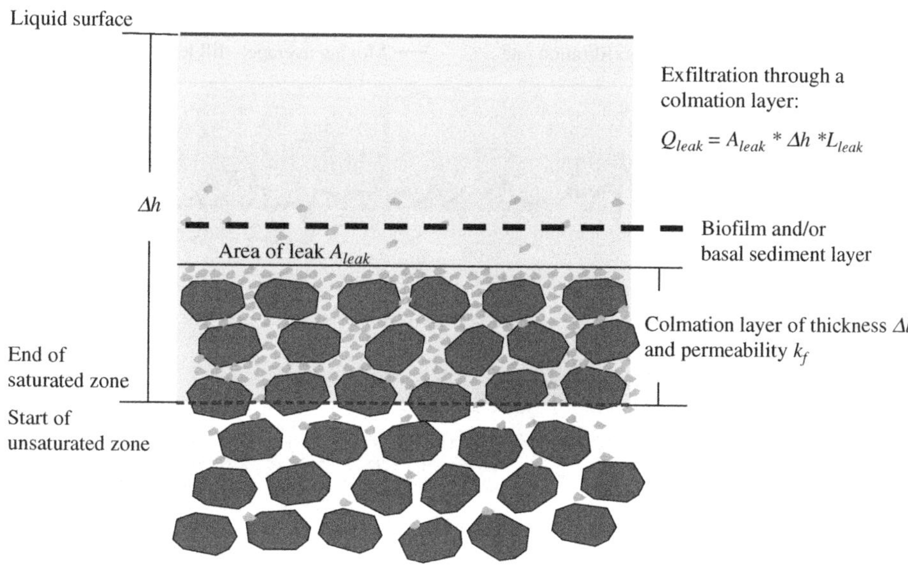

Figure 8: Representation of the structure and growth of the colmation layer (after Vollersten, 2007).

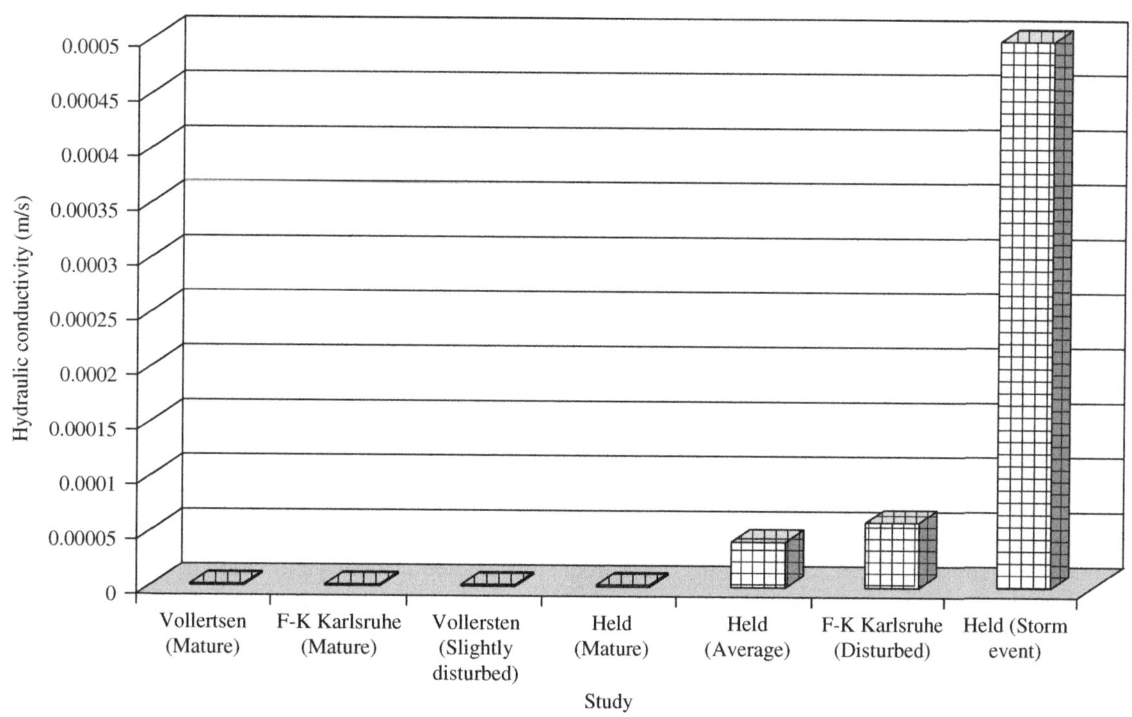

Figure 9: Results reported for exfiltration permeability measurements.

indicated by a solid black line in the lower plot of the figure (Wolf *et al.*, 2007). There is a continuous decrease in the exfiltration flux over time from the commencement of the experiment, reaching a quasi-stable equilibrium state which represents the average steady state exfiltration rate for the pipe defects. The long term behaviour illustrates a typical exponential decline over time; a typical decline pattern noted by many other workers. However, the period of stabilised exfiltration conditions which had been established under normal dry weather flow operational conditions, is subject to increases in leakage varying over 2 orders of magnitude above the equilibrium level. This is clearly seen from inspection of the variable pattern shown by the exfiltration rate (solid grey) in the lower part of Figure 10 and which suggest that variability in exfiltration under quasi-steady state conditions is a common and predictable occurrence.

Such leakage has been reported from many studies including those of Vollertsen and Hvitved-Jacobsen (2003), Blackwood *et al.* (2005b) and Held *et al.* (2005). It is considered that such breakthrough leakage is, in part, the result of

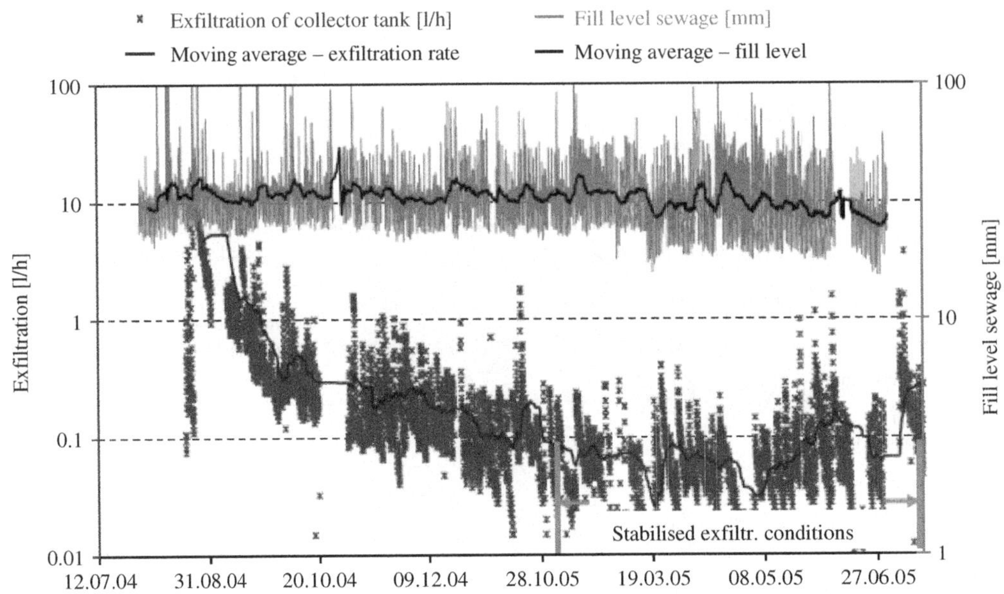

Figure 10: Time series of sewage flow depth and exfiltration rate. (From: Wolf *et al.*, 2007)

instability in the growth conditions and ageing process of the colmation and bridging layers. However, a major influence on such long term leakage losses is the effect of small scale, diurnal kinematic wave pulses which generate temporary local increases in bed shear stress which in turn cause scour and mobilisation of the clogging layers. Such pulses are exacerbated in small diameter sewers due to the shunting or "sliding dam" nature of the sediment material moving along the sewer bed (Littlewood and Butler, 2003) and the inclusion of cementitous toilet tissue exacerbates the dam building tendency (Memon *et al.*, 2007). Flow restriction following such in-sewer accumulation locally increases velocity, turbulence and bed shear and the increasing hydrostatic head following back-up and partial blockage, will increase flow rates as defined by Darcy's Law, through the accumulated layer. The shear rate through interstices in the retaining sediment subsequently increases, and can dislodge material allowing more flow to pass and increased scouring to occur on the sewer invert (Guzman *et al.*, 2007).

In the live wastewater rig studies undertaken in Dundee, Scotland, these flow pulses were noted to occur at quite low pressure heads in tests with a 6 mm half-vertical opening under 22–25 mm flow levels in a 150 mm diameter pipe with a gravel trench backfill (Blackwood *et al.*, 2005a). The short term pulses were believed to be due to temporary local increases in bed turbulence which were sufficiently strong to lift out the colmation fill material from the joint. In the Rastatt, German studies, the colmation layer appeared to be very vulnerable to changes in flow velocity both during peak diurnal flows as well as during storm events (Klinger *et al.*, 2007). During rig tests on live wastewater in the Dundee, Scotland experiments, flow depths at $Q_{70/75}$ in a 150 mm pipe were sufficient to generate velocities of 7–8 $l\ s^{-1}$ and generate bed shear stress in excess of 2.5 $N\ m^{-2}$; the frequently quoted critical yield strength for cohesive sewer bed sediment. Some workers have suggested that even lower bed shear stresses in the order of 1.1–1.4 $N\ m^{-2}$ are sufficient to prevent biofilm growth as well as to scour the colmation layer (Guzman *et al.*, 2007).

There is also considerable evidence of disturbance and removal of clogging layers during sewer pressure testing even at fairly low flushing pressures (~90 psi). Vollersten *et al.* (2002) have reported increased exfiltration rates by a factor of 4 to 5 times from steady-state levels following low pressure jetting. Some workers have also described "wash-out" cavities located beneath holes and gaps in sewer pipes attributed to the scouring effect of exfiltrating effluent (Vollertsen, 2007). Such cavities would also dislocate any transition zone such that subsequent blockage would only be comprised of the colmation layer.

None of the cited exfiltration investigations refer to any specific rupture pressure head values and there is little supporting evidence to underpin the legitimacy of elevated pressure conditions as a necessary or only precursor to the disturbance and mobilisation of the colmation layer, or indeed to disprove that flow levels and associated shear velocities below full hydraulic capacity are incapable of achieving the same "critical" flow conditions to initiate such disturbance. In their experimental sewer rig tests, Vollertsen *et al.* (2002) and Vollertsen and Hvitved-Jacobsen (2003) evaluated the effect of increase in water pressure without increasing the flow velocity on exfiltration loss from an open joint. The immediate effect of the pressure increase was an increase in exfiltration which increased from about 0.3 $l\ d^{-1}$ to approximately 55 $l\ d^{-1}$, i.e. by over a factor of 100. Assuming the validity of the Darcian equation, a large part of the rise in exfiltration was attributed to the increase in pressure, and the rest to an increase in leakage factor by approximately a factor of 7. After several repetitions of the test, the increase in leakage factor compared to the half-full flowing pipe became smaller (less than a factor of 2). Similar experiments conducted with holes in the pipe demonstrated a similar response to increased pressure, although the change in leakage factor was less pronounced for holes at the pipe invert compared to open joints.

The colmation layer, particularly when it is at an immature growth stage, can be easily damaged by diurnal effluent peak flows and operational sewer cleaning both of which can generate turbulent bed conditions. Thus the commonly held view that it is only sudden precipitous sewer backups and associated elevated pressure heads which severely disrupt the hydraulic regime, and consequently initiate exfiltration breakthrough, is not a fully justifiable assertion. What is generally not disputed is that sewer leakages never seal completely and that all sewer systems will be subject to leakage loss over extended time periods although severe sewer blockage can induce local pressurised conditions to an extent that catastrophic colmation rupture can also occur.

ACKNOWLEDGEMENTS

We would wish to acknowledge the support and discussion of Jes Vollersten (Department of Biotechnology, Chemistry & Environmental Engineering, Aalborg University, Denmark) and David Blackwood (Urban Wastewater Technology Centre, University of Abertay, Dundee, Scotland) in the preparation of this chapter.

REFERENCES

Ab-Wahab, M.A., Abdul-Talib, S., Baharom, B. and Marwi, M.S. (2004) Exfiltration from pipes: Effects of flow rates, different leakage areas and soil beddings. 138–146 in Bertrand-Krajewski, J-L., Almeida, M., Matos, J. and Abdul-Talib, S (2004; Edits): *Sewer Networks and Processes Within Urban Water Systems*. (Proc 18th European & 1st Asian Junior Scientists Workshop: Sewer Networks and Processes within Urban Areas). IWA Publishing, London, UK. ISBN 978 1843395065.

Amick, R.S. and Burgess, E.H. (2000) *Exfiltration in Sewer Systems*. Report EPA/600/R-01/034, US Environmental Protection Agency, National Risk Management Research Laboratory. Cincinnati, Ohio, US

Beal, C., Gardner, T., Menzies, N.W., Rassam, P. and Vieritz, A. (2004) Prediction of steady-state flux through variably saturated zones within a septic absorption trench. *Proc. SuperSoil2004, 3rd Australian & New Zealand Soils Conference*. 5–9 September 2004. University of Sydney, Sydney, Australia.

Blackwood, D.J., Ellis, J.B., Revitt, D.M., Gilmour, D.J. and Staines, A. (2005a) Factors influencing exfiltration processes in sewers. *Water Science & Tech.*, **51**(2), 147–154.

Blackwood, D.J, Ellis, J.B, Revitt, D.M., Gilmour. D.J. and Staines, A. (2005b) Exfiltration from sewers: Is it a serious problem? In: Mikkelsen, P. (Edit): *Proceedings IWA 10th International Conference on Urban Drainage (ICUD)*, 21–16 August 2005. Copenhagen, Denmark. IWA Publishing, London, UK. ISBN 978 1843395744.

Blackwood, D.J., Gilmour, D.J., Stainer, A., Ellis, J.B. and Revitt, D.M. (2008) *Riverside Dundee Sewer Exfiltration*. Final Report to UK Water Industry Consortium, January 2008. Urban Water Technology Centre, University of Abertay, Dundee, Scotland.

De Silva, D., Burn, S., Tjandraatmadja, G., Moglia, M., Davis, P., Wolf, L., Held, I., Vollertsen, J., Willaims, W. and Hafskjold, L. (2004) Sustainable management of leakage from wastewater pipelines. In: *4th World Water Congress: Water Supply and Water Services Management* (2005). Proc IWA Congress, 21–22 September 2004. Marrakesh, Morocco, IWA London. UK. ISBN 978 1843394709.

Ellis, J.B., Revitt, D.M., Vollertsen, J. and Blackwood, D.J. (2008) Factors influencing temporal exfiltration rates in sewer systems. *Proc. 11th.Int. Conf. on Urban Drainage (ICUD)*. August 2008 Edinburgh, Scotland. CD-ROM. IWA London, UK. ISBN 978 1899796 212.

Ellis, J.B., Revitt, D.M., Vollertsen, J. and Blackwood, D.J. (2009) Sewer exfiltration and the colmation layer. *Water Science & Tech.*, **59**(11), 227–2280.

Ellis, J.B., Revitt, D.M., Lister, P., Willgress, C. and Buckley, A. (2003) Experimental studies of sewer exfiltration. *Wat.Science & Tech.*, **47**(4), 61–67.

Guzman, K., La Motta, E.J., McCorquodale, J.A., Roiss, S. and Ermogenous, M. (2007) Effect of biofilm formation on roughness coefficient and solids deposition in small-diameter PVC sewer pipes. *ASCE, Journ Envir.Eng.*, **133**(4), 364–371.

Held, I., Klinger, J., Wolf, L. and Hotzl, H. (2005) Direct measurements of exfiltration at a sewer site under operating conditions. In: Howard, K.W.F. (Edit); *Proc. Mathias Eiswirth Memorial Volume*. Balkema, Rotterdam, The Netherlands.

Karpf, C., Traenckner, J. and Krebs, P. (2008) Hydraulic modelling of sewage exfiltration. *Proc. 11th Int.Conf. on Urban Drainage* (ICUD). August 2008, Edinburgh, Scotland. CD-ROM. IWA London, UK. ISBN 978 1899796 212.

Klinger, J., Turkovic, R., Wolf, L. and Hotzl, H. (2007) Long term investigations of the colmation processes at a real sewer defect. *Geophysical Research Abstracts*, **9**, 1–3.

Littlewood, K. and Butler, D. (2003) Movement mechanisms of gross solids in intermittent flow. *Water Science & Tech.*, **47**(4), 45–50.

Memon, F.A., Fidav, A., Littlewood, K., Butler, D., makropoulus, C. and Liu, S. (2007) A performance investigation of small-bore sewers. *Wat.Science & Tech.*, **55**(4), 85–91.

Norsker, N.H., Nielsen, P.H. and Hvitved-Jacobsen, T. (1994) Influence of oxygen on biofilm growth and potential sulphate reduction in gravity sewer biofilm. In *Proc. 2nd Int. Conf. Sewer Processes and Networks; The Sewer as a Physical, Chemical and Biological Reactor*. 25–28 May 1994. Aaalborg, Denmark. Also in Hvitved-Jacobsen, T. (1998; Edit): *The Sewer as a Physical, Chemical and Biological Reactor*. IWA Publishing, London, UK. ISBN 978 0080433790.

Oms, C., Gromaire-Mertz, M.C. and Chebbo, G. (2002) In-situ observations of the water-sediment interface in combined sewers using endoscopy. 5–12 in: *Proceedings 3rd Int.Conf. Sewer Processes and Networks.* April 2002. CEREVE, Ecole Nationale Pontes et Chaussees. Paris, France. Also in Bertrand-Krajewski, J-L and Chebbo, G. (Edits): *Proc. 3rd Int. Conf. Sewer Processes and Networks.* IWA Publishing, London, UK. ISBN 978 1843394365.

Rauch, W. and Stegner, T. (1994) The colmation of leaks in sewer systems during dry weather flow. *Water Science & Tech.*, **30**(1), 205–210.

Rice, R.C. (1974) Soil clogging during infiltration of secondary effluent. *Water Poll. Control Fed.*, **46**(4), 708–716.

Rocher, V., Azimi, S., Moilleroi, R. and Chebbo, G. (2002) Biofilm in combined sewers: Wet weather pollution sources and/or dry weather pollutant indicator. In: *Proceedings 3rd Int. Conf. on Sewer Processes & Networks*, April 2002. CEREVE, Ecole Nationale Pontes et Chaussees. Paris, France. Also in Bertrand-Krajewski, J-L and Chebbo, G. (Edits): *Proc. 3rd Int. Conf. Sewer Processes and Networks.* IWA Publishing, London, UK. ISBN 978 1843394365.

Rutsch, M., Rieckermann, J., Cullmann, J., Ellis, J.B., Vollertsen, J. and Krebs, P. (2008) Towards a better understanding of sewer exfiltration. *Water Research.* **42**, 2385–2394.

Selvakumar, A., Field, R., Burgess, E.H. and Amick, R.S. (2004) Exfiltration in sanitary sewer systems in the United States. *Urban Water Journal*, **1**(3), 227–234.

Velickovic, B. (2005) Colmation as one of the processes in interactions between groundwater and surface water. *Facta Universitates, Series Architecture & Civil Engineering*, **3**(2), 165–172.

Vollersten, J., Vorkamp, K., Hvitved-Jacobsen, T. and Jensen, M.A. (2002) *Udsiuning af spildevand fra aflohssystemer. (Exfiltration of wastewater from Sewers).* Miljoproject No.685. Danish Environmental Protection Agency, Danish Ministry of Envrionment, Copenhagen, Denmark.

Vollertsen, J. (2007) *The Phenomenon of Sewerage Leakage*. Unpublished Report. School of Environmental Engineering, Aalborg University, Denmark.

Vollertsen, J. and Hvitved-Jacobsen, T. (2003) Exfiltration from gravity sewers; A pilot scale study. *Wat.Science & Tech.*, **47**(4), 69–76.

Wolf, L., Morris, B. and Burn, S. (Edits). (2007) *Urban Water Resources Toolbox: Integrating Groundwater into Urban Water Management*. IWA Publishing, London. UK. ISBN 1843391384.

Chapter 7

Test and field applications of methods in Germany and France

Mandy Rutsch, Peter Krebs, Jérôme de Bénédittis and Jean-Luc Bertrand-Krajewski

INTRODUCTION

This chapter gives some brief examples of application of the methods presented in chapter 2 for various cities in Germany and France. These examples show:

(i) that the methods are applicable for very different geographic contexts and sewer systems,

(ii) the approaches yield some orders of differing magnitude of infiltration and exfiltration rates, and

(iii) they provide practical comments on the use and application of the methods.

Detailed reports and publications covering these field applications in more detail are available on the APUSS website (www.insa-lyon.fr/Laboratoires/URGC-HU/apuss)

APPLICATIONS IN GERMANY

Application in Dresden Salzb

Salzb is a small municipal catchment located in the eastern part of the city of Dresden. It has a total combined foul sewage sewer length of 9320 m, which corresponds to 350 individual sewer reaches. 4455 inhabitants are connected to this part of the combined sewer system which has an average annual wastewater discharge of 163,950 m^3/s. The sewer slope is 0.1% on average and is composed of concrete and egg-shaped sections, which were constructed in the period between 1900 to 1930. Sewer diameters range between 300 and 500/750 mm. The catchment is rather homogeneous in terms of its pipe and housing characteristics. Only 11 reaches have been surveyed and possess recorded CCTV data. However, even this small sample number of data information exhibits a poor condition for the sewer system with defect class representing the worst possible condition.

The catchment is very well defined by a piezometer coverage which extends from the head to the end of the catchment. A part of the catchment is situated within an old arm of the river Elbe causing a short time reaction to any subsurface groundwater movement. In addition, given a connection between storm water accumulation due to loamy soils, changes in groundwater levels are also quickly noticeable in the sewers as well. As shown in Figure 1, the sewers are situated within varying groundwater levels induced by the river Elbe, which could provide appropriate conditions for both infiltration and exfiltration.

Edited by Bryan Ellis and Jean-Luc Bertrand-Krajewski. ISBN: 9781843391494. Published by IWA Publishing, London, UK.

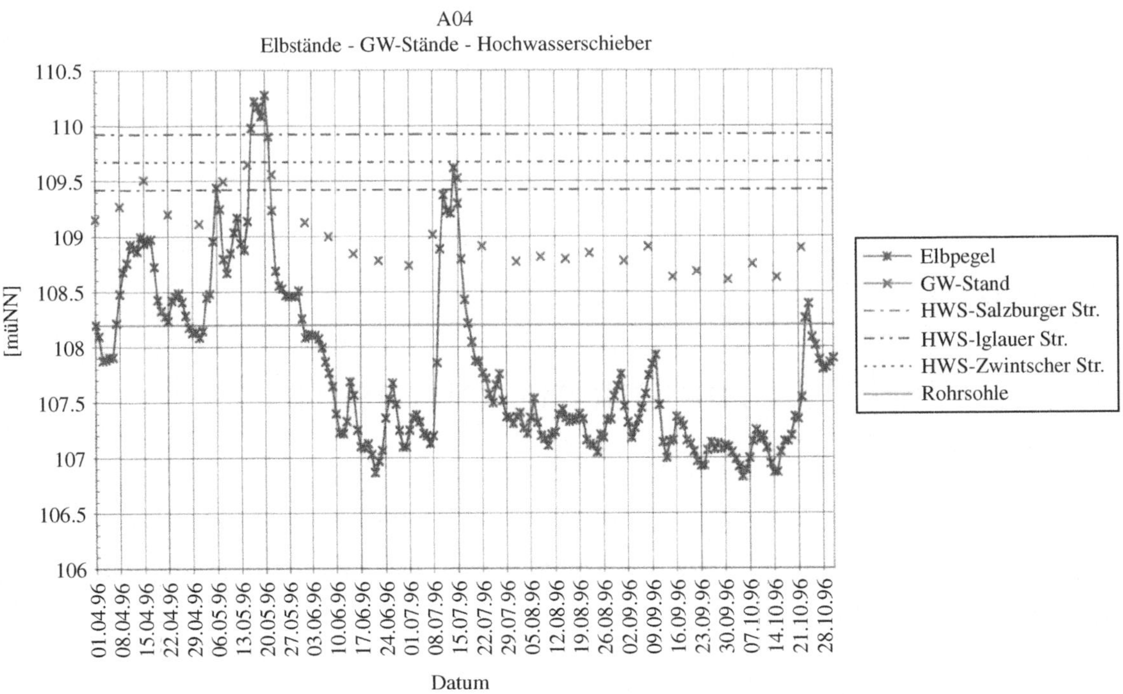

Bild C5.1: Darstellung der möglichen Einflußgrößen auf den Fremdwasseranfall an der Meßstelle A04

Figure 1: Comparison of sewer invert level, groundwater level and water level in the river elbe.

Methods applied and equipment used

The determination of infiltration was conducted by applying the pollution times series method (see Chapter 2). Initially a ultrasonic doppler flow sensor was installed in the sewer, and was accompanied later by an UV-VIS in-line spectrometer probe from s::can (www.s-can.at). To determine exfiltration, the QUEST-C methodology (see Chapter 2) was applied primarily due to the very low flow occurring at the head of the sewer system. Lithium chloride was used as the indicator tracer with Natrium bromide as the reference tracer. Background concentrations of the tracers remained at zero during tests as well as before and after the measurement test runs.

Infiltration: results and discussion

The measurement campaign was conducted from October 2003 through to March 2004, although very wet and snowy weather did not allow an evaluation of the whole data set. Given problems of measuring night-time low flow levels a weir was tested to raise water levels but this caused unsatisfactory values of suspended solids with sediments accumulating behind the weir and data from this period was abandoned. To calibrate the s:can probe a large sampling campaign was conducted over the whole period and Figure 2 shows the comparison of COD samples obtained from laboratory analysis compared to COD measured on-line.

The infiltration model was set up for several time periods as a plausibility test and to look at any seasonal variation in the time series. In Figure 3 the infiltration model is shown for the period of 18.02.04 until 27.02.04 with the calcluated infiltration rate being zero, which was also the result for other recorded periods.

Thus, the initial idea to evaluate a time series under varying groundwater conditions provided negative outcomes and comparison of the groundwater levels in the area with the water level in the sewer did not result in any seasonal variation. The very dry weather which occurred in 2003 resulted in a correspondingly low groundwater table, which did not rise to reach the sewer invert level until March 2004.

Exfiltration: results and discussion

The exfiltration rate for the sewer system was measured in the lower part of the catchment. At this point in the main sewer, the wastewater pipe has a diameter of 500/570 mm and extends over a reach length of 840 m. Records indicate that the main part of this sewer distance was undisturbed and undamaged with only some small upstream inflows. The output graphs from the analysed tracer concentrations are shown in Figure 4, which compare tracer concentrations with discharge as recorded throughout the experimental runs.

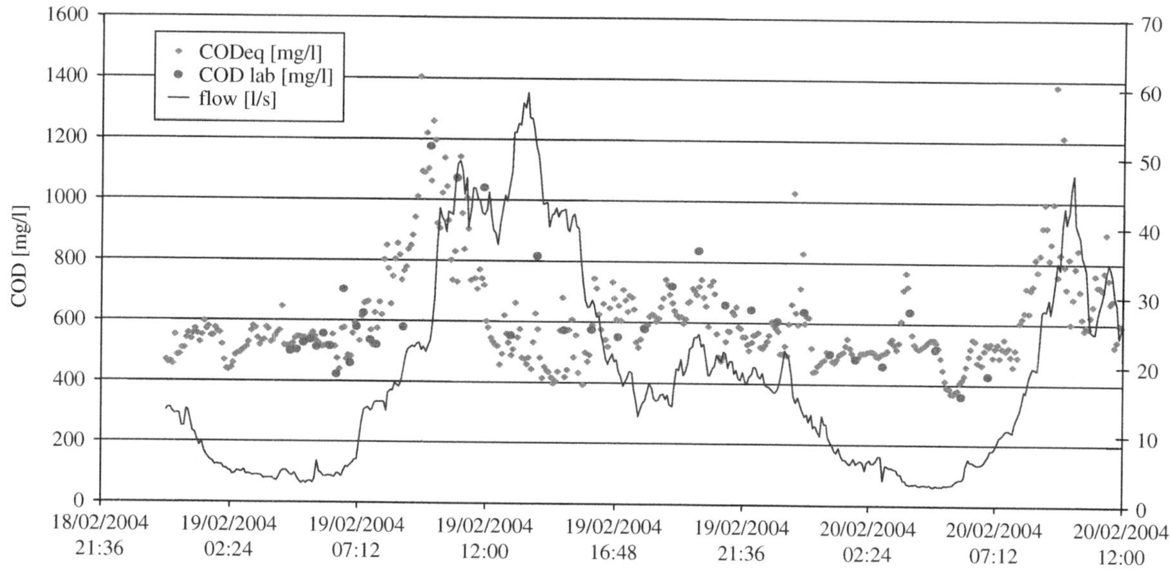

Figure 2: Sampling campaign in Dresden Salzb (19/20.02.04).

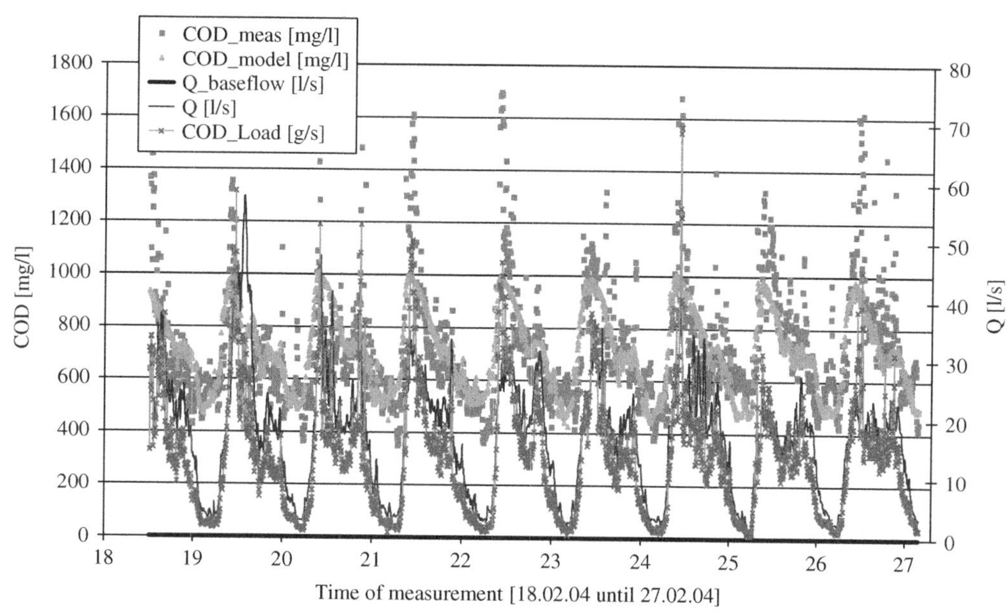

Figure 3: Infiltration Model Measuring Campaign for Dresden Salzb (18.02.04 until 27.02.04).

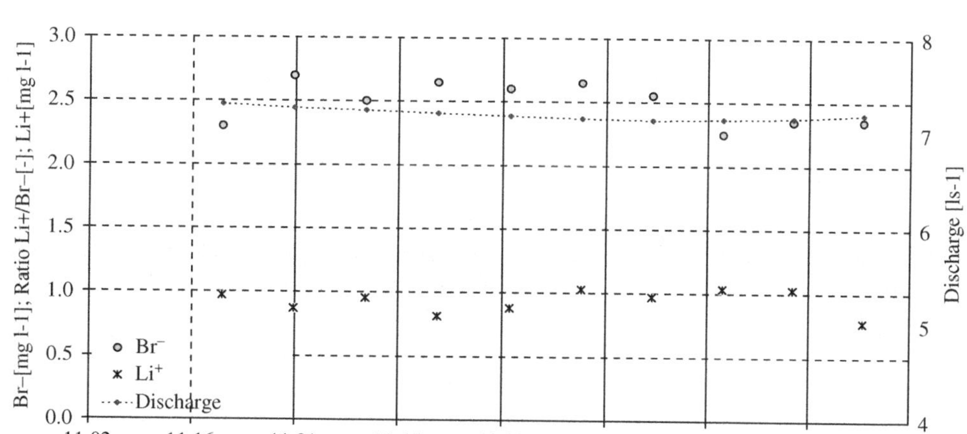

Figure 4: QUEST-C experiment Dresden Salzb (Li and Br comparison with flow data).

The flow during the experiment was very stable with the Lithium and Bromide tracer concentrations not fluctuating very much (Figure 4). The exfiltration rate was calculated to be 14% of the dry weather flow, which is an average of 0.56 l/s and ranging between 0.51 to 0.62 l/s. (or 48 m^3/d).

APPLICATION IN DRESDEN LOCKWITZ AND MUEVA

Mueva is a small municipal catchment composed of low density suburban housing in the vicinity of Dresden with 1300 inhabitants. The sewered catchment extends over 200 ha and has an average dry weather flow rate of 7.4 l/s. Both catchment and sewer system are very homogenous in their characteristics which facilitates the transfer of sewer reach information to the whole catchment. The infiltration rate is influenced by groundwater and a small stream located adjacent to the sewer. CCTV records were only available for a small number of sewers but indicate the system to be in poor condition. The Lockwitz catchment includes the sewered discharges of the Mueva catchment and serves a total of 2850 inhabitants covering a 400 ha connected area. The dry weather flow for this larger catchment adds up to 12 l/s.

Investigations of sewer flow in 2002 (Figure 5) exhibited a strong seasonal variation in the dry weather flow underlining the poor condition of the sewers and the varying effluent level in the sewer due to groundwater variations and variations of the water level in the nearby stream.

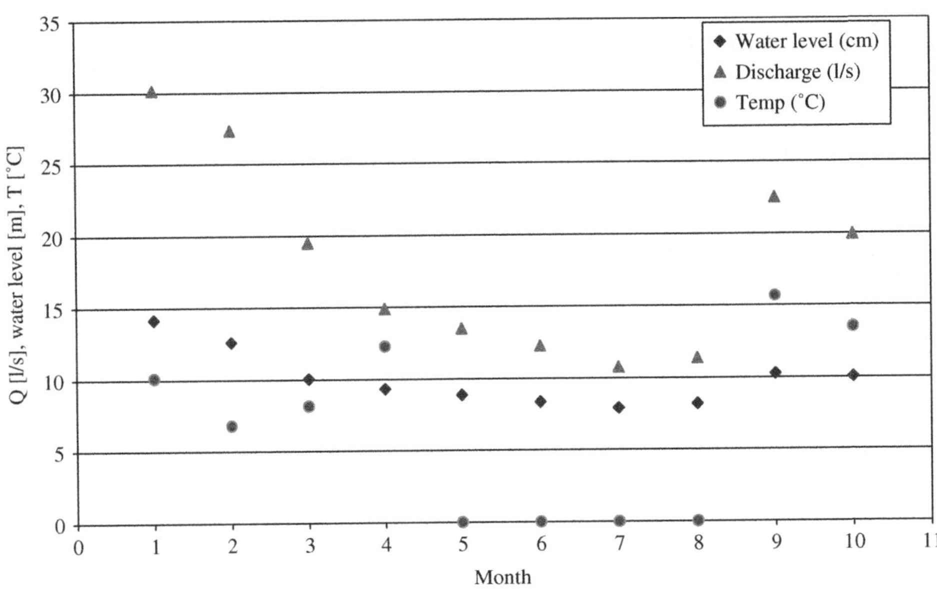

Figure 5: Investigation of dry weather flow in Lockwitz. (2002).

Methods applied and equipment used

The determination of infiltration was conducted by applying the pollution times series method (see Chapter 2). As in the Dresden Salzb investigations, an ultrasonic doppler flow sensor was initially installed, later accompanied by an UV-VIS inline probe from s::can (www.s-can.at). Additionally in Lockwitz, single reaches were blocked; at the upstream side of the sewer the reach was blocked using a balloon and at the downstream side a weir was installed (shown in Figure 6), the rising or falling water level observed at the weir, indicating exfiltration and infiltration, respectively.

Infiltration: results and discussion

At the Mueva catchment good measuring conditions were obtained with the measurement campaign lasting from October 2003 to December 2003 accompanied by several sampling campaigns to calibrate the s:can probe. As expected a very high infiltration rate of 4 l/s was observed, which is some 46% of the mean dry weather flow (Figure 7).

The Lockwitz catchment also provided good measuring conditions with the probe being tested the first time without power supply, which required more effort in terms of data readout and maintenance. The infiltration rate was calculated to be 7 l/s or 57% of the dry weather flow. Unfortunately, the probe could not be calibrated, thus the obtained measuring results were not considered in the final evaluation. The sewer was blocked over a length of approximately 513 m probably due to the very low 0.2% slope. However, the blocking exhibited very good and reproducible results and the infiltration rate was calculated as being 0.08 l/s or 0.57 l/(h*m).

Figure 6: Weir on the downstream side of the sewer.

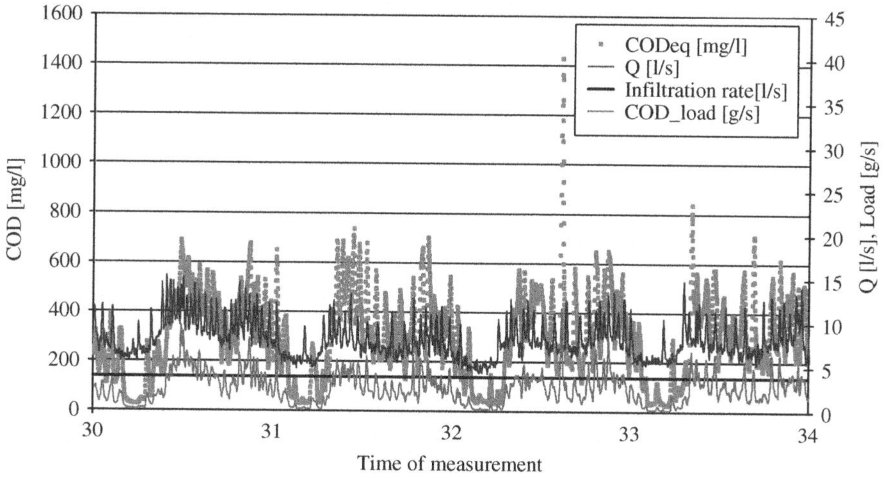

Figure 7: Measurement Campaign in Mueva (30.11.03–4.12.03).

APPLICATION IN GLADBECK

A test site was chosen in Gladbeck-Brauck with the connected area amounting to 67,9 ha and consisting of 22,1 ha impervious area. The sewered catchment serves 3670 inhabitants and the wastewater pipe has a dimension of 1400 mm at the measurement station. Data were collected during February 2004 after a long rainy period with approximately 140 mm of rainfall. The measurement campaign was conducted over a period of 2 weeks, with flow rate recorded using an ultrasonic-doppler probe and with COD and TSS concentrations measured using the s::can UV-VIS spectrometer, which was carefully calibrated on-site. The determination of infiltration was conducted by applying the pollution times series method (see chapter 2).

Infiltration: results and discussion

Figure 8 illustrates the recorded hydrographs and pollutographs and the estimated infiltration rate for the Gladbeck-Brauck site. A strong exponential decrease of the infiltration rate is evident, induced by the preceding rain event. The decrease in the infiltration rate causes a slight increase in the amplitude of the pollution concentration, further signifying the decrease of the infiltration rate. The obtained results are shown in Table 1 differentiated after the exponential decrease noted in the infiltration rate.

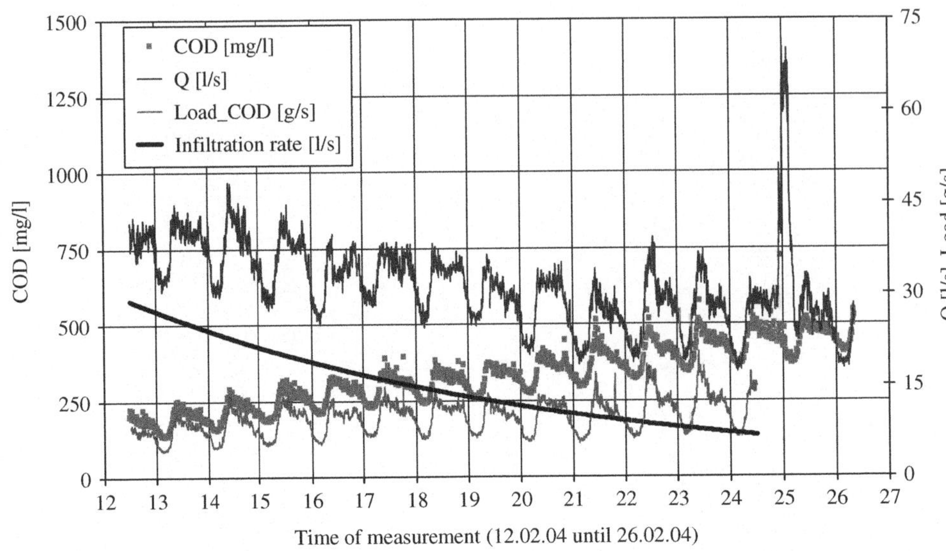

Figure 8: Measurement campaign in gladbeck.

Table 1: Characterisation of diurnal sewage flow for the investigated days.

Hydrograph		**12./13.02.04**	**20./21.02.04**	**23./24.02.04**
Mean COD conc.	mg/l	184	361	436
Mean flow rate Q	l/s	38	27	26
Relative infiltration rate Q_{inf}/Q_{sewage}	%	72	38	27.6
Characteristics	–	High infiltration rate	Medium infiltration rate	Low infiltration rate
	–	Day after a storm event	Fading out interflow	Dry weather day

APPLICATION IN BOTTROP

A test site with 4080 inhabitants was chosen in a subcatchment in the north of the city of Bottrop. The catchment has a connected area of 61.6 ha with a 26.5 ha impervious area and a sewer diameter of 1500 mm at the measurement station.

Methods applied and equipment used

The data were collected during April 2004 with the measurement campaign being conducted over a period of 2 weeks. Flow rate was recorded using an ultrasonic-doppler probe with COD and TSS concentrations measured with the s::can UV-VIS spectrometer, which was carefully calibrated on-site. The determination of infiltration was conducted by applying the pollution times series method (see chapter 2).

Infiltration: results and discussion

Figure 9 illustrates the recorded hydrographs and pollutographs and the estimated infiltration rate. The investigation in Bottrop was not successful due to adverse weather condition and untrained staff as well as laboratory difficulties. The infiltration rate was calculated as being 14.6% (Figure 9) although this result carries a high degree of uncertainty, particularly as the modelled COD time series does not comply with the measured time series.

APPLICATION IN DRESDEN-HELLERAU

Dresden-Hellerau is located on the upland slopes of the city with the sewer system being very homogeneous in terms of both structure and location. The sewers were constructed in 1920 using clay as the construction material. The geomorphic situation at the site showed a disruption from sandy soils to mountainous forms and occurence of floating groundwater. The catchment is illustrated in Figure 10 as viewed on the Aquabase software (see Chapter 10).

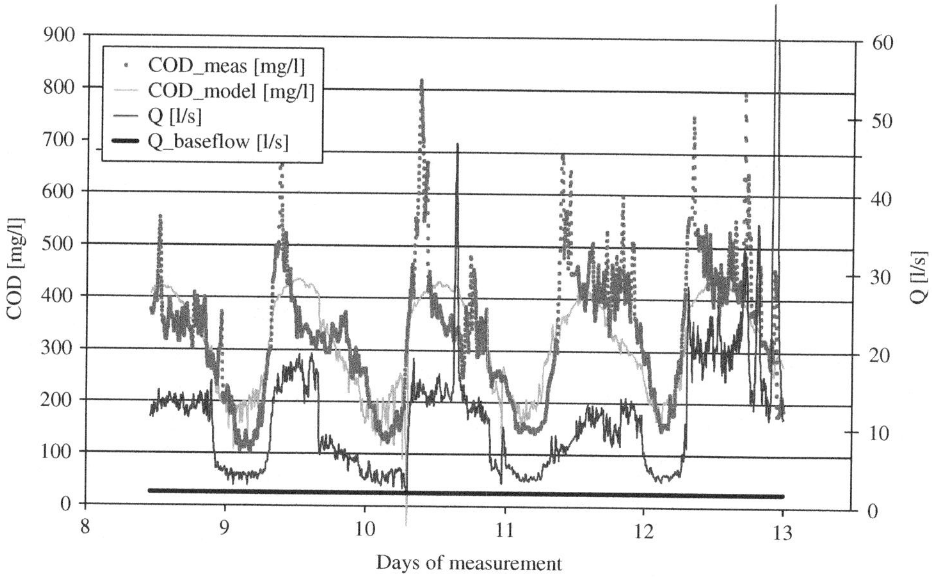

Figure 9: Measurement campaign in bottrop.

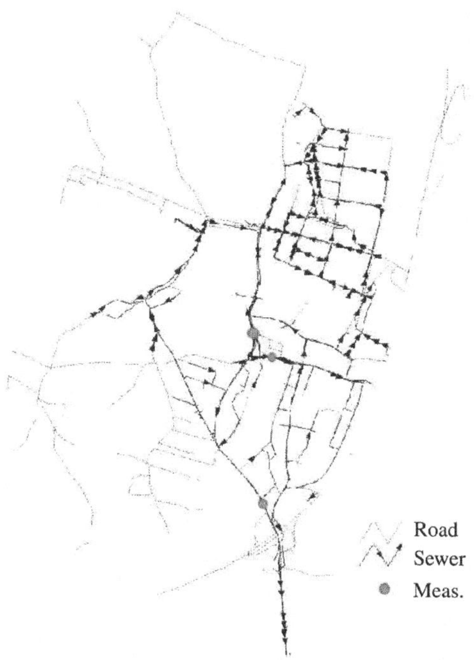

Figure 10: Sewer system in Dresden Hellerau.

The good conditions allowed a testing of four different sewers within the one sub-catchment to find possible similarities. It was assumed that the sewer line did not have any groundwater influence as far as the performance of exfiltration measurements were concerned.

Methods applied and equipment used

To determine infiltration, flow measurements and mass balance calculations were performed with exfiltration measurements being conducted using the QUEST method as described in Chapter 2.

Infiltration: results and discussion

Flow measurements at Mor showed an influence of groundwater, with an infiltration rate of approximately 40% being estimated. Due to little flow in the other sewer lines, flow measurement was only possible for this site. The initial assumption that there was little if any groundwater influence could thus not be proven.

Exfiltration: results and discussion

Exfiltration could not be determined at any of the investigated sites as indicated in the final column of Table 2.

Table 2: Characteristics and exfiltration rates in dresden-hellerau.

Name	Length [m]	Date of constr.	Material	DN [mm]	Slope [%]	$Q_{Ex,mean}$ [%]
Doer	660	1920	Clay	250	2–6	0
Pfarr	560	1920	Clay	250	2–6	0
Mars	1170	1920	Clay	250	3–6	0
Mor	730	1920	concrete	250/375	0.40	0

After conducting the infiltration measurements, the influence of groundwater over the whole catchment area cannot be excluded. The flow in the Doer, Pfarr and Mars catchments was too low to conduct meaningful flow measurements and thus these data could not be used in the data evaluation. The conductivity of the wastewater at the test sites was very low and stable, and a good recognition of the tracer pulse shapes was possible. As these measurements were the first QUEST measurements to be conducted within any of the German cities, a lack of working experience might have led to unreliabilities in the experimental test runs. However, due to lack of valuable CCTV records, a detailed test of plausibility was not possible.

APPLICATION IN BERLIN-MAL

The Berlin-Mal area was built in the 1990s and shows a homogeneous appearance with multi-storey buildings and a grid-like street order. It is a high density development with approximately 60,000 inhabitants. The catchment is situated on the Barnim-Plateau with boulder clay as aquitard and an occurrence of floating groundwater. The distance to the groundwater table varies between 10 to 20 m. Structural data and CCTV records were available for the sewer system together with groundwater data. However, groundwater influences were excluded from consideration based on evidence of soil maps and work experience of local sewer operators. This assumption was underlined from examination of hydrographs from the last few years which showed that the discharge decreased from 1995 to 2002 by about 30% due to changes in water consumption; a significant drop in summer is attributed to property vacancies during this summer vacation period. In addition, the hydrographs are very stable and do not show any seasonal variation, which might be expected for groundwater-influenced areas.

In addition to the APUSS exfiltration tracer measurement, flow measurements at the end of the catchment were conducted to observe the minimum night flow. These flow measurements showed a water deficit compared to the expected night flow for an area having 60,000 inhabitants. This could be attributed to people's nightly behaviour or to a loss of water due to exfiltration.

Methods applied and equipment used

The determination of infiltration was conducted by applying the pollution times series method with exfiltration applying the QUEST and QUEST-C methodology (See Chapter 2). Determined flow measurements were compared to measurements conducted by the local sewer operators and to the drinking water consumption. The initially observed wastewater deficit could not be proven in any substantial magnitude although a water deficit was observed, which might be reasonably be connected to exfiltration.

Infiltration: results and discussion

The probe was installed at the lower end of the catchment. Given the excellent working conditions (power supply, a locked territory, and support from the local operators), the probe delivered good quality results with Figure 11 showing an example of the measured hydrographs and the modelled concentrations.

The hydrograph shapes deviate from those of other German catchments and the COD concentration do not follow the flow over the day. While the pollutant concentration has only one peak in the morning, the flow has a second peak in the evening. This was attributed to local people's daily behaviour. However, using an enhancement of the pollutant times series approach, the infiltration rate was calculated to be approximately 5%.

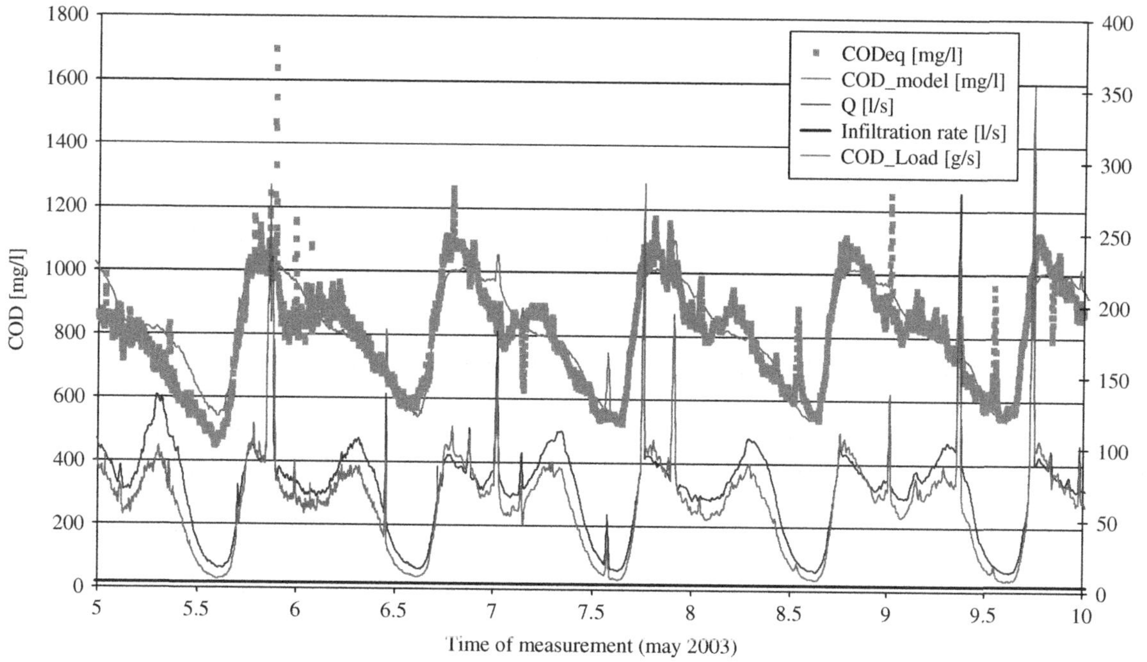

Figure 11: Flow and CODeq for sub-catchment berlin-MAL.

Exfiltration: results and discussion

Nine exfiltration measurements were carried out in Berlin-MAL although two of them did not succeed due to a disturbed baseline and unsatisfactory data quality. The characteristics and results of the remaining valid tests are shown in Table 3. According to information given by the local sewer operator, the quality of workmanship in the 1980s, when these sewers were built, was rather poor which is also demonstrated by the different materials used within one sewer section. The exfiltration rate in Table 3 is given in volume per leakage area and per time interval in order to compare it to other results. The leakage area was obtained by a summary of all leakages occurring in the wetted perimeter and which are visible from the CCTV records.

Table 3: Characteristics and exfiltration rate of investigation sites in berlin-MAL.

Name	Length [m]	Material	DN [mm]	A_{Leak} [cm^2]	$Q_{Ex,mean}$ [%]	$Q_{Ex,mean}$ [m^3/cm^2/d]	$Q_{Ex,5\%}$ [m^3/cm^2/d]	$Q_{Ex,95\%}$ [m^3/cm^2/d]
Fal	1035	Various	600	464	3.5	0.17	0.03	0.815
Bies	1060	clay, concrete	200–400	848	15.5	0.073	0.051	0.094
Kroe/Cri	640	Clay	300–400	416	0.5	0.007	0.009	0.25
Kroe/Rib	2000	BGM, Clay	300–600	917	2.5	0.033	0.02	0.053
Schw/Ki	706	BGM, PVC	200–400	246	9.6	0.114	0.106	0.127
Schw/ Rib	1490	Various	200–600	454	17.6	0.54	0.42	0.48
Hag/Ki	780	PVC, clay	200–300	227	6.7	0.078	0.016	0.14

As there is no groundwater intrusion, the wastewater is not diluted and therefore shows a very high and unstable background conductivity. Confirmatory measurements at night were not possible, as the flow was too low but the conductivity did not reduce that much as evidenced by the conductivity records over 24 hours. It is known that high background conductivity can cause serious difficulties for the APUSS tracer measurement when employing NaCl as the tracer substance. Disturbed background conductivity can also affect the recognition of the pulses. The 10% quantile gives an idea about the uncertainty of the measurement. For example, the exfiltration rate in the Fal sub-catchment ranges from 0.03 to 0.815 $m^3/cm^2/d$, which actually requires more measurements to reduce the high observed range. The Schw/Rib sub-catchment on the other hand, has only a small uncertainty, although the determined high exfiltration rate of 0.54 $m^3/cm^2/d$ is questionable.

APPLICATION IN BERLIN-LR

The catchment is situated in the north of the city of Berlin on the Teltow plateau. It is a residential area with a high fraction of green and unpaved surface and single family housing. It is characterised by boulder clay as an aquitard, whilst confined groundwater and floating groundwater occurs. The groundwater level is approximately 10 m below the ground surface. Surface conditions vary between landfill, built-up areas, with soils being characterised by medium sand to medium loamy sand to loamy sand. The clay-built sewer system has two construction periods, 1930 and 1965 with the main sewer dimension being 200 to 450 mm. Data availability for the catchment was good and CCTV data of the observed sewer section also existed. Initial observations of the night flow at the lower end of the catchment exhibited a water deficit, when comparing the measured night flow with the expected night flow for an area having 50,700 inhabitants. This could be attributed to people's nightly behaviour or to a loss of water due to exfiltration. The water consumption and wastewater discharge respectively remained rather constant over the years of observation. A distinct decrease of discharge in July and August is attributed to vacation time. The missing seasonal variation suggested that the sewers were not likely to be influenced by groundwater. Unfortunately it was not possible to employ the infiltration s::can probe at this site to either confirm or negate this suggestion.

Exfiltration: results and discussion

Table 4 shows the measured exfiltration rates together with site characteristics and standard deviatiosn for the Berlin-LR catchment. The parallel application of QUEST and QUEST-C in the Stein sub-catchment yielded similar results, although the standard deviation for the QUEST experiment is much larger than that of the QUEST-C experiment. The second parallel measurement had to be interrupted as the site was not suitable for installing the required equipment. The third parallel measurement (Ldamm) exhibited two completely different results. It was decided to choose the result obtained with the QUEST-C measurement, as the result is similar to the exfiltration rate of the Scabi sub-catchment. The Ldamm and Scabi sub-catchments share the same sewer and most of the defects occurred within this part of the system. A similar exfiltration for both sites was thus expected. A very high exfiltration rate was observed in the Ekzlo sub-catchment which averaged 0.26 $m^3/cm^2/d$ (18.8% of the dry weather flow). Whilst measurement errors should be minimal, it is nevertheless questionable as to whether such a high exfiltration rate can actually occur. The Locke sub-catchment also exhibited a rather high exfiltration rate with 42% of the dry weather flow but here the leakage area is rather large, which could explain the recorded exfiltration rate of 0.11 $m^3/cm^2/d$.

Table 4: Exfiltration rates and sewer characteristics in Berlin-LR.

Name	Length [m]	Material	DN [mm]	A_{Leak} [cm^2]	$Q_{Ex,mean}$ [%]	$Q_{Ex,mean}$ [$m^3/cm^2/d$]	$Q_{Ex,5\%}$ [$m^3/cm^2/d$]	$Q_{Ex,95\%}$ [$m^3/cm^2/d$]
Stein	1220	Clay	200–350	530	8.8	0.10	0.093	0.108
Stein QUEST	1220	Clay	200–350	530	14.6	0.143	0.025	0.164
Altlt QUEST	1450	Clay	200–350	585	8.6	0.052	0.019	0.062
Ldamm	1460	Clay	350–500	990	5.6	0.066	0.031	0.107
Scabi	1320	Concrete	300–500	1072	5.0	0.046	0.045	0.046
Gerst	1870	Clay	250–350	568	20.5	0.07	0.057	0.082
Locke	1475	Clay	200–450	1623	42	0.11	0.096	0.125
Ekzlo	1350	Reinf. concrete	450–600	639	18.8	0.26	0.185	0.333
Ketti	1100	Asb. cementos	300–450	108	2.3	0.043	0.04	0.046

EXFILTRATION SUMMARY

In total, 29 exfiltration measurements have been carried out in Dresden and Berlin by applying the QUEST and/or QUEST-C methods previously described in chapter 2. Some sewer lines have been investigated using both methods. In Figure 12, the exfiltration rates for measurements undertaken in the 19 catchments within these two cities are shown in association with the mean flow rate recorded during the particular measurement. The measured exfiltration rates average to approximately 1.0 l/s, which equals 86 m^3/d. After analysis of the CCTV records, the exfiltration rates were also referred to the pipe leakage area (Figure 13). However, the number of useful measurements further decreased since no CCTV data were available for the sewer lines in Dresden.

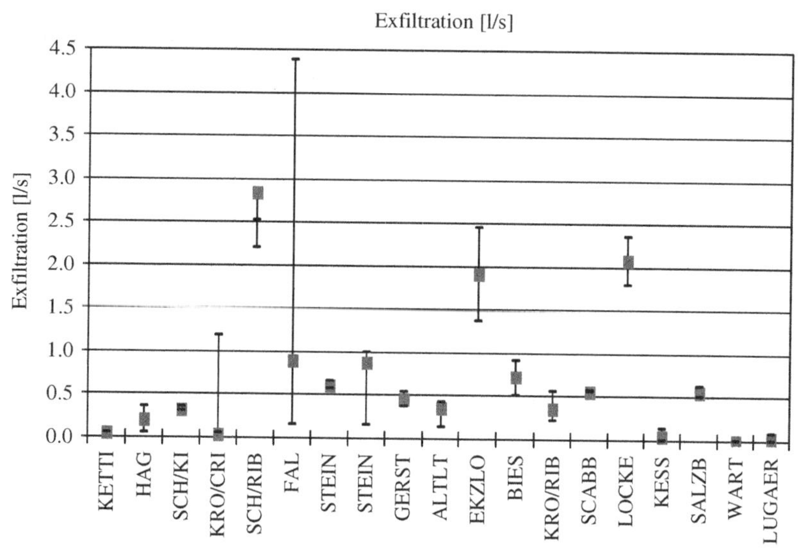

Figure 12: Measured exfiltration rates [l/s] related to a mean flow rate.

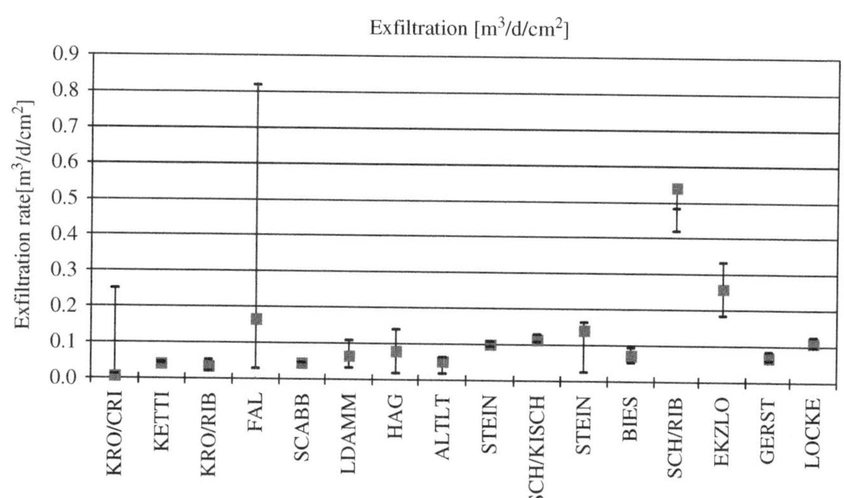

Figure 13: Measured exfiltration rates [$m^3/m^2/d$] depending on the leakage area [m^2].

As a plausibility test, the exfiltration rates obtained in real sewers were compared to the results of various laboratory experiments described in the literature. Field tracer measurements exhibited exfiltration rates only some 10% of the exfiltration rates obtained in the reported laboratory experiments. This deviation can be attributed to a concentration on single defects in the laboratory experiments, to the steady state condition obtained during one experiment (pressure test) and in some cases may also reflect the enveloping soil conditions in the laboratory experiments.

The practical applicability of the QUEST procedure appears to be good and is not complicated. The provided standard operation procedure (SOP) provides a sound basis to perform the measurements, but needs to be adapted to specific individual and local requirements. Preliminary campaigns are needed before carrying out the experiments, such as tracer baseline definition and flow recording. High background conductivity can cause serious difficulties for the measurement when using sodium chloride as the tracer. The data analysis is rather sophisticated but provides a reliable source of both exfiltration rate and measurement uncertainties. The application of the QUEST-C method is relatively simple by comparison. However, the set-up is sophisticated and expensive (at least during the prototype phase of the APUSS project). As the utilised laboratory equipment is not necessarily adapted or appropriate for direct field application, a very careful handling is required. Furthermore, the confirming analysis of the samples demands the use of high accuracy instrumentation in the laboratory. The uncertainty in the lab was calculated to approximate 10% in case of low concentrations of Br^- and Li^+ (background concentration and concentration less than 2 mg/l) and 5% in cases of higher concentrations or when analysing the standards. A double or triple analysis of the samples is necessary.

In order to obtain reliable results, detailed information is required, especially on flow conditions and on natural tracer baseline concentrations. To diminish the measurement errors inherent in both methods, a large number of tests needs to be performed. It is highly recommended i) to closely follow the standard operation procedure (SOP) and ii) to undertake some preliminary training on the method and on the equipment applied in the field.

INFILTRATION SUMMARY

The pollution time series method as described in Chapter 2 was applied in six German sub-catchments:

- 2 measurements in collaboration with Emschergenossenschaft in Gladbeck and Bottrop,
- 1 measurement in collaboration with BWB (Berlinerwasserbetriebe) in Berlin,
- 3 measurements in collaboration with Stadtentwässerung Dresden in Dresden.

Site characteristics and results are summarised in Table 5. The measurements show very good and in particular, reproducible results although the success of the method is subject to some restrictions. Flooding of the s:can probe due to backwater must be excluded and a certain minimum infiltration rate is necessary to keep the uncertainty of the measurement small (the lower the rising velocity of the water level, the higher becomes the uncertainty of reading the scale). The installation of the UV-VIS s:can probe is restricted to a water level in the sewer pipe higher than 10 cm.

Table 5: Sub-catchments for infiltration measurements in Germany.

Site	Gladbeck	Bottrop	Berlin	Sal-zb_Street	Mueva	Lockwitz
Average dry weather flow	12 l/s	10 l/s	50 l/s	7 l/s	7.4 l/s	12 l/s
Characteristics	Suburban catchments with well described ground-water and sewer conditions		Homogenous, densely populated area with a low groundwater table	Varying ground-water levels induced by the river Elbe	Homogenous, suburban catchment, influenced by groundwater and a small stream next to the sewer	
Inhabitants	3 670	4 080	60 000	4 455	1 300	2 850
Infiltration rate	27.6%	14.6%	5%	0%	46%	57%
Measuring period	2 weeks	4 weeks	4 weeks	6 months	3 months	5 days
Comment	Exponential decrease of infiltration rate induced by preceding rain	Poor data quality	Validation of exfiltration measurements	Long time investigation including investigation of exfiltration		Long time series of flow measure-ments; probe was not calibrated

To obtain satisfactory results, the following prerequisites must be fulfilled for the infiltration measurements:

- good supporting conditions such as a power supply and support by the local authorities reduce the otherwise very high effort to maintain the probe at the measuring station.
- well trained staff on site for reading out data and sampling as well as in the laboratory for sample analysis is highly recommended.
- a sampling campaign is mandatory to assure a suitable calibration of the on-line instruments. A huge deviation was recorded in Gladbeck when comparing the results for a calibrated and a non-calibrated probe.
- small alterations in the hydrograph or pollutograph might already indicate measurement errors.

APPLICATIONS IN LYON, FRANCE

The results presented in this section have been obtained in the Ecully experimental catchment, located in the western suburbs of Lyon in southern France. This is a residential catchment of 245 ha having an ovoid-shaped combined sewer system which follows the natural drainage lines and is underlain by modern alluvium lying over a granitic basement. A permanent flow monitoring station has been established at the outlet to the sewer system since 2001 and is capable of volume and rate recording at 2 minute intervals. Average flow rates vary between 60 to 140 m^3/hour with an average conductivity of 800 µS/cm. Analysis of daily hydrographs indicated that large amounts of infiltration water contributed to the dry weather flow.

Exfiltration

After preliminary tests had been carried out between April 2003 and June 2003 to train the monitoring team and to obtain adequate information on the proposed site, the QUEST protocol was adapted to the specificity of the investigation reach. The utilized tracer solution had a NaCl concentration varying between 170 and 180 g/L. The volumes of solution used for the indicator pulse and the reference pulse were respectively 25 L and 4 L. The time interval between the two indicator pulses was 20 minutes and the first reference was injected 10 minutes after the first indicator, in order to obtain fully separated peaks. Every measurement campaign had a duration of two hours and was performed with four reference and three indicator pulses.

An example of conductivity time series obtained during an experiment is illustrated in Figure 14. The reference and indicator signals present peaks of conductivity in about the same concentration range, between 3500 and 4200 µS/cm; the conductivity baseline being stable during the experiment at 700 µS/cm. The exfiltration ratio (%) was calculated for each indicator pulse by taking into account the two reference peaks directly before and after its passage. An average exfiltration ratio with an associated uncertainty was calculated for each of the six measurement campaigns realised between July 2003 and July 2004.

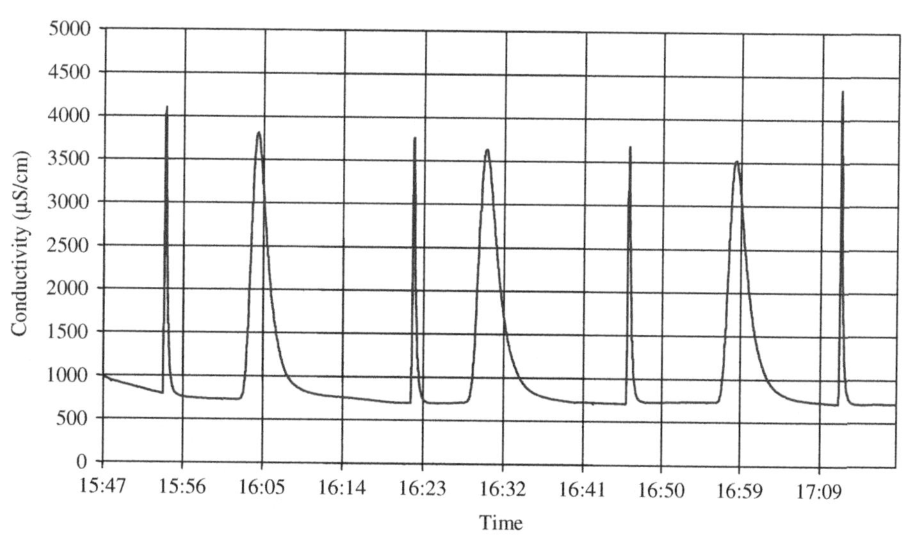

Figure 14: Example of experimental results during a measurement campaign of exfiltration (July 2003).

The uncertainty calculations were performed with a preliminary version of the data analysis routine, in which the uncertainty in the conductivity baseline was not considered. However, as this baseline was rather stable during the experiments, the associated uncertainty is assumed here to be low and negligible compared to the other uncertainties. The uncertainty calculation considers the analytical uncertainty on one single value of exfiltration ratio close to 1% and the uncertainty linked to the distribution of the values of exfiltration ratio which depends essentially on the variability of the wastewater flow rate during the experiment. The results obtained during the six measurements campaigns carried out in Ecully are illustrated in Figure 15.

The exfiltration ratio varies between 2% and 16%. It can also be observed that the rate of exfiltration exhibits some variability. It is hypothesised that it i) decreases during periods of high groundwater table (GWT) and ii) increases

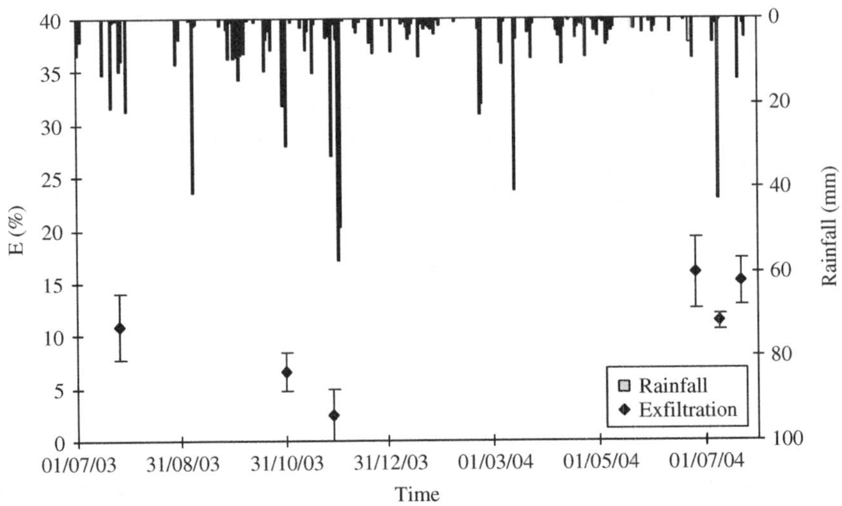

Figure 15: Results of the six measurement campaigns of exfiltration (July 2003–July 2004).

during periods of low GWT. In July 2004, an event-based variability is assumed. After a strong rainfall event (40 mm), the exfiltration ratio decreased from 15% to 10%. Therefore, it can be argued that exfiltration follows a temporal pattern which is similar to that of infiltration and that the variability of exfiltration is also influenced by rainfall frequency and intensity. Additional experiments will be necessary to confirm these hypotheses in order to develop a better understanding of the phenomena.

The application of the QUEST procedure is not straightforward as various parameters influence the measurement campaign design and can affect the determination of a reliable exfiltration ratio. These parameters include the choice of sewer reach as well as variability in sewage characteristics and the wastewater flow rates. The Ecully experiments suggest that the method should be restricted to sewer lengths less than 1 km with a significant wastewater flow rate occurring upstream of the measuring reach to allow for good mixing and transit time for the tracer. It is also recommended that a reference injection be applied upstream of the measurement reach in order to minimise the potential variability and associated uncertainties as advocated in the field tests undertaken in the UK (see Chapter 9).

Infiltration

The most frequently used method employed in the Ecully catchment was the stable isotope method, as favourable local conditions were observed, with clearly distinct origins of drinking water and of groundwater having different isotopic signatures. The oxygen isotope composition of Rhone river water is roughly 1.5‰ to 3‰ lighter than that of the Saone and as a consequence this difference in isotopic composition between wastewater resulting from household discharges and that of local groundwater, it can be reliably used to evaluate sewer infiltration. In addition, the measurement of infiltration rates based on the pollutant time series method using UV-VIS spectrometry could also be applied as measured in a collector tank housing pumped wastewater in a continuous through-flow stream. Numerous measurement campaigns have been carried out, at hourly, daily and seasonal scales. Application of the times series method based on local COD calibration of the s:can probe and comparison with laboratory analysis, indicated a matrix effect for night time samples which made comparison with the isotopic method impossible.

Usually, infiltration flow was assumed to be constant at a daily scale whilst strict wastewater flow is characterised by strong variations due to human activities, with two peaks occurring respectively in the morning and in the evening. The daily cycle of strict wastewater (i.e. drinking water) can be observed in Figure 16 with a strong decrease of contributions during the night period but with strong flow peaks during the evening (20:00) and the morning (07:00–09:00). This figure also reveals that the infiltration rate (in m3/h) presents important hourly variations, with infiltration varying by up to a factor of 1 to 2 during peak periods, and which are partly explained by private groundwater pumping installations used by some inhabitants in the Ecully catchment.

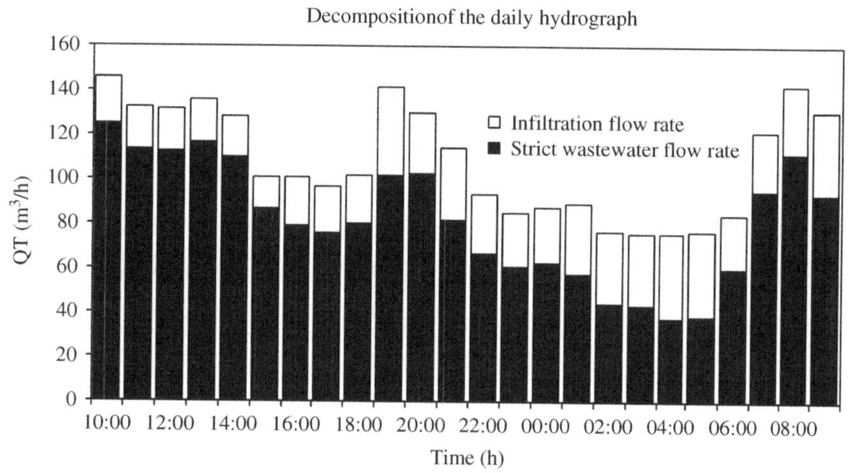

Figure 16: Composition of a daily total hydrograph using the stable isotopes method.

Table 6: Estimated values of infiltration and their 95% confidence interval uncertainties on 17/12/2003.

	Infiltration ratio (% of dry weather flow)	Total dry weather flow Q_T (m^3/h)	Infiltration flow Q_{INF}(m3/h)
Day period	38.0 ± 12.4	167.2 ± 2.4	63.5 ± 20.8
Night period (02h00–06h00)	56.0 ± 15.4	74.6 ± 4.1	41.7 ± 11.7

A measurement campaign was carried out in December 2003 with several samples in order to study the spatial variability of the reference values for drinking water and for groundwater and to estimate the impact of uncertainty linked to the spatial variability in infiltration ratios. The main results of this analysis are given in Table 6. The infiltration flow (in m^3/h) appears more important in the day time than during the night, when the infiltration ratio (in% of total dry weather flow) is higher. These values are due to the fact that what is named here as constituting infiltration, is actually both real infiltration (i.e. groundwater entering the sewer pipes through defects) and groundwater discharged into the sewers by the private groundwater pumping installations.

Chapter 8

Test and field applications of methods in Italy

Valentina Prigiobbe and Mario Giulianelli

SITE DESCRIPTION

The APUSS methods for assessing infiltration and exfiltration have been applied in two catchments in Rome, namely Infernetto and Torraccia (Figure 1). The first catchment is essentially a residential area having a twenty years old separated sewer system with an underlying geology mainly characterized by sand and gravel. The second catchment is also a residential area but having a ten years old combined sewer system whose geology is mainly characterized by volcanic tuff. A third catchment, named "Mostacciano", was also initially considered but was not finally included in the case study work due to problems of sewer access.

The Torraccia catchment

The experimental Torraccia catchment is about 85 ha, with 55 ha residential and 30 ha rural occupancy and is located on a pyroclastic plateau 40 m above sealevel. Permeability of the volcanic material is variable because of included clay lenses that cause the formation of a multi-layered groundwater. At 30–32 m below ground there is a confined aquifer. The urban area is characterized by residential dwellings with courtyards and gardens. The sewer network comprises an ovoid-shaped combined system (1 × 1.2 m, 1.5 × 1.8 m and 1.5 × 2.1 m laid between 4–9 m below ground level) and is some 14 years old; the pipe material is concrete. The sewer network has been implemented in the AquaBase software (see Chapter 9) as illustrated in Figure 2.

The Infernetto catchment

The experimental "Infernetto" catchment is about 5.5 km^2 and characterized by residential properties with private gardens. The north-east part is located on coastal deposits at 12–14 m above sealevel; the south-west part is located on alluvium of the Tiber river delta at 2–3 m above sealevel. Thus the subsoil is characterized by silt-sand sediments, with clay and gravel in the south-west possessing poor technical characteristics. The shallow groundwater varies between 6–7 m and 3 m. The sewer network is a circular-shaped separated system (0.3 m, 0.6 m, 0.8 m and 1 m, laid at 2–6 m below ground level) and is some 19 years old. The composite pipe material includes concrete (size 1.0 m), ceramic (sizes 0.6 m and 0.8 m) and PVC (size 0.3 m). The sewer network has also been implemented in the AquaBase software (Figure 3).

Edited by Bryan Ellis and Jean-Luc Bertrand-Krajewski. ISBN: 9781843391494. Published by IWA Publishing, London, UK.

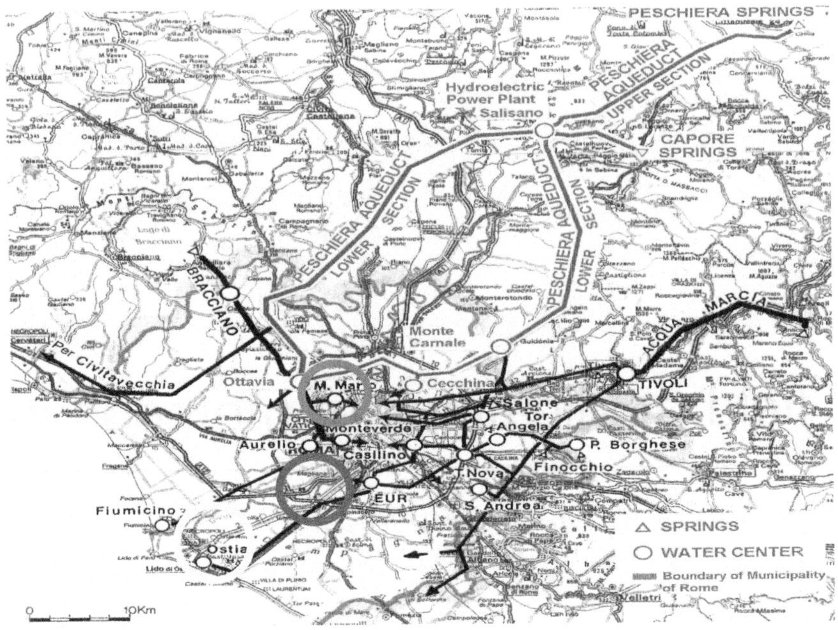

Figure 1: Map of Rome catchments. (circled in white the "Infernetto" catchment, circled in light grey the "Torraccia" catchment and circled in dark grey the "Mostacciano" catchment).

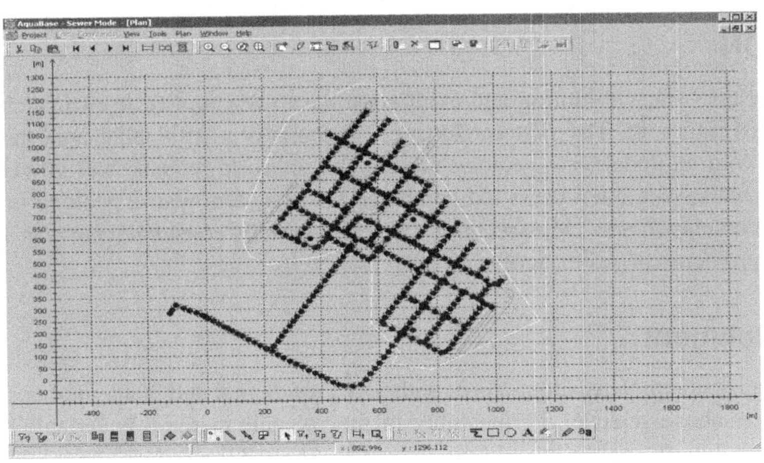

Figure 2: The sewer network in the Torraccia catchment.

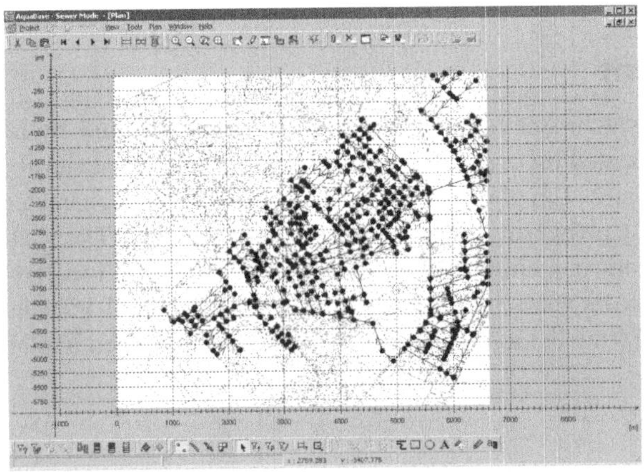

Figure 3: The sewer network in the Infernetto catchment.

METHODS APPLIED AND EQUIPMENT USED

The methods applied in the case study included the QUEST and QUEST-C tracer methodologies (see Chapter 2) as well as the pollutograph and isotopic methods:

- in applying the QUEST method, the equipment used consisted of a submerged probe (Sigma 900max) for level and velocity measurements and three WTW conductivity probes (device LF 197 with sensor TetraCon 325). The conductivity data were recorded on a datalogger (GRANT SQ400) with a resolution of one second. The chemical tracer used was NaCl with a purity of 97%.
- in applying the QUEST-C method a submerged probe (Sigma 900MAX) was used for level and velocity measurements, two pumps (Velp, model SO311) for dosing the tracers with a further two pumps (Watson & Marlow, model SCIQ 323) used for taking wastewater samples. The ion concentrations were determined by means of a Dionex Dx100 analyser (with anion column AS14 and cation column CS12).
- in applying the pollutograph method, the submerged probe (Sigma 900max) was also used for level and velocity measurements, and an on-line spectrolyser (s::can) for spectrographic analysis with the measured data being directly stored into a portable computer (con::vert, s::can).
- in applying the isotopic method a submerged probe (Sigma 900max) was employed for the level and velocity measurements, and an automatic sampler (SIGMA 900MAX) for taking wastewater samples from the sewer stream. All effluent samples were filtered through 0.45 μm Watman filters and analyzed (Finnigan MAT spectrometer) in the laboratory for $\delta^{18}O$ abundance determination (Epstein and Myeda, 1953).

Infiltration: results and discussion

Infiltration was assessed in the catchments of Torraccia and Infernetto, with the isotopic method used in the former catchment and both isotopic and pollutograph methods in the latter catchment.

Isotopic method

The isotopic method applied in the Rome case studies was based on the hydrograph separation method (Chapter 2) to identify the two source components of groundwater and drinking water with the sampling points in the two catchments shown in Figures 4a and b. No springs or other sources were found in the investigated areas following systematic hydrogeological surveys.

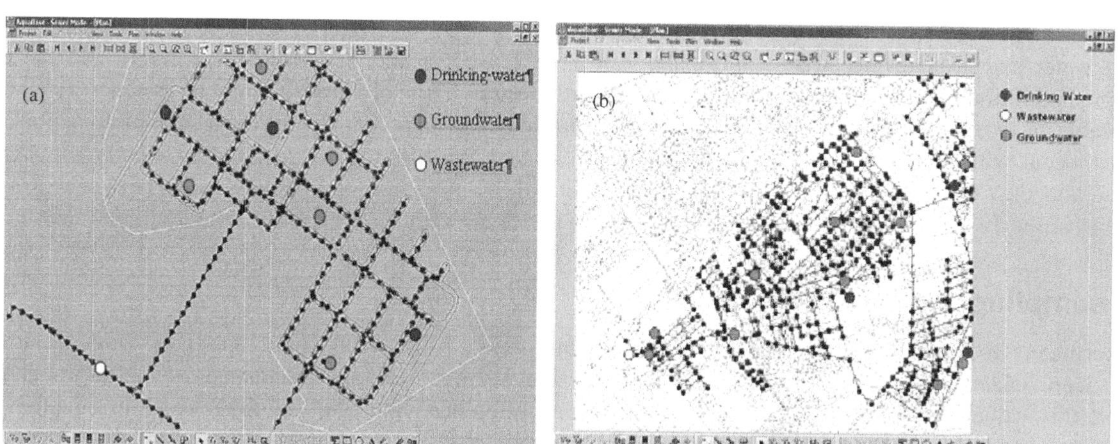

Figure 4: Sampling points in the Torraccia and Infernetto catchments.

Before applying the isotopic method the following hypotheses were verified:

- the drinking water source was unique;
- the differences between $\delta^{18}O_{foulwater}$ and $\delta^{18}O_{infiltration}$ were more than a certain percentage based on the expected infiltration ratio (de Benèdittis and Bertrand-Krajewski, 2004);
- the infiltration contributions had a low spatial variability over the investigated area with a random standard deviation of the isotopic values of groundwater samples being less than the laboratory analysis random error of ±0.2‰.

The second point was confirmed only after field investigations, even though available former data supported such a hypothesis (Longinelli and Selmo, 2003). In order to quantify the uncertainty in the application of the isotopic method in Rome, the principal sources of error were firstly identified and quantified:

Zero order errors

The isotopic characterization of the samples followed the procedure suggested by Epstein and Mayeda (1953) with manual preparation of the sample and determination using the mass-spectrometer. Errors here could be due to both the low accuracy of the operator and the inefficiency of the equipment; the precision assessed by the laboratory is ±0.2‰. If the equation for assessing the infiltration rate Q_{inf}(m/s) is:

$$Q_{\text{inf}} = Q_{ww} \frac{{}^{18}O_{ww} - {}^{18}O_{dw}}{{}^{18}O_{gw} - {}^{18}O_{dw}} \tag{1}$$

the wastewater flow rate (Q_{ww}) must firstly be measured, and the infiltration rate (Q_{inf}) estimate might then be affected by the error deriving from this measurement. In the current application, the flow meter used was an area-velocity Doppler system and from technical data, the level can be measured with an accuracy of ±0.01156 m and a precision of ±0.00346 m and the velocity with an accuracy of ±2.0% of the measured value; the precision of the $\delta^{18}O$ analysis was ±0.2%.

First order error

An important hypothesis of the isotopic method is the spatial and the temporal homogeneity of the two model components (i.e., drinking water, ${}^{18}O_{dw}$ and parasitical water, ${}^{18}O_{gw}$). The spatial variability over the Infernetto and Torraccia catchments was estimated separately for drinking water and infiltration contributions. As the drinking water supply is supplied from a unique source, the spatial variability should be low and this was confirmed from isotopic composition and conductivity studies. The temporal variability of $\delta^{18}O$ (as determined by the error propagation equation of linear models) was equal to ±0.03‰ and ±0.07‰ in Torraccia and Infernetto, respectively.

Nth order error

The mixing model can be applied only when the different sources have sensible different isotopic compositions. de Benèdittis and Bertrand-Krajewski (2004) have determined the relationship between infiltration ratio uncertainty and the isotopic difference between the two sources (drinking water and groundwater) The 18-oxygen differences between drinking water and groundwater were 2.63‰ and 2.36% in Torraccia and Infernetto respectively and the expected uncertainty in infiltration is around 5% for both areas. Nevertheless, as a rule of thumb it might be assumed that the method is applicable in those parts of Rome where the shallow groundwater is recharged by local precipitation (unless $\delta^{18}O$ variations occur within the unsaturated and saturated zones), and the drinking water is independently supplied by the Peschiera aqueduct (see Figure 1). The isotopic composition of the precipitation in Rome varies between −6‰–−5‰, whilst that of the Peschiera aqueduct potable water supply varies between −9‰–−8‰ (Longinelli and Selmo, 2003).

Field sampling

The experiments in Torraccia were carried out over a 10 hour period, from 11.00 a.m. to 8.00 p.m. Three drinking water samples were taken from different public fountains and the average isotopic abundance was determined as being −8.14‰; this value has been considered as the 18-oxygen reference value for drinking water in equation 1. Four groundwater samples were taken from piezometers spread evenly over the catchment area, and the determined isotopic average abundance was −5.74‰; this value has been considered as the 18-oxygen reference value for groundwater. At the outlet of the Torraccia sewer system five grab wastewater samples were taken from the sewer effluent and the flow rate was measured. Figure 5a shows the results obtained from laboratory analyses of the three water matrixes.

The $\delta^{18}O$ abundance in wastewater is slightly higher than that in drinking water but much lower than that found in groundwater, which means the wastewater is a mix of the two sources, but the groundwater contributions are negligible in comparison to those due to the aqueduct. Thus the infiltration rate must be low.

The experiments in Infernetto were carried out over a 24 hour period. Five drinking water samples were taken from different fountains and the isotopic average abundance was −8.11‰; this value has been considered as the 18-oxygen reference value for drinking water. Twelve groundwater samples were taken from piezometers and the isotopic average abundance was −5.75‰; this value has been considered as the 18-oxygen reference value for groundwater. At the outlet of the Infernetto sewer system, twenty three wastewater samples were taken from the sewer stream and the flow rate

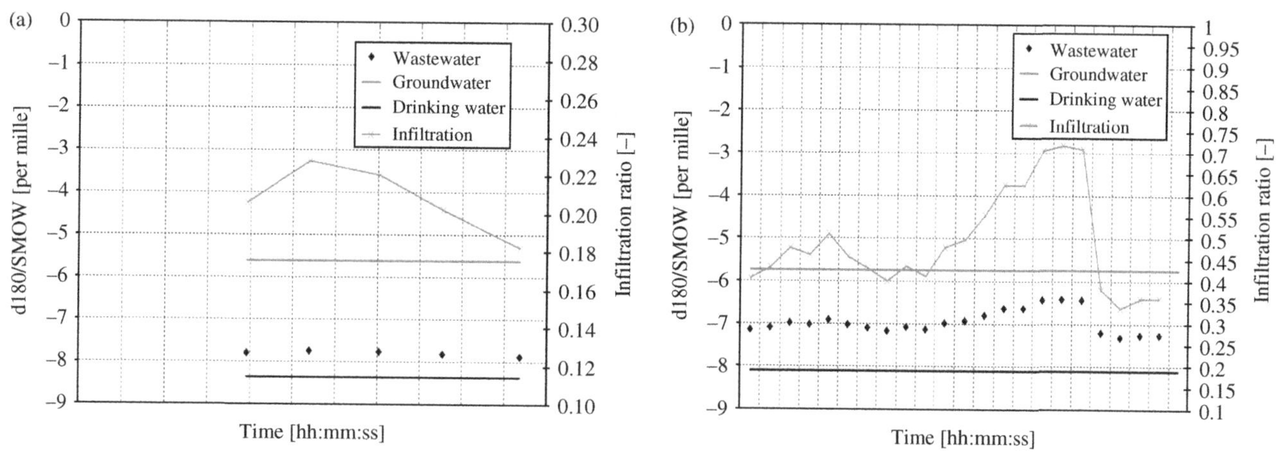

Figure 5: a) $\delta^{18}O$ in drinking water, groundwater and wastewater in Torraccia. b) $\delta^{18}O$ in drinking water, groundwater and wastewater in Infernetto.

measured. Figure 5b shows the results from laboratory analyses of the three water matrixes. The $\delta^{18}O$ abundance in wastewater is higher than that in drinking water but lower than that in groundwater. This is taken to mean, that like Torraccia, the wastewater is a mix of the two sources, although in the Infernetto case the expected infiltration ratio is much higher than in the Torraccia area.

The individual systematic and random errors have been quantified (Table 1) and used for error propagation analysis carried out on the data using a random sampling routine based on the Monte Carlo method. The statistical uncertainty affecting the infiltration ratio and infiltration rate estimated by the isotopic method has been calculated with a sample size equal to 10,000.

Table 1: Errors affecting the infiltration ratio.

Error	Parameters	Random error	Systematic error	Estimation
Spatial variability of groundwater composition [‰]	$\delta^{18}O_{GW}$	I.: 0.09 T.: -	–	From field investigations
Temporal variability of drinking water composition [‰]	$\delta^{18}O_{DW}$	I.: 0.07 T.: 0.03	–	From conductivity measurements
Conservation of $\delta^{18}O$ during the water use [‰]	$\delta^{18}O_{WW}$	I.: - T.: -	–	From field investigations
Level measurement [m]	Water depth	0.00346	0.01156	Technical data sheets
Velocity measurement [%]	Flow velocity	2.0	–	Technical data sheets
$\delta^{18}O$ variation during the storage [‰]	$\delta^{18}O_{WW}$	I.: - T.: -	–	From field investigations
Laboratory analysis [‰]	$\delta^{18}O_{GW,DW,WW}$	0.20	–	From laboratory data

-Note: "I." indicates the Infernetto area; "T." indicates the Torraccia area.

The daily average infiltration percentage in the Torraccia catchment has been estimated at 14.20%±1.80% with a precision of ±21%. The daily average infiltration percentage in the Infernetto catchment has been estimated at 49.90%±11.90% with a precision of ±20%. It can be assumed that the precision of the method is 20%, and thus the infiltration rate in Torraccia is not significant. From Figure 6, it is possible to see, as in Infernetto, that the infiltration ratio changes over the experimental period, increasing at night and decreasing during daylight hours. The variability in infiltration rate is negligible (±7.82 L s^{-1}) in comparison to the wastewater flowrate variability (±29.93 L s^{-1}). The infiltration rate is quite steady throughout the day, whereas it increases during the night (±11.90%), when the drinking water consumption decreases.

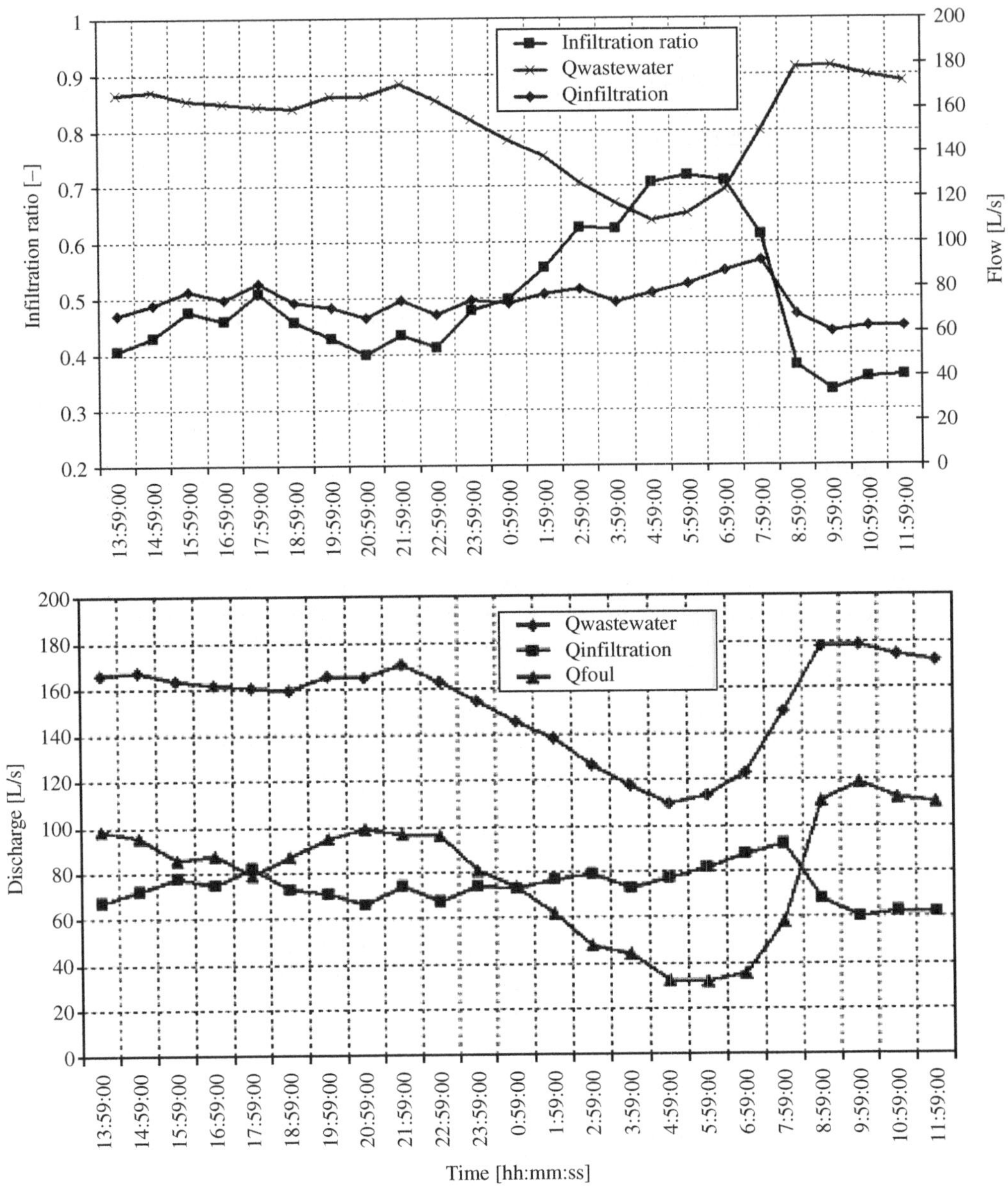

Figure 6. Infiltration ratio, infiltration rate and wastewater flow rate in the Infernetto catchment.

Pollutograph method

The equation for assessing the exfiltration ratio (R_{inf}) by the pollutograph method using COD as the parameter of interest is:

$$R_{\text{inf}}(t) = \frac{Q_{Infiltraion}(t)}{Q_{wastewater}(t)} = \frac{COD_{foul}(t) - COD_{ww}(t)}{COD_{ww}(t)} \tag{2}$$

where the wastewater COD (COD_{ww}) is measured by the on-line spectrometer probe. $COD_{foul}(t)$ has been considered as a Fourier function of the time because of its diurnal and weekly periodicity:

$$COD_{foul} = \sum_{i=1}^{N} a_i \sin(\omega_i t + \varphi_i) \tag{3}$$

where a_i, ω_i and ϕ_i are respectively amplitude, frequency and phase of the function and N is the number of the function used for the COD_{foul} model. These various parameters can be estimated mathematically with a non-linear regression model.

Preliminary experimental campaigns

The s::can probe was mounted in the sewer at the outlet pumping station of the urban network draining the Infernetto area as shown in Figure 7 with a protective rubber surround to the analyser. Figure 8 shows the COD concentrations measured, reflecting a high degree of disturbance with signal drift being evident after two days following cleaning by compressed air. The disturbances can be related to turbulence from the weir discharging the wastewater into the pumping station with the drift being caused by biofilm growth over the UV-VIS window of the s::can probe. As a consequence the probe was moved upstream where the flow was less turbulent and the probe windows were cleaned with alcohol solution (10% in weight) every two days. After these precautions the measurements improved considerably as indicated in Figure 9. Figure 10 shows the results of laboratory calibration of the COD probe with a much better correlation (0.98) being observed for night-time readings. The COD data has been modified, prior to final calculation of the infiltration ratio using equation 3, by taking into considering the calibration curve for both diurnal and nocturnal periods.

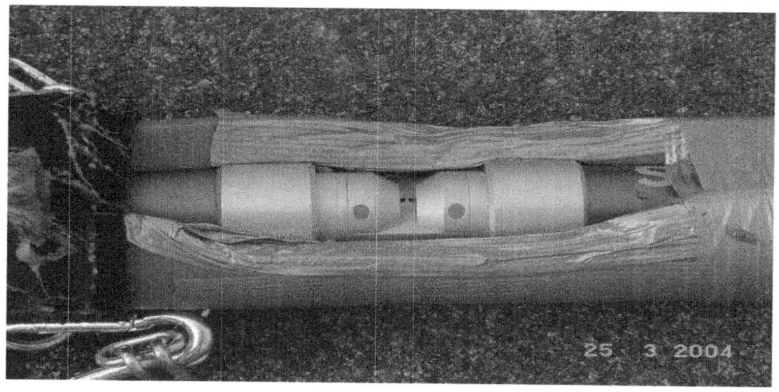

Figure 7: The s::can probe and its cover.

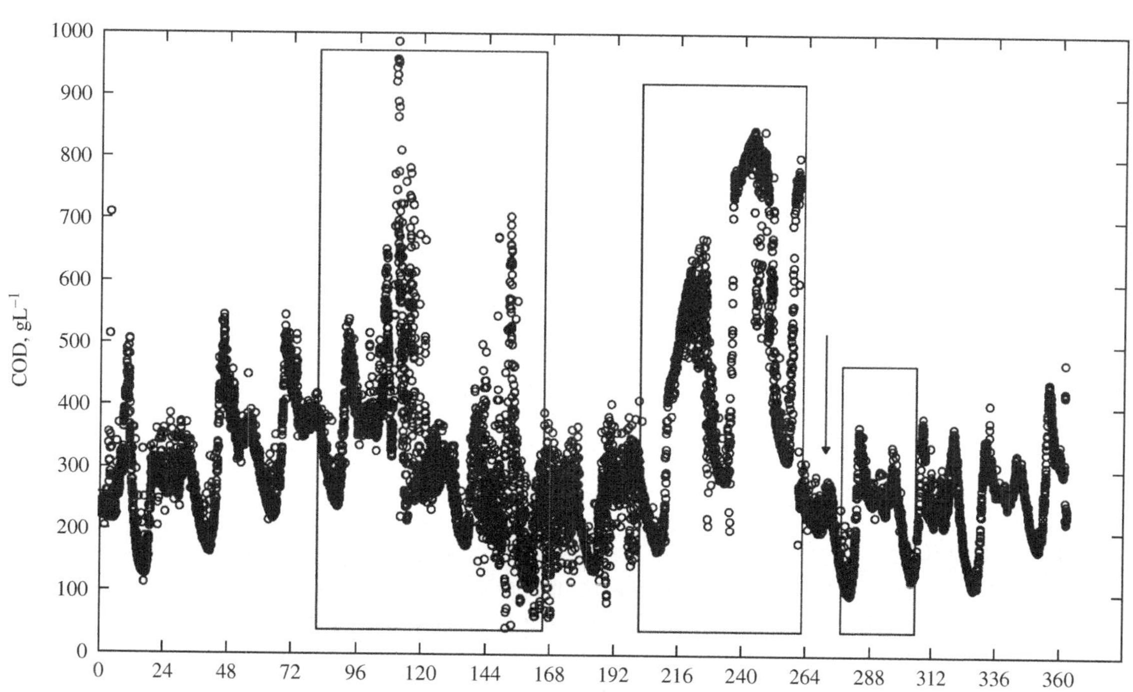

Figure 8: Variability of COD measurements over time. (Data were collected every 2 minutes, 24 March – 14 May 2004. The two large boxes highlight where the COD measurement pattern is disturbed. The smaller box refers to a 1 day pattern following cleaning of the sensor as indicated by the arrow).

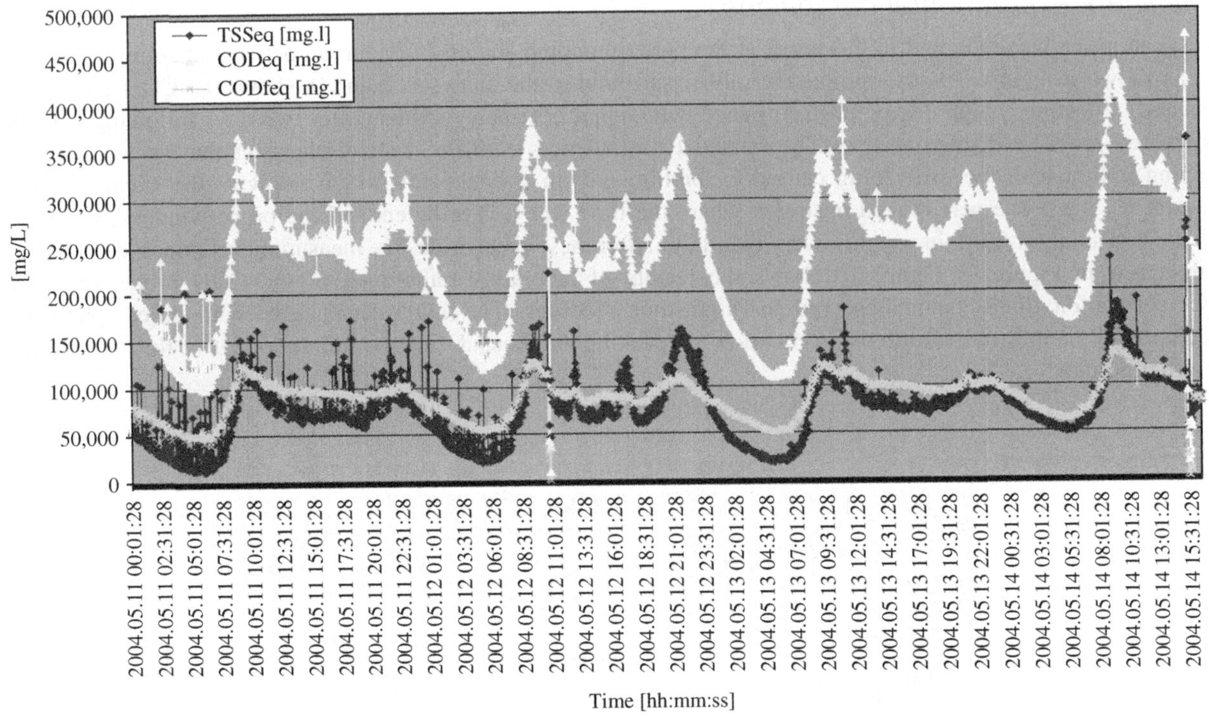

Figure 9: COD measurements after changing installation location and cleaning.

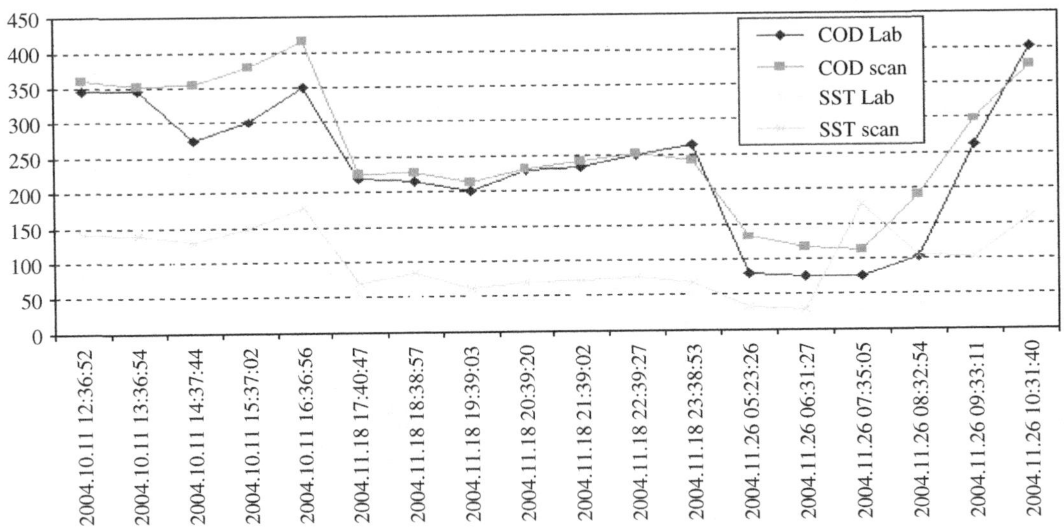

Figure 10: COD calibration.

Infiltration calculation

An algorithm written by Kracht (2003) in R-script has been applied to calculate the infiltration ratio and rate with the results displayed in Figures 11 and 12 with Table 2 providing an overall summary.

EXFILTRATION: RESULTS AND DISCUSSION

Exfiltration was assessed using the QUEST and QUEST-C methods in the Torraccia catchment but only the former method was applied in the Infernetto catchment.

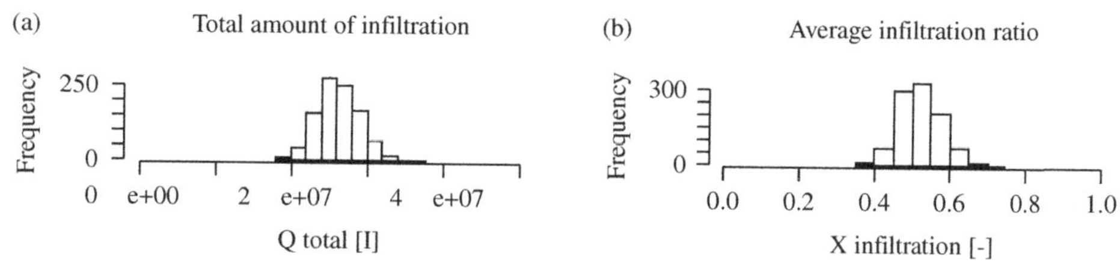

Figure 11: Infiltration rate (a) and infiltration ratio (b).

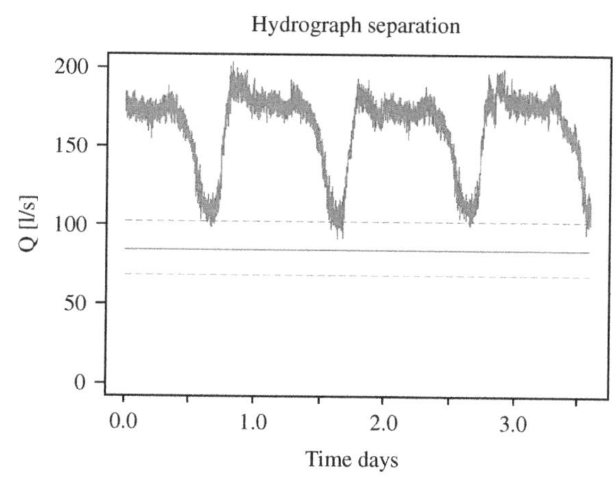

Figure 12: Hydrograph separation of the flows in infernetto. (wastewater flow rate, upper plot; infiltration flow rate lower horizontal line).

Table 2: Infiltration ratio [–] and rate [L s^{-1}] calculated after calibration of the measured COD data.

	Min	1st Quantile	Median	Mean	3th Quantile	Max
Infiltration ratio [–]	0.3629	0.4842	0.5189	0.5227	0.5601	0.7198
Infiltration rate [Ls^{-1}]	58.62	78.21	83.81	84.42	90.46	116.26

QUEST-C

The equation applied for exfiltration ratio ($Exfr_c$, %) assessment is:

$$Exfr_c = \left[1 - \left(\frac{\frac{c_{Br} q_{Br}}{C_{Br}}}{\frac{c_{Li} q_{Li}}{C_{Li}}} \right) \right] 100 \tag{4}$$

where C_{Li} and C_{Br} are the concentration of Li^+ and Br^- ions respectively in the outlet samples and c_{Li}, c_{Br} represent concentrations added to the wastewater flow with q_{Li} and q_{Br} being the flow reates of the injector solution pump.

Before the application of the QUEST-C method a general uncertainty analysis was carried out in order to plan the experiments accurately and to investigate if the tests were feasible with the proposed model (i.e. equation 2). Thus, for each variable in equation 2, the UMCs (Uncertainty Magnification Factors) and UPCs (Uncertainty Percentage Factors) were calculated. The first set of factors indicate the influence of the uncertainty in that variable on the uncertainty in the result, and the second (UPFs) give the percentage contribution of the uncertainty in that variable to the squared uncertainty in the result (Coleman and Steele, 1998). Table 3 shows the results of the uncertainty analysis for the different sources of error and Figure 13 the distribution of the UPC and UMF factors.

From the UPC calculation two different results were obtained:

(1) C_{Br} is the principal source of uncertainty on the exfiltration ratio when the adsorption of Br^- on the solid matter is considered equal to the estimated value, consequently the exfiltration by equation 2 could be underestimated. Thus, the distance between Manhole 2 and 3 should be calculated for an accurate tracer mixing;

(2) C_{Li} is the principal source of uncertainty on the exfiltration ratio when the adsorption of Br^- on the solid matter is neglected. This means the errors affecting C_{Li} (e.g.: transport and IC analysis) have to be carefully reduced. In order to reduce the effect of transport, the exfiltration ratio should be evaluated from the average concentrations of the tracers.

Table 3: Standard Deviation of Each Source of Uncertainty.

	Preparation	Field application	Laboratory analysis					
Parameters	**Weight chemical tracer [gr]**	**Dilution [mL]**	**Dosing of tracer solution**	**Adsorption on solid matter in sewer [mg/L]**	**Transport [mg/L]**	**Sampling**	**Storage of samples**	**Ion – Chromato-graph [%]**
c_{Li}	0.1	0.4	0					10
c_{Br}	0.1	0.4	0					10
q_{Li}			0					
q_{Br}			0					
C_{Li}				0.0	0.108	0	0	10
C_{Br}				0.0–2.0	0.016	0	0	10

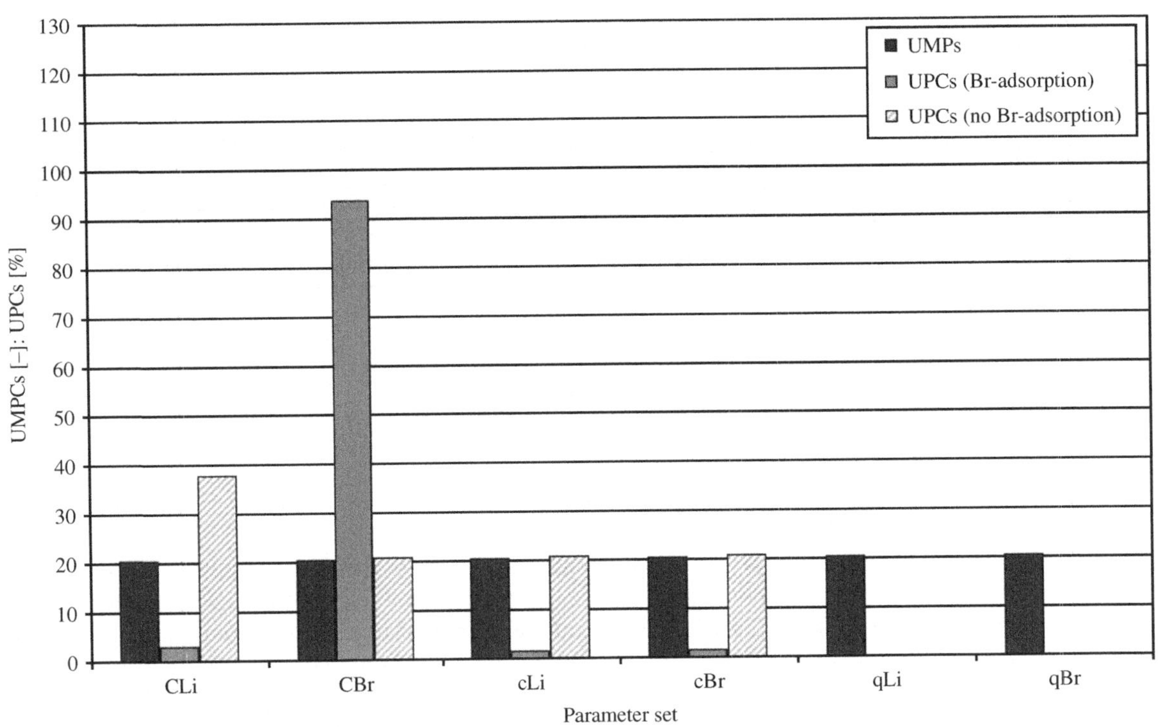

Figure 13: UMF and UPC values.

Three experiments were carried out in the Torraccia catchment to test the watertightness of three sewer reaches. The investigated reaches were 4–9 m below ground level with a slope of 0.9%. Reach 1 is 400 m long and egg-shaped (120 × 210 cm), Reach 2 is 482 m long and egg-shaped (120 × 210 cm) and Reach 3 is 680 m long and egg-shaped (180 × 120 cm). All the pipes are in subsoil characterized by cracked tuff and pozzolan and the backfilling is concrete. The groundwater submerges the sewer pipes only during wet weather and is rapidly drained towards a deeper aquifer due to the high permeability of the cracked tuff. The principal differences between the investigated reaches are: the flow rate, the number of house connection and nodes, the traffic load and the vegetation as summarised in Table 4.

Table 4: Different features of the investigated pipes.

	Reach 1	Reach 2	Reach 3
Average Flow rate [L/s]	28.20±1.97	19.97±2.04	9.50±1.35
Water depth [cm]	13.35±1.97	7.59±1.97	7.49±1.97
Number of HC	0	0	23
Number of nodes	0	0	7
Traffic	No	No	Yes
Vegetation	Grass	Grass	Eucalyptus

Table 5 indicates the dosing concentrations used and the exfiltration ratio percentage computed from the average concentration of the tracers. The computed exfiltration ratios in the table show two negative values. The negative exfiltration recorded for Reach 1 may be due to illicit lithium discharges in the sewer during the experiment because of the strong deviation from zero that is more than the estimated uncertainty (Rieckermann and Bares, 2005). The absolute negative value of exfiltration ratio for Reach 2 is within the uncertainty value of the QUEST-C results, and it can be assumed that the exfiltration from this pipe was negligible during the investigation which confirms the good structural state of the sewer. The high exfiltration ratio computed for Reach 3 implies that there are significant damages along this pipe reach, but it is difficult to say if the calculated value is representative of the structural damage that might occur across the catchment as a whole.

Table 5: Tracer dosing concentrations and exfiltration rates.

	Reach 1	Reach 2	Reach 3
cLi [gr/L]	18.04	20.10	24.22
cBr [gr/L]	50.35	58.86	50.98
qLi [mL/min]	62.61	45.43	28.44
qBr [mL/min]	62.91	36.32	23.75
Exfiltration [%]	−8.69±0.62	−1.73±4.69	20.80±4.46

QUEST

The equation applied for exfiltration ratio ($exfr_c$, %) assessment using the QUEST methodology with NaCl as the selected chemical tracer is:

$$exf = 1 - \frac{M_2 \int_{t=startIpeak}^{t=endIpeak} e'(C_{Ind}(t) - C_{bl}(t))dt}{M_1 \int_{t=startRpeak}^{t=endRpeak} e(C_{Ref}(t) - C_{bl}(t))dt} * 100 \tag{5}$$

where M_2 is the dosed mass of tracer for the reference signal [kg]; e' is the calibration coefficient [gr NaCl cm/μS]; $C_{ind}(t)$ is the conductivity of wastewater stream during the reference passage at the measuring manhole [μS cm^{-1}]; C_{bl} is the natural conductivity of the wastewater stream during the reference passage which has to be estimated by a linear or nonlinear regression [μS cm^{-1}]; $t_{end(start)}R_{peak}$ is the time of the end (beginning) of the reference peak; $t_{end(start)}I_{peak}$ is time of the end (beginning) of the indicator peak. The *e'* and the e values are equal if the conductivities of the reference and the indicator peaks vary in the same interval over the calibration line.

The QUEST method can be applied in two different ways:

- not allowing the reference and the indicator conductivity peaks to overlap each other;
- allowing overlapping.

The QUEST method was applied without overlapping the reference and indicator peaks because the peaks were very distinguishable and no peak fitting is needed before integrating the peak function. A disadvantage of this approach is that the distance between the reference and the indicator peaks might prejudice the flow value estimations over the dosing time and thus travel times need to be carefully determined. A further potential disadvantage is that the model is non-linear ($C_{bl} = f(t)$, $C_{ind} = f(M_1, t, C_{bl})$ and $C_{ref} = f(M_2, t, C_{bl}, C_{ind})$) and reference and indicator peaks must be distinguished using known peak formulae. The advantage of reference and indicator peaks being very close to each other, is that it allows the flow value estimated by the reference peak to be more representative of the flow corresponding to the passage of the indicator peak.

Equation 5 gives a more accurate exfiltration ratio if:

- the same equipment is used for measuring the conductivity of both reference and indicator peaks;
- the values of the calibration coefficients (*e* and *e'*) are the same;
- the reference and indicator tracer solutions are prepared with the same equipment;
- the reference and the indicator peaks are as close as possible to each other.

Given these conditions, the systematic errors affecting the numerator and the denominator are the same and the propagated systematic errors in the results are minimized (Chrakroun *et al.*, 1993).

The set-up of the experimental field dosage procedure needs to consider the following questions:

(1) where should the tracer for the indicator and reference signals be dosed along the investigated reaches?
(2) where should the conductivity be measured?
(3) how much tracer should be dosed?
(4) when is the optimum time for the experiment to be carried out during a dry weather day?
(5) how can the overlapping of the reference and the indicator peaks at the measuring manhole be avoided?

The manholes selected for tracer dosing and downpipe concentration measurements were chosen on the basis of the mixing length, which for the Torraccia catchment varied between 75 m and 130 m. A model was developed for the manhole configurations based on three continuity equations, three mass balances and three exfiltration ratio equations based on the simple network scheme as shown in Figure 14.

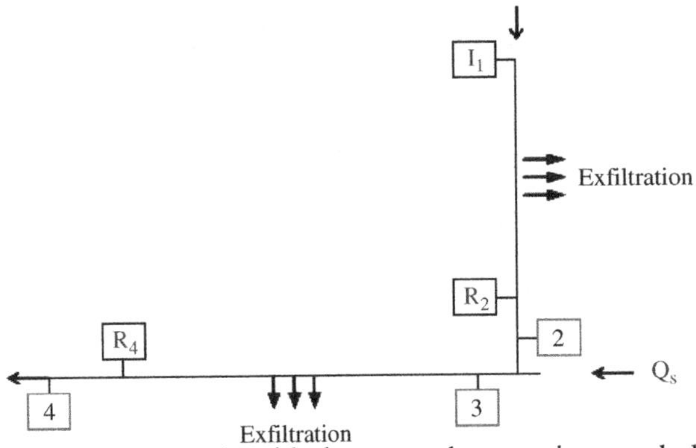

Figure 14: Scheme of investigated sewer network, with dosage and measuring manhole indicated. (I_1 = tracer dosage for indicator signal; R_2 = tracer dosage for reference signal; 2,3 and 4 = measuring points, Q_5 = lateral inflow).

The tracer dosage was determined considering that the peak-baseline ratio needs to be as high as possible whilst at the same time avoiding potential precipitation of salt. The natural conductivity of wastewater and the flow rate were recorded over a three day dry weather period in order to define the time when they are as steady as possible (a maximum variability in conductivity was fixed at 100 $\mu S\ cm^{-1}$ and an acceptable variability of flow rate fixed at $\pm 5\ L\ s^{-1}$). Over the experimental dosing period it was possible to ensure that the peaks did not overlap because:

- the baseline can be modelled satisfactorily with a linear function although the number of regression points influences the distance between the peaks, allowing the maximum number of data points to be fixed at 200 in order to reduce the duration of the experiment as much as possible;
- the unknown flow during the tracer indicator passage can be obtained from the reference peaks measured just after and before that of the indicator.

Preliminary measurements were carried out in order to determine the shape of the peak at the end of each investigated reach in order to evaluate the time interval between two subsequent dosages. Furthermore, from these tests the duration of the peak can be roughly determined in order to avoid peak overlap during the investigation and to decide correctly when the tracer solution can be dosed. The duration of the investigation has to be as short as possible in order to minimise variability in natural conductivity and flow rate during the quasi-steadystate measurement period.

The method was applied in both the Infernetto and Torraccia catchments and a practical form of equation 5 is:

$$exf = 1 - \frac{M_2 * AreaPeak_1}{M_1 * \overline{AreaPeaks_2}} \tag{6}$$

where $AreaPeak_I$ is the area under the Indicator peak and $\overline{AreaPeaks_2}$ is the average area under the four references peaks (shown in grey on Figure 15) placed immediately before and after the indicator peak (solid line in Figure 15).

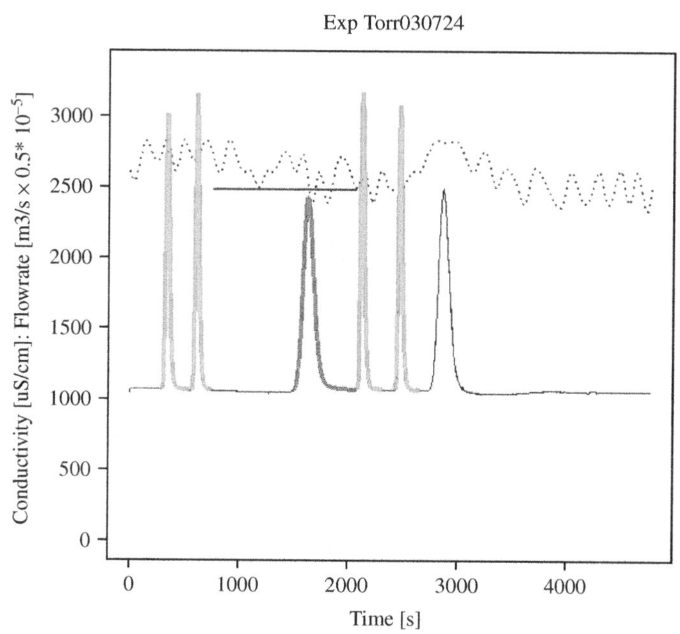

Figure 15: Examples of conductivity peaks recorded during the field experiments. (light grey show reference peaks and indicators are in solid black).

The results for five of the experimental runs are shown in Table 6 with the error propagation carried out using a routine written in R-script by Rieckermann and Bares (2005). The table also shows characteristics of the sewer system. The first three listed in the table refer to dosing runs conducted over the same reach in the Torraccia catchment which had no house connections (HCs) but did contain one node in the middle of the sewer reach. The fourth line listed in Table 6 refers to an experimental run conducted in the same sewer reach but above the confluence node in the sewer pipe. The last line refers to an experimental run conducted within Torraccia on a sewer reach having five nodal confluences and several house connections.

Table 6: Characteristics of QUEST experiments and results for the Torraccia catchment.
(T: experiment carried out in Torraccia catchment; I: experiment carried out in Infernetto catchment).

Experiment code	Investigated length [m]	Duration [s]	Flowrate [L s-1]	Mean Exf. Ratio [%]	Coverage of the ground and backfilling
030710T	800	2000	10.63±0.50	7.11±0.02	Grass, concrete backfilling
030724T	800	2000	12.50±0.51	0.09±0.19	Idem
030730T	800	2000	12.48±0.45	5.94±0.41	Idem
040616T	400	4000	8.48±0.45	−1.73±4.69	Idem
040408I	1500	9000	22.59±0.45	1.66±0.11	Road, gravel backfilling

Table 7 shows the exfiltration ratios calculated by means of the above equations without considering the sources of uncertainty and the exfiltration ratio before and after a confluence node. The low exfiltration ratio indicated in Table 7 for the Torraccia experiments (T) was expected given that the tested sewer pipes were quite new and, principally, that the backfilling of the sewer trench was concrete. The low exfiltration ratio shown for the Infernetto experiments (I) was also expected given that a large part of the sewer system in the catchment is submerged by infiltration. Finally, from inspection of Table 7 the first part of the investigated reach seems to be less damaged in comparison to the second lower part. The negative values of the exfiltration ratio are due to errors occurring during the measuring and data analysis and do not indicate any addition of salt along the pipe run, as there are no house connections along this reach that might have accidentally discharged salt. As the expected value of exfiltration in all investigated reaches was zero, it can be assumed that the accuracy of the method is between 2%–3%.

Table 7: Exfiltration ratio [%] evaluated for two tested reaches.

	030710T	030724T	030730T	Mean	St.Dev.
First reach	−0.12	−0.24	−0.21	−0.19	0.06
Second reach	2.96	2.16	2.14	2.42	0.47

REFERENCES

Chakroun, W., Taylor, R.P., Steele, W.G. and Coleman, H.W. (1993) Bias Error Reduction Using Ratio to Baseline Experiments-Heat Transfer Case Study, *Journal of Thermophisics and Heat Transfer*, **117**(4), 754–757.

Coleman, W.H. and Steele J.R. (1998) *Experimentation and Uncertainty Analysis for Engineers*. 2nd Edition. John Wiley & Sons Inc., New York, US. ISBN 0471121460.

de Benèdittis, J. and Bertrand-Krajewski, J.-L. (2004) *Measurement of infiltration rates in urban sewer systems: Use of oxygen isotopes*. WEM, Book.

Epstein, S. and Mayeda, T. (1953) Variation of ^{18}O content of water from natural sources. *Geochim.Cosmochim. Acta* **4**, 213–224.

Kracht, O. (2003) *Guidelines: The stable isotope composition of water as a natural tracer for the quantification of extraneous infiltration into sewer systems*. Internal APUSS protocol, EAWAG, Zurich, Switzerland.

Longinelli, A. and Selmo, E. (2003) Isotopic composition of the precipitation in Italy: a first overall map. *Journal of Hydrology*, **270**, 75–88.

Rieckermann, J. and Bares, V. (2005) *Standard Operation Procedures (SOP): Quantification of Exfiltration from Sewers with Artificial Tracers. (QUEST)*. EAWAG, Zurich, Switzerland: 48pp.

Chapter 9

Test and field applications of methods in the UK

D. Michael Revitt, J. Bryan Ellis and Nick Paterakis

INTRODUCTION

The traditional approaches for determining exfiltration have been based on rather inaccurate methods of flow measurement, pressure testing, analysis of diurnal flow and load variation, water balancing and expensive CCTV inspections (Rutsch *et al.*, 2005). As described in Chapter 2 of this book, two novel tracer methods identified as the QUEST (Quantification of Exfiltration from Sewers with artificial Tracers) and the QUEST-C techniques have been developed within the APUSS project. The former involves mass balance comparisons over an investigated sewer reach and a leak free reference section following the pulse injection of a single tracer (Rieckermann and Gujer, 2002). The QUEST-C method requires access to similar sewer stretches but because two different tracers are dosed continuously, in-line monitoring equipment possessing a low response time is not essential (Rieckermann, 2005).

The standard tracer for use in the QUEST technique is sodium chloride because of its proven inertness in the presence of sewage with regard to adsorption, chemical and microbiological degradation, and photolysis. Previous workers have used conductivity probes as the in-sewer detection technique for this tracer (Rieckermann and Gujer, 2002; Guillanelli *et al.*, 2003) despite the fact that this mode of detection is not specific to sodium chloride and that high dosing concentrations are required to overcome the natural background salinity of sewage. Typical dosing concentrations are 200 g/L and when combined with the large dosing volumes these create problems relating to efficient mixing and dispersion of the tracer downstream of the dosing point. Alternative ionic tracers, which can be utilised at lower dosing concentrations due to their negligible presence in natural waters and sewage, are lithium (usually as lithium chloride) and bromide (usually as sodium or potassium bromide). However, for these tracers there are currently no direct monitoring probes available, which can tolerate the hostile sewer environment, and therefore they are not appropriate for use in the pulse dosing method (QUEST) but can be utilised in the continuous dosing approach (QUEST-C).

Synthetic dyestuffs are extremely sensitive tracers because of their ease of detection at very low concentrations. This is particularly true for fluorescent dyes such as Rhodamine WT (RWT), Sulphorhodamine B and Sodium Fluorescein, and the former has been widely used as a tracer in both surface waters (Edwardson *et al.*, 2003) and groundwaters (Flury and Wai, 2003). The continuous monitoring of Rhodamine WT can be achieved using a dedicated submersible fluorimeter probe but, prior to this research study, the use of such a probe had not been thoroughly tested in a sewer environment where interferences due to high turbidity are to be expected. The behaviour of Rhodamine WT in sewage could also be complicated by the existence of two isomers in the commercially available product. The *para*-isomer has been shown to demonstrate conservative behaviour under laboratory test conditions whereas the *meta*-isomer exhibited efficient sorption to mineral surfaces (Vasudevan *et al.*, 2001).

A series of batch tests using fresh raw sewage, carried out as part of the research study, have shown that all three fluorescent dyes exhibit adsorption, up to an extent of 10%, over exposure periods of up to 24 hours (Chapter 3). However, these laboratory

 Assessing infiltration and exfiltration on the Performance of Urban Sewer Systems (APUSS). Edited by Bryan Ellis and Jean-Luc Bertrand-Krajewski. ISBN: 9781843391494. Published by IWA Publishing, London, UK.

tests were not able to fully duplicate the conditions to which a tracer will be subjected during the determination of sewer exfiltration rates. This chapter describes a series of field experiments which have been devised to test the behaviour of Rhodamine WT during sewer transport and hence assess its applicability for use as a tracer in the determination of exfiltration. The performances of Rhodamine WT in both pulse dosing (QUEST) and continuous dosing (QUEST-C) experiments has been investigated and compared to that of Li and Br ion tracers in the latter methodology. The results, which have previously been partially reported (Revitt *et al.*, 2006) enable the two approaches to be compared in terms of their feasibility and efficiency for the determination of exfiltration rates and also provide further evidence of the suitability of tracer studies for this purpose.

SITE DETAILS AND TRACER METHODOLOGIES

Site description

The pulse dosing (QUEST) and continuous dosing (QUEST-C) experiments were conducted on a 660 m length of combined sewer located at Potters Bar in North London, UK. The monitored section had an average slope of 0.00173 m/m and consisted of a 510 m length of circular 450 mm concrete pipe passing into a 150 m length of 600 mm steel piping. The last section is located above ground on a sewer 'bridge' and therefore could be observed to be watertight thus representing an ideal reference section against which to measure exfiltration in the up-sewer stretches using different tracer methodologies. The test sewer section was also free from sedimentation which might cause inhibition of tracer recovery due to potential absorption in dead zones.

Access to the sewer for dosing and monitoring purposes was by means of existing manholes. Two up-sewer manholes (identified as U1 and U2) were used for the introduction of the indicator tracer and were separated from the down-sewer monitoring manhole (identified as manhole A) by distances of 660 m and 480 m, respectively. Manhole B, located 150 m up-sewer of manhole A was consistently used for the introduction of the reference tracer. A submersible fluorimeter (Turner Designs SCUFA III) was positioned at manhole A to obtain direct in-sewer readings of the Rhodamine WT concentrations. In addition, this monitoring point also included equipment for depth measurement (pre-calibrated STS Model 8370 pressure transducer) and 30 second time-averaged flow velocity measurement (Ott Nautilus C200 ultrasonic probe) to facilitate the calculation of sewer flow rates.

Experimental procedures

QUEST dosing technique

This technique requires *in-situ* monitoring, with short time resolution measurements, to enable good peak definition to be obtained. The single tracer used throughout the QUEST experiments was RWT however, two different strategies (identified as 1 and 2) were employed with subsequent downstream detection at 1 second intervals.

Strategy 1 QUEST experiments

In this strategy, a single pulse dose of accurately measured tracer solution (Rhodamine WT; 20 g/L) was introduced at either manhole U1 or U2 accompanied by two earlier or later pulse doses of the fluorescent dye at manhole B. The dosing arrangements employed for Strategy 1, in which a constant ratio between indicator and reference of 2.5 to 1 was maintained, are shown in Table 1.

Table 1: Dosing arrangements for the Strategy 1 QUEST experiments.

Reference code	Upstream indicator dosing point	Indicator dosing volume (20 g/L RWT)	Reference dosing volume (20 g/L RWT)
QUEST 1a	U1	50 mL	20 mL
QUEST 1b	U1	25 mL	10 mL
QUEST 1c	U1	100 mL	40 mL
QUEST 1d	U2	50 mL	20 mL
QUEST 1e	U2	25 mL	10 mL

The optimum dosing volumes for efficient downstream detection were found to be 50 mL and 20 mL for the indicator and reference doses, respectively. The low volumes of tracer used and the absence of a significant density difference between the dosing solution and the sewage facilitated rapid and efficient mixing of the two components with no evidence of tracer filaments.

The timings of the dosing introductions were critically important in order to ensure that the downstream indicator and reference signals were monitored as discrete peaks which were also sufficiently close to minimize the impact of influencing factors such as changes in the flow regime and possible changes in sewage quality, particularly turbidity. To assist the achievement of this scenario, the times of travel between the dosing points and manhole A were determined prior to the commencement of the exfiltration experiments.

Strategy 2 QUEST experiments

In these experiments alternate indicator and reference doses were introduced continuously at dosing points U1 and B, respectively over time periods of at least 1 hour. The objective was to obtain a series of situations in which a reference peak could be clearly identified as being associated with an indicator peak. Flow rates were monitored consecutively to assess the impact of any changes in this parameter on the measured peak areas.

QUEST-C dosing technique

These experiments involved the prolonged introductions of tracers at constant rates (using a peristaltic pump) at dosing points U1 and B with subsequent monitoring at the downstream manhole A. Different combinations of three tracers (RWT, Li (as lithium chloride) and Br (as potassium bromide)) were used. The peristaltic pumps were adjusted to deliver nominal dosing rates of 1 L/minute and accurate values determined volumetrically. The optimum dosing periods were for up to 10 minutes with careful synchronisation to ensure that the respective tracers reached the monitoring point (manhole A) simultaneously and were therefore subjected to similar conditions during their transport through the monitored sewer sections. The dosing concentrations were 1.0 g/L for RWT and Li and 2.5 g/L for Br. RWT tracer concentrations reaching the monitoring point were measured *in-situ* whereas Li and/or Br concentrations were determined by laboratory analyses following the collection of grab samples, at 30 second intervals. Flow measurements were obtained during all QUEST-C experiments.

Sample collection and laboratory analysis

The fluorescent signal produced by RWT can be strongly influenced by variations in turbidity and therefore the fluorimeter was calibrated with raw sewage and RWT solutions (0 to 1000 μg/L) prior to and on completion of each set of field experiments. Repeated analyses of the same RWT dosed sewage samples indicate standard deviations in the measured values not exceeding 0.6% in the concentration range up to 600 μg/L. The sewage samples collected at manhole A during the field experiments were returned to the laboratory for the analysis of Li (by flame emission spectrophotometry) and Br (by ion chromatography). Spectrophotometric analysis for Li at 670.0 nm was performed in triplicate on supernatant solutions following centrifugation of the sewage samples at 3000 rpm for 5 minutes. Li standards at concentrations of 0, 250 μg/L, 500 μg/L and 1000 μg/L were prepared in sewage samples which were collected immediately prior to each monitoring exercise and these were also clarified by centrifugation. This calibration approach compensated for any interference with the Li emission signal by components of the sewage sample, particularly the high levels of sodium. The variability in the measurement of Li under these conditions was $\pm$ 0.7% within the peak concentration monitoring range of 200 to 600 μg/L and the detection limit (expressed as 3 times the standard deviation of the blank) was 3.4 μg/L. Bromide determinations were carried out using ion chromatography (Dionex ICS-1500) with Br standards of 0, 0.625 mg/L, 1.25 mg/L and 2.5 mg/L prepared in collected raw sewage as described for the Li analysis. Following centrifugation, the standards and collected samples were filtered at 0.45 μm and passed through a Sep-Pak C-18 column prior to analysis to protect the analytical column. Samples (25.0 μL) were automatically injected on to a high capacity anion-exchange column (2×250 mm) and isocratically eluted using a carbonate/bicarbonate eluent at a flow rate of 1.5 mL/minute. The anion peaks were detected using ECD and under the applied conditions, the bromide peak was efficiently resolved from the adjacent nitrite and nitrate peaks. The analytical precision for bromide was lower than for the other tracers with a typical standard deviation of $\pm$ 3.1% within the peak concentration monitoring range of 0.5 to 1.7 mg/L. The procedure detection limit for Br was 11.9 μg/L.

RESULTS AND DISCUSSION–QUEST EXPERIMENTS

Strategy 1 Quest experiments

The results obtained for the Strategy 1 experiments are represented in Figure 1 by the temporal variation in the recorded Rhodamine WT concentrations (readings taken at 1 second intervals) at the monitoring man-hole A following the introduction of tracer pulses at dosing points B and U1. The first two peaks are reference peaks (dosing at B) followed by the indicator peak (dosed at U1) and two more reference peaks. The dosing strategy was designed to achieve peak separation but also to ensure that the reference peaks immediately before and after the indicator peak were as close as possible to ensure that the tracer was exposed to the same sewer flow and sewage quality conditions during its passage through the sewer. This particular situation can be difficult to obtain and in Figure 1 it can be seen that the post-indicator reference peak is too close preventing complete resolution whereas in the case of the pre-indicator peak, there is too much separation.

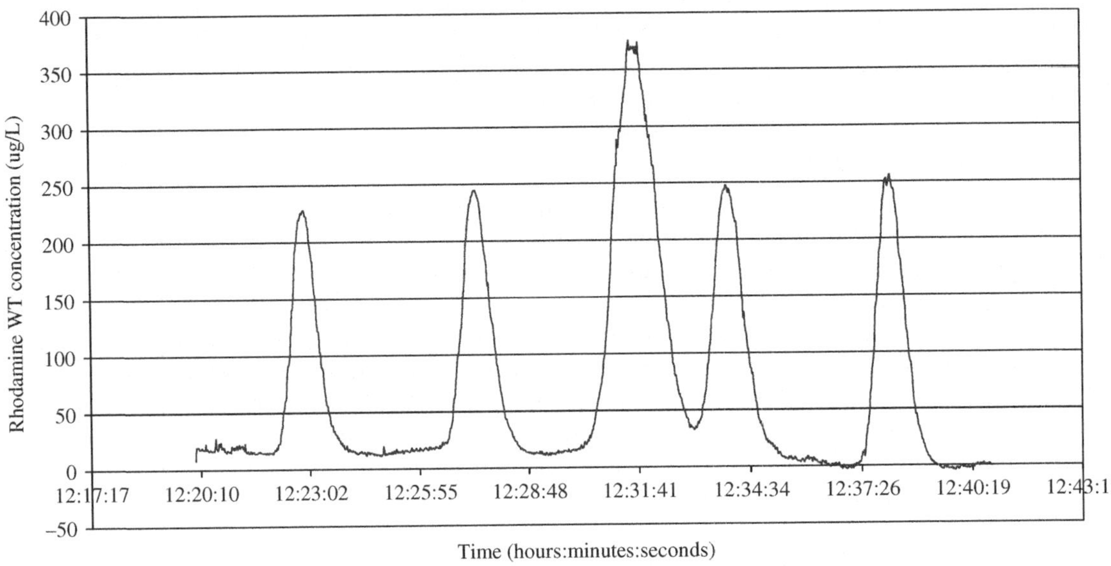

Figure 1: Separation of Rhodamine WT peak concentrations during QUEST 1a experiment.

Good peak separation is essential to enable accurate estimation of the peak areas which are required for the calculation of the amount of tracer passing the monitoring point. The adopted peak fitting process involved separate baseline adjustments for each of the tracer peaks followed by an area estimation. The crude results for each of five experiments are shown in Table 2 with the ratios representing the indicator:reference peak area values. There is clearly substantial variability around the expected value of 2.5 assuming the absence of any exfiltration in the indicator stretch (U1 to B) of the sewer. This variability is evident within single experimental runs as well as between the different experiments and the range of predicted exfiltration rates (−6.0% to 13.2%; Table 2) clearly question the viability of this measuring approach. Possible errors in the technique include failure to introduce consistent tracer doses and the problems associated with the time lapse between many of the indicator and reference doses. These problems were addressed by carrying out another series of QUEST experiments (Strategy 2) which are described below.

Table 2: Estimated peak area ratios for Strategy 1 QUEST experiments.

	Peak area ratios				Average peak area ratio	Average equivalent exfiltration
	Peak 3:Peak 1	Peak 3:Peak 2	Peak 3:Peak 4	Peak 3:Peak 5		
QUEST 1a	2.65	2.33	2.31	2.06	2.34 ± 0.21	6.4%
QUEST 1b	2.99	2.59	2.37	–	2.06 ± 0.26	−6.0%
QUEST 1c	1.86	1.95	2.55	2.32	2.17 ± 0.28	13.2%
QUEST 1d	1.97	2.24	2.93	–	2.38 ± 0.40	4.8%
QUEST 1e	2.05	2.10	2.65	1.98	2.20 ± 0.27	12.0%

3 = indicator peak; 1, 2, 4, and 5 = reference peaks

Strategy 2 QUEST Experiments

The temporal variations in the recorded RWT concentrations at the monitoring manhole A following the repeated introductions of tracer pulses of 50 mL and 20 mL, respectively at dosing points U1 and B are shown in Figure 2 for a 1 hour duration field experiment. Also shown are the flow rates which were monitored at 30 second intervals within the sewer at man-hole A. The smaller reference peaks associated with each indicator peak can be readily identified and application of the previously described peak fitting approach gives calculated ratios of indicator to reference peaks between 2.21 and 2.71. This is a similar range to that derived for the average values for the Strategy 1 QUEST experiments and is not considered sufficiently reliable to support this technique as a practical method for the determination of exfiltration rates.

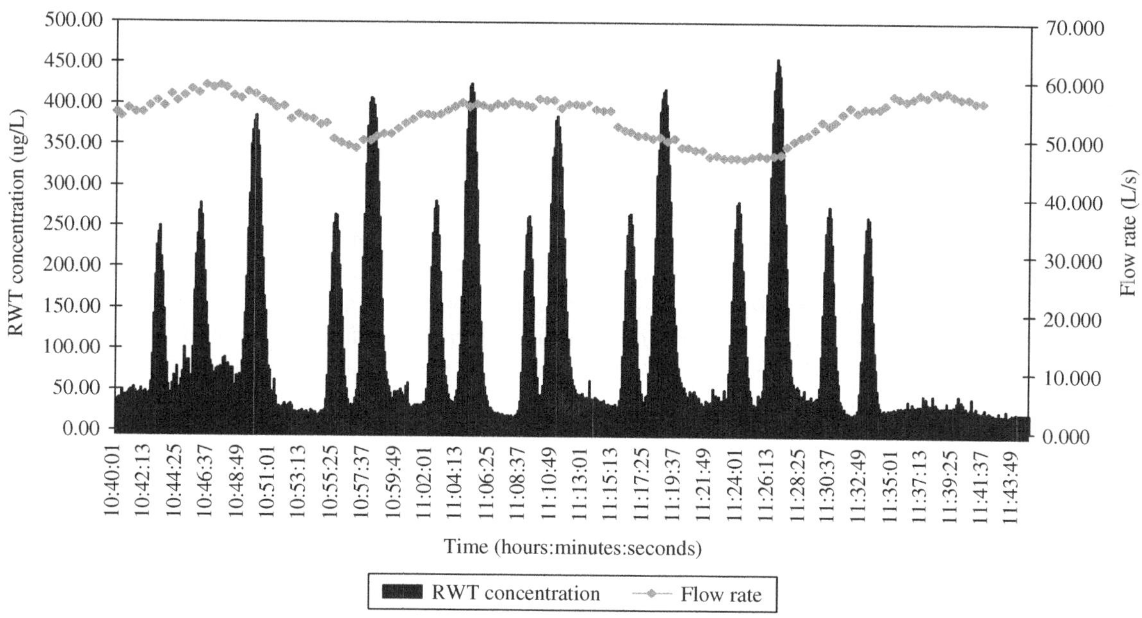

Figure 2: Separation of Rhodamine WT peak concentrations and variation of flow rates during a Strategy 2 QUEST experiment.

The average ratio of the indicator peaks to the adjacent reference peaks was 2.47 ± 0.17. The expected ratio, based on the dosed volumes and concentrations, and assuming that RWT behaves conservatively and there is no exfiltration within the indicator stretch of the sewer was 2.49. Therefore, the average monitored value predicts an exfiltration rate of 0.8% but the variability about the mean value and the range of the measured ratios limit the confidence which can be placed on this technique as a realistic overall approach for the determination of exfiltration. Rieckermann *et al.* (2005) have developed a sophisticated curve fitting approach for accurately assessing the peak areas but visual observation of Figure 2 shows that there is a basic underlying non-conformity in the different reference peaks as well as intermittent baseline noise which inevitably leads to variability in the calculated result.

It is evident that the observed signal intensities will be influenced by fluctuations in the background turbidity. Every attempt was made to counteract this by calibrating the fluorimeter using sewage samples collected immediately before the exfiltration experiments commenced. One possible error identified in the Strategy 1 experiments was that variations in sewer flow rates could influence the monitored concentrations. In the experiment illustrated in Figure 2, the flow rates can be seen to follow a shallow sinusoidal pattern between a maximum of 59.1 L/s and a minimum of 49.5 L/s. This variability can be expected to influence the RWT peak heights and areas and emphasises the need for careful control of the respective dosing times to achieve indicator and reference peaks which are subjected to similar flow conditions but are still resolvable for accurate quantitative analysis. The simultaneous determination of flow rates enables the comparison of indicator and reference peaks based on RWT loadings as well as concentrations. Inspection of the trends in Figure 2 show some evidence that tracer concentrations do increase at lower flow rates and vice versa but this is not a consistent trend. In an attempt to compensate for variations in flow, Rhodamine WT loadings were calculated and the temporal trends plotted. The ratios of indicator to reference peak areas based on tracer loadings (2.32–2.51) were less extreme than those calculated using concentrations but still too variable to have full confidence in the reliability of the QUEST technique, based on pulse dosing. The measurement of a pulse dose is made more difficult by the possible occurrence of intermittent tracer interferences which may be magnified during the short time duration of the pulse. A way to overcome this problem is to replace pulse dosing by continuous dosing (over time periods of up to 10 minutes) as this then allows an average signal due to the tracer to be derived and subsequently used to calculate the exfiltration rate.

RESULTS AND DISCUSSION–QUEST-C EXPERIMENTS

The QUEST-C experiments involve the prolonged introductions of two different tracers at constant rates (using a peristaltic pump set to a nominal delivery rate of 1 L/s) at dosing points U1 and B with subsequent monitoring at the downstream man-hole A. In addition to the continued use of Rhodamine WT, lithium (as lithium chloride) and bromide (as potassium bromide) were also employed and their sewer concentrations determined by collecting timed grab samples and returning these to the laboratory for analysis by flame emission spectrophotometry (Li) and ion chromatography (Br). An important advantage of these multi-tracer experiments over the pulse dosing technique is that they are less susceptible to short term variations in sewage characteristics (eg flow rate) which may influence the tracer signal.

Details of the dosing strategies used to test the QUEST-C technique are provided in Table 3. The QUEST-C1 and QUEST-C4 experiments were designed to assess the behaviours of the different tracers either along the full length of the monitored sewer (U1 to A) or the reference section (B to A), respectively. The comparison of two experiments involving continuous dosing with Rhodamine WT over the indicator section of the sewer (QUEST-C2) and continuous dosing with Li over the reference section of the sewer (QUEST-C3) enables an estimation of the exfiltration rate in the upstream indicator stretch of the sewer to be made. However, these were not performed simultaneously and therefore complete overlap of the tracer signals at the downstream monitoring site was not achieved. QUEST-C5 and QUEST-C6 experiments were carried out to achieve this requirement and hence eliminate any variability in the results due to changes in sewer conditions. Similarly, the QUEST-C7 experiment achieved this situation but with the tracers reversed relative to the indicator and reference doses. Flow rates within the sewer were monitored at the outlet manhole site A for all the QUEST-C experiments.

Table 3: Dosing arrangements for the QUEST-C experiments.

Reference code	Indicator dosing details at U1	Reference dosing details at B
QUEST-C1	Mixed tracer containing Rhodamine WT (1 g/L), Li (1 g/L) and Br (2.5 g/L) at 1 L/minute for 9 minutes	–
QUEST-C2	Rhodamine WT tracer (1 g/L) at 1 L/minute for 9 minutes	–
QUEST-C3	–	Li tracer (1 g/L) at 1 L/minute for 9 minutes
QUEST-C4	–	Mixed tracer containing Rhodamine WT (1 g/L), Li (1 g/L) and Br (2.5 g/L) at 1 L/minute for 9 minutes
QUEST-C5	Rhodamine WT tracer (1 g/L) at 1 L/minute for 5–6 minutes	Mixed tracer containing Li (1 g/L) and Br (2.5 g/L) at 1 L/minute for 5–6 minutes
QUEST-C6	Rhodamine WT tracer (1 g/L) at 1 L/minute for 5–6 minutes	Mixed tracer containing Li (1 g/L) and Br (2.5 g/L) at 1 L/minute for 5–6 minutes
QUEST-C7	Mixed tracer containing Li (1 g/L) and Br (2.5 g/L) at 1 L/minute for 5–6 minutes	Rhodamine WT tracer (1 g/L) at 1 L/minute for 5–6 minutes

QUEST-C1 and QUEST-C4 experiments

Monitoring of the Rhodamine WT and lithium concentrations at the downstream manhole A during these experiments resulted in concentration profiles for both tracers which mirrored each other exactly (see Figure 3 for the results for the QUEST-C1 experiment), indicating that they behaved consistently within the sewer environment regardless of whether they were introduced at dosing locations U1 (QUEST-C1) or B (QUEST-C4). These tracers have also demonstrated similar behaviours within wetlands at dosing concentrations of 100 μg/L for RWT and 1000 μg/L for Li (Dierberg and DeBusk, 2005). There are few reports of the use of RWT as a hydraulic tracer in sewage contaminated environments but Adams *et al.* (1998) found negligible interactions with suspended sediments during residence time studies of a harbour heavily impacted by combined sewer overflows. Lithium has been successfully used as a tracer in an engineered wetland receiving municipal wastewater effluent (Gray and Sedlak, 2005) and in a submerged biofilter for the tertiary nitrification of treated domestic sewage (Seguret and Racault, 1998).

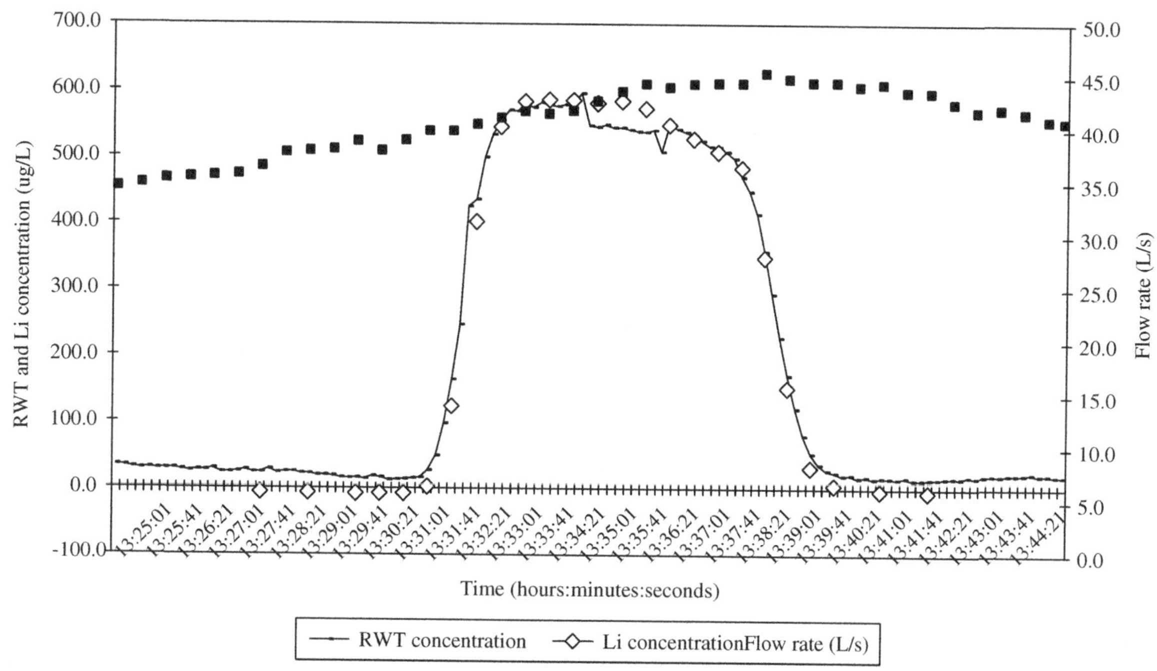

Figure 3: Temporal variations in RWT and Li concentrations and flow rates during QUEST-C1 experiment.

The viability of a tracer for the detection of any losses within the sewer environment is determined by its ability to behave conservatively. The QUEST-C1 and QUEST-C4 experiments enabled this characteristic to be tested for RWT and Li by comparing the monitored downstream tracer loadings with those introduced at the dosing point. The loading rates of both tracers at monitoring station A, calculated from the measured concentrations and the flow rates, are plotted for the QUEST-C1 experiment in Figure 4. The total amounts of each tracer passing the downstream sewer manhole can be derived from these loading diagrams and the results are compared with the introduced amounts for these dosing experiments to deduce the recovery efficiencies given in Table 4; this table also gives the tracer recovery efficiencies for the QUEST-C2 and QUEST-C3 experiments.

The percentage tracer recoveries for RWT and Li were consistently within the range 97.3 % to 101.4% indicating that both are relatively inert and demonstrate little removal during sewer transport for periods of up to 20 minutes. The best agreement between these two tracers was for the QUEST-C4 experiment involving the shorter reference section of the sewer (150 m). Over the full length of the monitored sewer (660 m), there were greater discrepancies but no clear indications as to which of the two tracers demonstrated the most conservative behaviour. Previous work has shown that both tracers were returned at more than 95% of the injected amount during passage through submerged aquatic vegetation dominated mesocosms (Dierberg and DeBusk, 2005). Other studies in wetlands showed recoveries of only 29% of the introduced RWT over 24 hour residence times with the principal loss mechanism being due to irreversible adsorption (Lin *et al.*, 2003) whereas Keefe *et al.* (2004) found overall RWT mass losses of 21% during passage through an impermeable wastewater–dependent wetland due to a combination of photolysis (10%) and sorption (11%) reactions. In a different environmental context, Pang *et al.* (2004) observed that between 4% and 14% of RWT was lost during transport through pumice sand columns due to irreversible adsorption of the *meta*-isomer although this isomer constitutes approximately 50% of commercial grade RWT (Vasudevan, 2001). In this study, the minor observed differences between RWT and Li can be seen to be within the experimental error associated with the determinations of the flow rates and peak tracer concentrations, which were used to calculate the loading rates.

The results for Br over both the longer and shorter sewer sections predict recovery rates above 100% and this is believed to be primarily due to the greater analytical problems associated with the determination of Br. Bromide has previously been reported to be a conservative tracer in a number of different environmental situations involving exposure to sewage including transport through the sewage effluent contaminated zone of a sandy aquifer (Blanford, 2005; Harman *et al.*, 1996) and a sand and gravel aquifer (Hess *et al.*, 2002) and passage through a subsurface flow

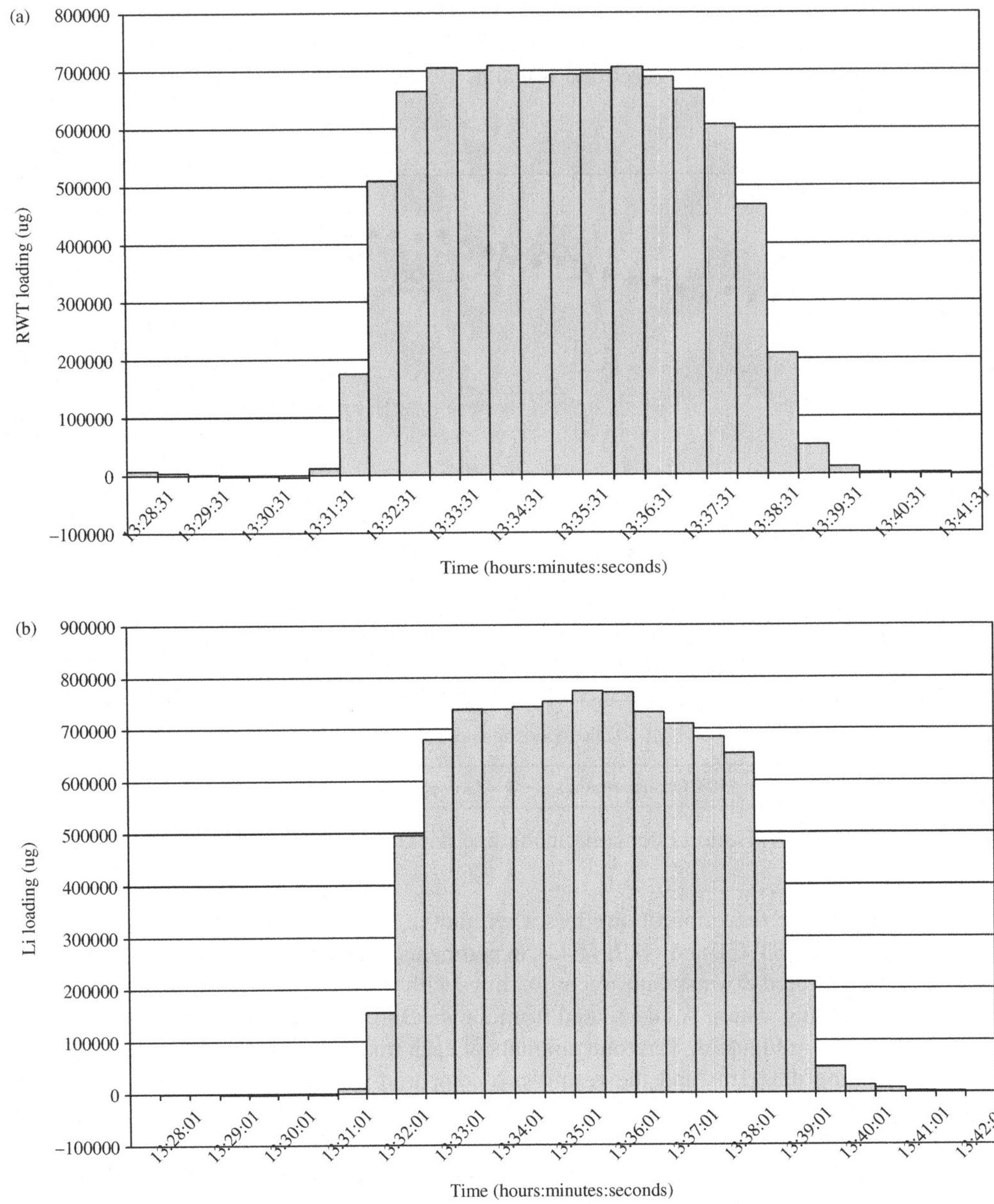

Figure 4: Tracer loading diagrams for a) RWT and b) Li for the QUEST-C1 experiments.

Table 4: Percentage recoveries of different tracers during QUEST-C1 to QUEST-C4 experiments.

	Tracer	Dosing amount (mg)	Monitored amount (mg)	Average percentage recovery
QUEST-C1	RWT	9270 ± 40	9050 ± 80	97.6
	Li	9270 ± 30	9400 ± 90	101.4
	Br	$23{,}170 \pm 60$	$25{,}300 \pm 900$	109.2
QUEST-C4	RWT	8560 ± 40	8390 ± 80	98.0
	Li	8560 ± 30	8410 ± 90	98.2
	Br	$21{,}400 \pm 60$	$23{,}200 \pm 800$	108.4
QUEST-C2 and QUEST-C3	RWT	8690 ± 40	8740 ± 80	100.6
	Li	8500 ± 30	8270 ± 90	97.3

constructed wetland receiving wastewater (Matamoros *et al.*, 2005). In this study bromide was determined in the laboratory using an ion chromatographic technique with an analytical precision of ±3.1% in the measured concentration range. This was considerably less precise than the corresponding techniques used for the determination of Rhodamine WT (in situ fluorimetry; ±0.6%) and lithium (flame emission spectrophotometry; ±0.7%) resulting in less reliable results for the bromide tracer.

QUEST-C2 and QUEST-C3 experiments

The combined tracer curves for these experiments together with the relevant flow rate changes are shown in Figure 4. Rhodamine WT represents the indicator tracer having been introduced at dosing point U1 and lithium is the reference tracer. Because the relative timings for the introductions of the two tracers were not correctly judged, the prolonged tracer peaks do not directly overlap. The slopes associated with both the Rhodamine WT and Li peak concentrations (Figure 5) identify the dangers with taking non-coincident average peak values, particularly where influencing changes (such as flow rate increases and decreases) occur within the sewer.

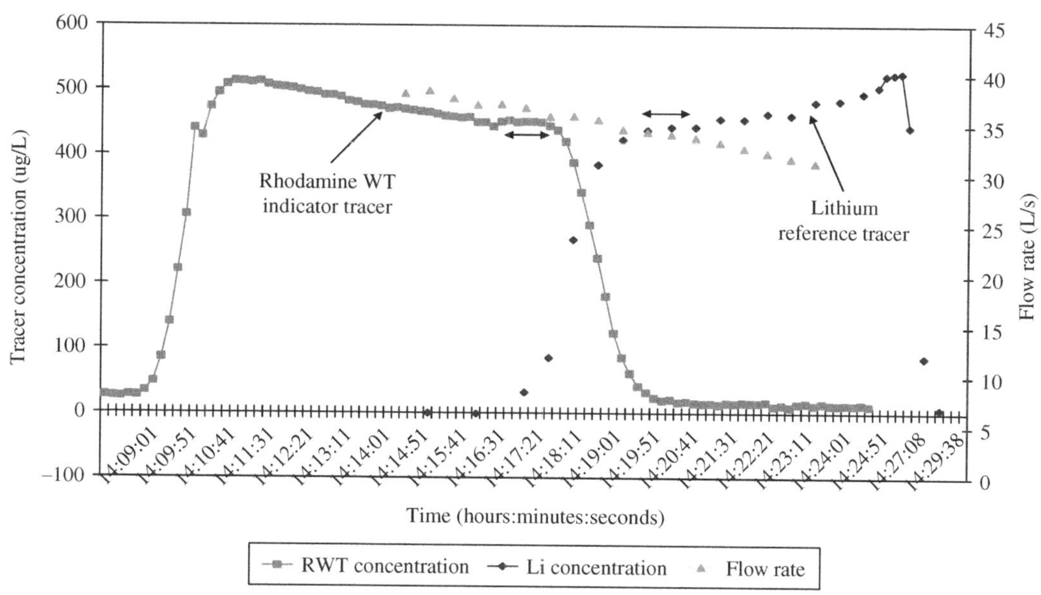

Figure 5: Tracer concentration curves and flow rates for QUEST-C2 and QUEST-C3 experiments.

The extent of exfiltration, E, within the indicator stretch of the monitored sewer can be expressed as:

$$E = 1 - \left[\frac{(Total.tracer.loading)_{ind}}{(Total.tracer.loading)_{ref}}\right]$$

from which standardisation of the measured in-sewer values relative to the dosed values gives:

$$E = 1 - \left[\frac{\left(\frac{Q \times C_M}{q \times C_D}\right)_{ind}}{\left(\frac{Q \times C_M}{q \times C_D}\right)_{ref}}\right]$$

where Q = sewer flow rate; q = dosing rate; C_M = measured concentration; C_D = dosed concentration.

For concurrent measurements of indicator and reference signals, the sewage flow rate (Q) is constant for both tracers and if the tracer dosing rates are identical the overall expression simplifies to:

$$E = 1 - \left[\frac{\left(\frac{C_M}{C_D}\right)_{ind}}{\left(\frac{C_M}{C_D}\right)_{ref}}\right]$$

Using the average tracer concentrations and associated flow rates, from the short plateaux regions identified in Figure 5, an exfiltration value of −0.53% can be calculated from the results obtained from the QUEST-C2 and QUEST-C3 experiments. Although this is consistent with the expected performance of the tested sewer, greater confidence in the derived result would be obtained if the experimental conditions were amended in order to achieve maximum overlap of the tracer curves. The QUEST-C5, QUEST-C6 and QUEST-C7 experiments were carried out to satisfy this requirement as well as to further test the versatility of the three different tracers.

QUEST-C5, QUEST-C6 and QUEST-C7 experiments

In the QUEST-C5 and QUEST-C6 experiments, Rhodamine WT was used as the indicator tracer with a mixed tracer of lithium and bromide being used as the reference. These tracer identities were reversed for the QUEST-C7 experiment. The measured concentration profiles for Rhodamine WT, lithium and bromide for the QUEST-C5 experiment are plotted in Figure 6. The results obtained for each experiment in terms of the average peak concentrations and their variability in the region of overlap are shown in Table 5 together with the calculated percentage exfiltration in the indicator stretch of the monitored sewer. The greater variability in the peak Br signal, despite its expected conservative behaviour, is indicative of the lower analytical precision with which this anion could be determined in sewage compared to RWT and Li. Therefore, for Br there is a reduced confidence concerning the use of the obtained results to estimate the degree of exfiltration. However, the study of Br behaviour provides a useful back-up confirmation of the in-sewer performances of the other tracers, particularly Li. The dosing concentration for Br was 2.5 g/L compared to 1.0 g/L for Li and the monitored ratio of the average peak concentrations was 2.51 which indicates that both tracers behaved similarly in terms of interactions during their passage through the sewer.

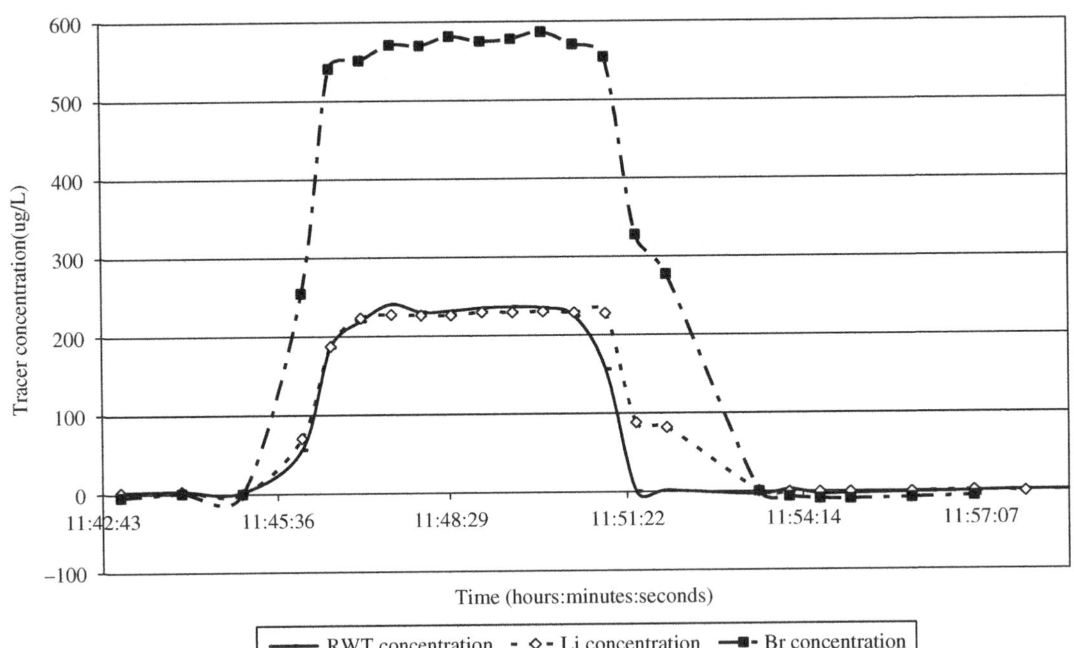

Figure 6: Temporal variations in monitored RWT, Li and Br concentrations during QUEST-C5 experiment.

The exfiltration results obtained from the QUEST-C5 and 7 experiments are in good agreement and predict an overall loss due to exfiltration of between 0.38% and 0.86%. The conformity of these results would suggest that all three tested tracers are interchangeable in terms of their use as either the indicator or reference tracer although only experiment IV examined the use of RWT as the reference tracer. The average exfiltration value of 0.62% predicted by these experiments is below the 1% to 2% of average sewer flow determined for an aquifer located in Nottingham, UK as a result of water balance calculations (Yang *et al.*, 1999) and equates to 0.92 L/s/km based on the monitored average sewer flow rate of 84 L/s. This is also below the range of values (1.0–9.0 L/s/km) listed by Gruenfeld (2000) based on an analysis of data provided by USEPA funded exfiltration monitoring projects.

The most tested tracer combination was that using RWT as the indicator and Li as the reference (QUEST-C2, QUEST-C3, QUEST-C5, and QUEST-C6 experiments). For the QUEST-C2 and QUEST-C3 experiments, this

Table 5: Peak tracer concentrations and predicted exfiltration rates using different combinations of tracers for the overlapping indicator and reference signals.

	Tracer combination and peak concentration (μg/l)		Predicted percentage exfiltration
	Indicator	Reference	
QUEST-C5	RWT; 234.4 ± 3.0	Li; 228.5 ± 1.6	−0.381 ± 0.002
	RWT; 234.4 ± 3.0	Br; 573.0 ± 10.2	−0.685 ± 0.005
QUEST-C6	RWT; 243.5 ± 7.2	Li; 254.2 ± 8.9	+1.078 ± 0.009
	RWT; 243.5 ± 7.2	Br; 659.3 ± 5.6	−2.571 ± 0.021
QUEST-C7	Li; 212.3 ± 8.0	RWT; 213.8 ± 1.2	−0.697 ± 0.023
	Br; 529.9 ± 15.5	RWT; 213.8 ± 1.2	−0.860 ± 0.059

combination predicted a positive value for the exfiltration (+0.53%), which can only be true if the indicator tracer was entering the sewer during the experiment. This is unlikely and hence the positive value has to be taken as a measure of the overall experimental error associated with this technique. A similar result was obtained for the RWT/Li tracer combination during the QUEST-C6 experiment with the value of +1.08% again indicating a positive deviation in the overall experimental error associated with the technique. Based on the positive and negative results in Table 5, it can be optimistically deduced that the QUEST-C technique is capable of determining the percentage exfiltration within a sewer system to within ± 1.0%. However, the exfiltration result obtained using Br as tracer in the QUEST-C6 experiment (−2.57%) is outside this accuracy range. This can be partly explained by the difficulties encountered with accurately analyzing Br in the collected sewage samples and this is supported by the large difference between the determined value (−2.57%) and that obtained when Li was used as the reference tracer (+1.08%) under identical field measurement conditions. Therefore it is concluded that the experimental results indicate that the QUEST-C technique is capable of measuring the percentage of sewer exfiltration to an accuracy of ±1.0% when RWT and Li are used as either the indicator or reference tracers for, at least, sewer lengths less than 1 km.

CONCLUSION

The two described traced techniques (QUEST) and (QUEST-C) are shown to represent feasible practical methods for the determination of exfiltration in sewer systems although with different degrees of reliability. Although the QUEST technique is, in principle, easier to use there can be major problems with the determination of the amount of tracer passing the monitoring point represented by the peak area. The nature of pulse dosing means that any measured interferences occurring within the sewer system are focused on the tracer plug and hence the impact on the determined peak height or area may be considerable. An additional problem involves compensating for baseline variability particularly at low dosing concentrations. The use of higher dosing concentrations/volumes of RWT would partially solve this problem but it would be important to ensure that this did not affect the mixing efficiency or pose possible problems for the downstream environment *e.g.* the receiving sewage treatment works.

The methodological approach adopted in the QUEST-C experiments eliminates any errors due to flow variations and the background tracer concentrations can be more accurately subtracted from the average measured peak concentrations. In the described series of experiments the downstream recorded peak tracer concentrations were coincident for between 6 and 9 minutes and it is envisaged that longer peak overlap times (*e.g.* typically employing dosing times of approximately 20 minutes) will provide increased confidence in the accuracy of the calculated peak average concentration. The analytical problems associated with determining Rhodamine WT (turbidity interference) and Li (interference from Na) in sewage were overcome by calibrating the relevant analytical instruments with sewage samples collected prior to and after the completion of each QUEST-C experiment. However, this approach would still be vulnerable to changes in sewage composition during the progress of an individual experiment. In such circumstances the overall accuracy could be increased by carrying out more frequent calibrations supported by further studies of the impact of sewage component variability, such as increases in organic sewage solids, on tracer performance. Nevertheless, the results obtained from the reported series of QUEST-C experiments indicate that the use of Rhodamine WT (as indicator) and Li (as reference) are the most reliable tracer combination with the ability to estimate the occurrence of exfiltration over a sewer length of less than 1 km to ±1%.

ASSESSMENT OF INFILTRATION

Introduction and field site

The submersible s:can UV-VIS spectrophotometer for the determination of COD time series, as described in chapter 2, was employed at a Thames Water sewage treatment works (STW) in order to estimate sewer infiltration rates. The application of this technique was undertaken within the context of a combined sewer system which was known to be the source of continuing problems due to excessive infiltration gains particularily following winter wet weather conditions and seasonal rises in the underlying water table. The investigated urban catchment at Berkhamsted in W Hertfordshire (SE England) which covers a total of nearly 8 km^2 and represents a commuting suburb of 25,000 population located some 25 km north west of Central London (Figure 7 and inset). The catchment possesses a mixed urban land use with dwellings located on relatively steep side slopes with retail, commercial and main communication facilities chiefly located along the valley floor parallel to the River Bulbourne and the Grand Union Canal. The strongly reticulate combined sewer system is comprised of circular ceramic pipes ranging in diameter from 0.15 m to 0.525 m with more than 60% of the network being greater than 100 years old. The smaller diameter pipes run directly down the steep valley sides and are intercepted by larger parallel pipes running at right angles along the valley floor. The sewered drainage from the south side of the valley crosses the river in the town centre (Figure 8) to meet the parallel sewer lines on the north side of the catchment which then divert the wastewater flows in an easterly direction to the main sewage treatment works (Figure 7).

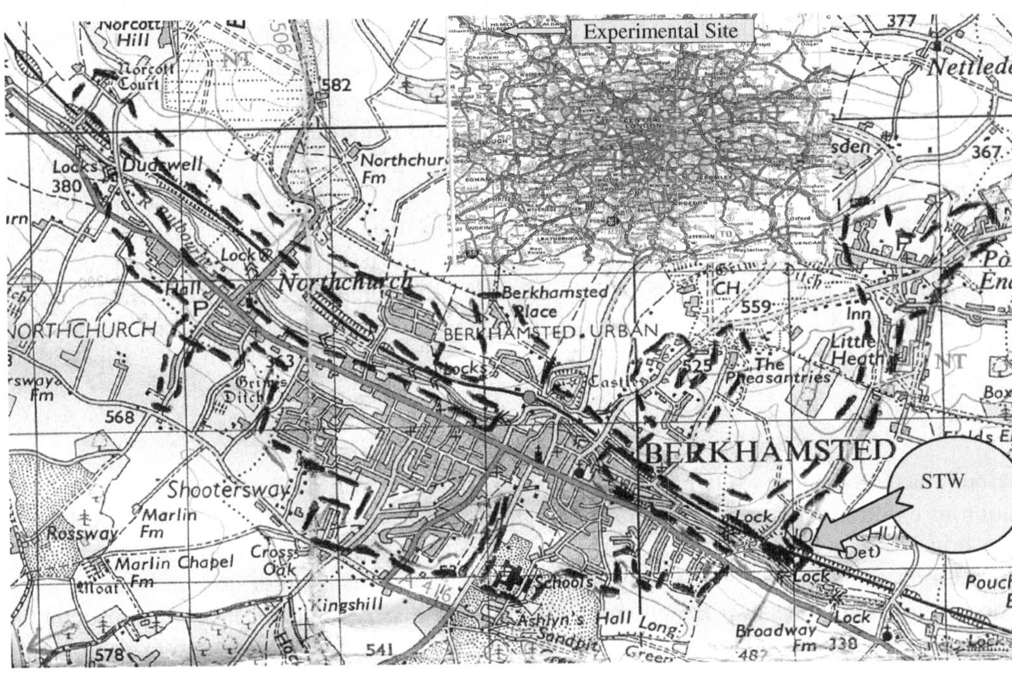

Figure 7: The Berkhamsted catchment.

Figure 8 shows the sewer inlet at the receiving sewage treatment works (Figure 8a) and the variation in daily depths and flows recorded over a typical week (Figure 8b) with the minimum night-time baseflow level varying between 70 and 120 l/s indicating an endemic infiltration gain to the system. COD concentrations observed in the immediate upstream reaches of the sewer pipe are also shown (Figure 8c), indicating very low values in the main trunk line running along the valley floor and confirming the dilution caused by infiltrating groundwater.

The underlying geology is relatively simple comprising permeable and fissured Chalk overlain by a thin and inconsistent layer of Boulder Clay with a narrow alluvial valley floor. The regional Chalk aquifer which is supplied directly by rainfall, and is supplemented to the south-west by artificial recharge, represents a valuable resource of drinking water for the Greater London region. The underlying regional Chalk aquifer is subject to considerable seasonal variation in groundwater level (Figure 9) and the Berkhamsted catchment has long suffered from substantial infiltration into the sewer system with out-of-sewer property flooding occurring during storm events concurrent with elevated groundwater levels, mainly in the winter/early spring periods. The elevated groundwater levels following the extreme winter 2001 and 2002/03 rainfall events are evident from Figure 9 resulting in a considerable field gradient to the trunk sewer line running at depths of between 0.5 m–2.5 m below the valley floor, causing basement and property flooding as well as extensive infiltration into the sewer system.

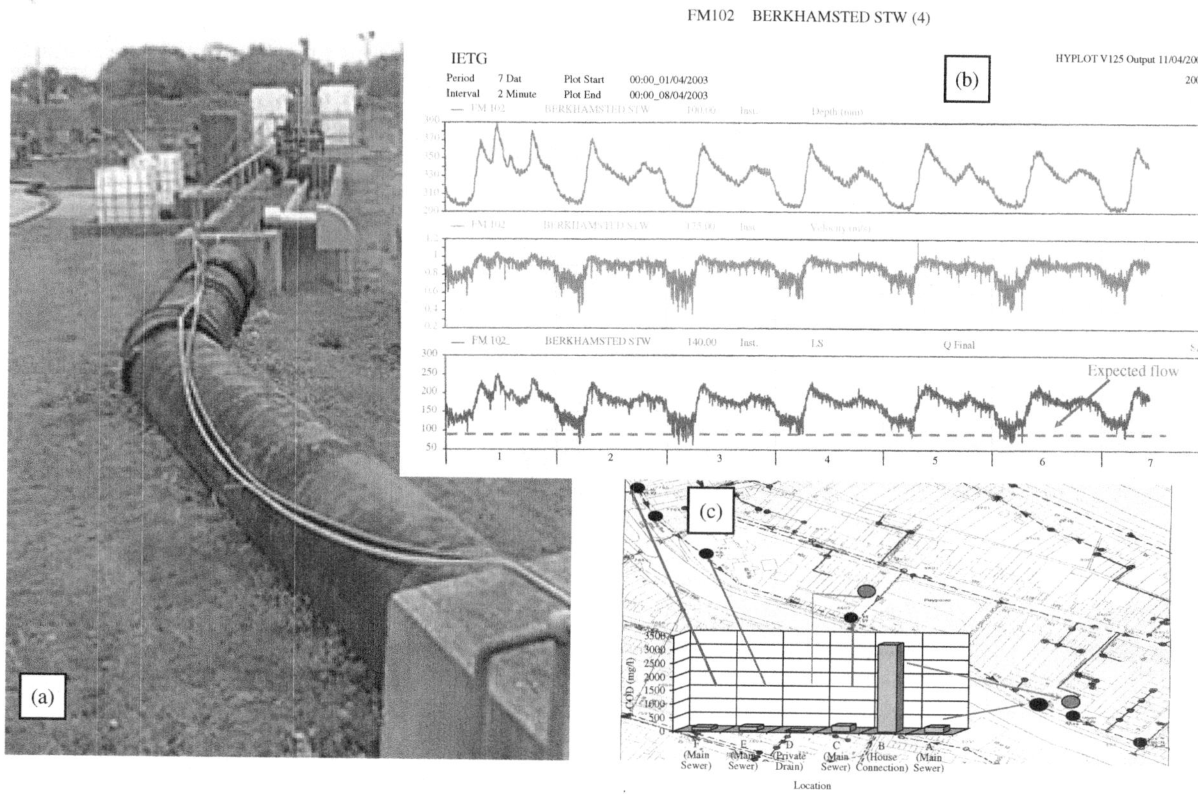

Figure 8: Berkhamsted STW sewage flows.

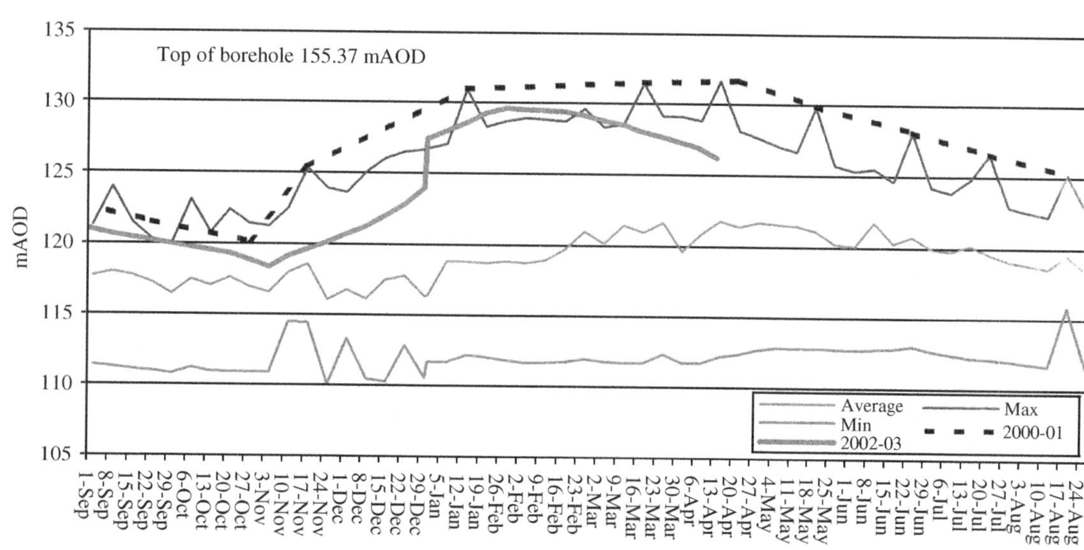

Figure 9: Groundwater detail for the Berkhamsted area.

The geomorphology of the catchment exacerbates the flooding problem as the head pipes can surcharge due to their small diameter and steep gradients. During the excessively wet winter period (January–April) of 2001, the groundwater levels attained 132 mAOD at nearby boreholes compared to an average height of the valley floor of 100–110 mAOD. This led to substantial and widespread infiltration into the sewer system and the STW flow records suggest that the average infiltration rate during this period was around 3000%. This compares to a winter season average of about 280–300% as estimated by Thames Water given that calculated average dry weather flow (DWF) based on the recorded flow time series for the catchment should be in the order of 50 l/s.

Field monitoring

Figure 10 illustrates the instrumental set-up at the inlet of the Berkhamsted sewage treatment works for the infiltration test runs. Both field observation and inspection of the COD time series suggests that the return humus mixer pipe located at the inlet did not interfere with the instrumental readings for the inflow wastewater which were timed to coincide with non-operational periods of the return mixer pipe. In addition, both the sampler inlet rose and the spectrophotometer were located near the invert of the inlet flume which was largely outside the essentially surface mixing zone of the return pipe.

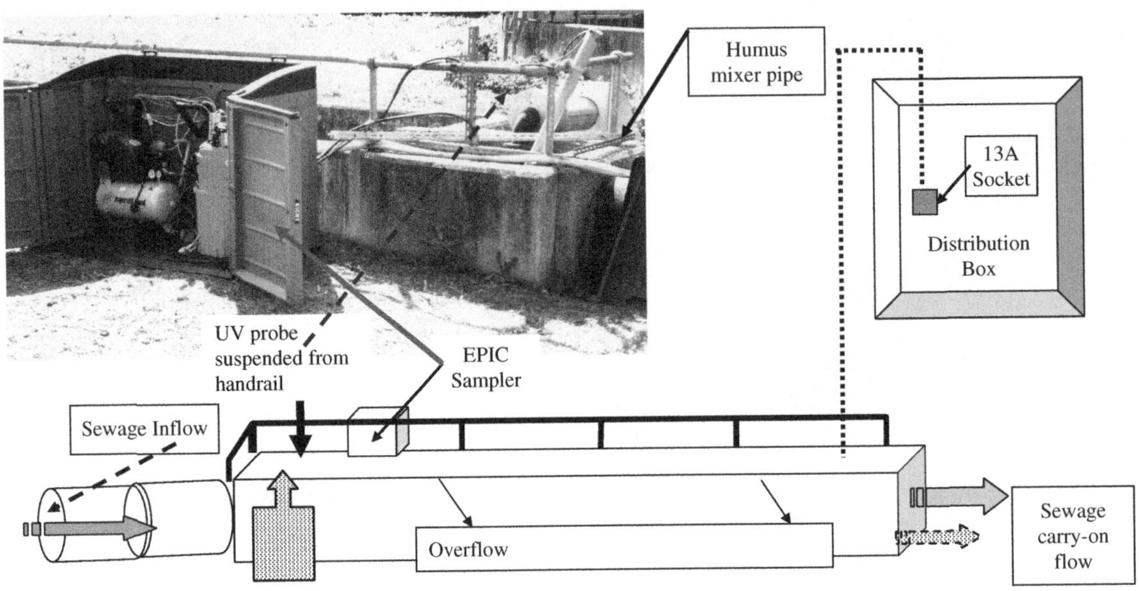

Figure 10: Instrumental Set-up at Berkhamsted Sewage Treatment Works

The COD-equivalents in the incoming wastewater were measured at 3 minute intervals with a submersible s::can UV-VIS spectrophotometer (200–750 nm range) installed in the inlet works. The optical window of the spectrophotometer was automatically cleaned with a 5 second blast of compressed air 20 seconds before the commencement of the measuring cycle. In-situ calibration of the instrumental readings was undertaken using conventional laboratory COD analysis of automatic samples taken at hourly intervals at the inlet. These were supplemented by grab samples. Wastewater flow readings obtained at 15 minute intervals from the site ultrasonic flow recorder were provided by Thames Water. Two measurement runs were conducted at the Berkhamsted site. The first test run operated over the period 16–23 September 2004 and the second run from 23–30 September 2004 but only the results from the first operational run are presented and discussed here.

Results and discussion

The fraction of infiltrating water is determined from a combined analysis of the time series of the discharged wastewater and the monitored COD concentrations. The data analysis is based on the use of a mixing model (outlined in Chapter 2), that describes the pollutant concentration in the wastewater in terms of its dependence on flow (Q) and time (t). The employed parameter set thus contains time dependent variables describing the infiltration rate and pollutant concentration in the foul sewage. It is assumed that the COD of the infiltrating water is negligible.

Figure 11 presents the full seven day COD time series as recorded at the wastewater inlet which shows evidence of possible sensor fouling (and/or other interference) during the first three days of the test run. The instrumental cleaning cycle was adjusted (to provide a 10 second air blast) in an attempt to overcome this problem and the last 4 days of the record appear to be reasonably clear of such interference. Figure 12 therefore displays a reduced data set for the last 4 days of the test run as derived from the R-Script procedure (Rieckermann and Bares, 2005). The graphical displays of Figure 12 show the modelled discharge (Q; l/s) and the data fit COD (mg/l) time series with the latter comprising the measured COD-equivalents (in black) and the calculated model COD time series (in red). The series provides a good reproduction of the measured data and satisfactorily explains the major elements of the wastewater pollutograph. The central hydrograph plot illustrates the constant baseflow separation and the diurnal wastewater variations. There is no evidence in the time series data for any interflow component over the time period of the test runs. The baseflow separation was calculated from the mixing equation by means of a Monte Carlo optimisation procedure and the accumulated infiltration results are listed in Table 6.

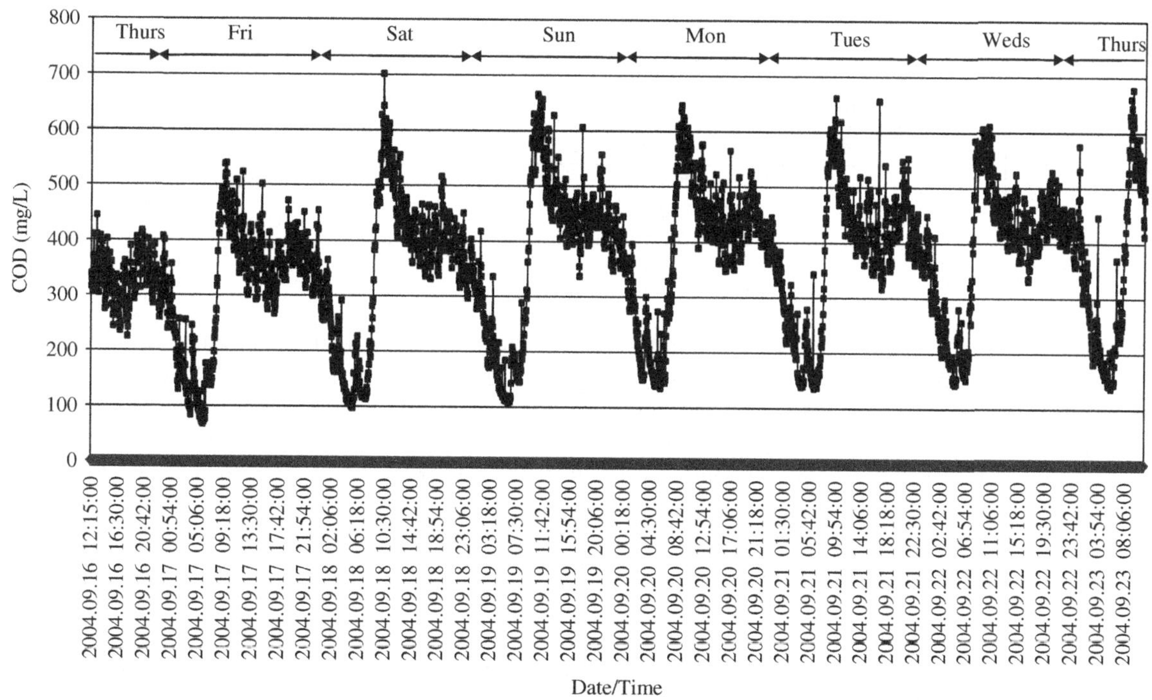

Figure 11: COD Time Series for Berkhamsted STW 16–23 September 2004.

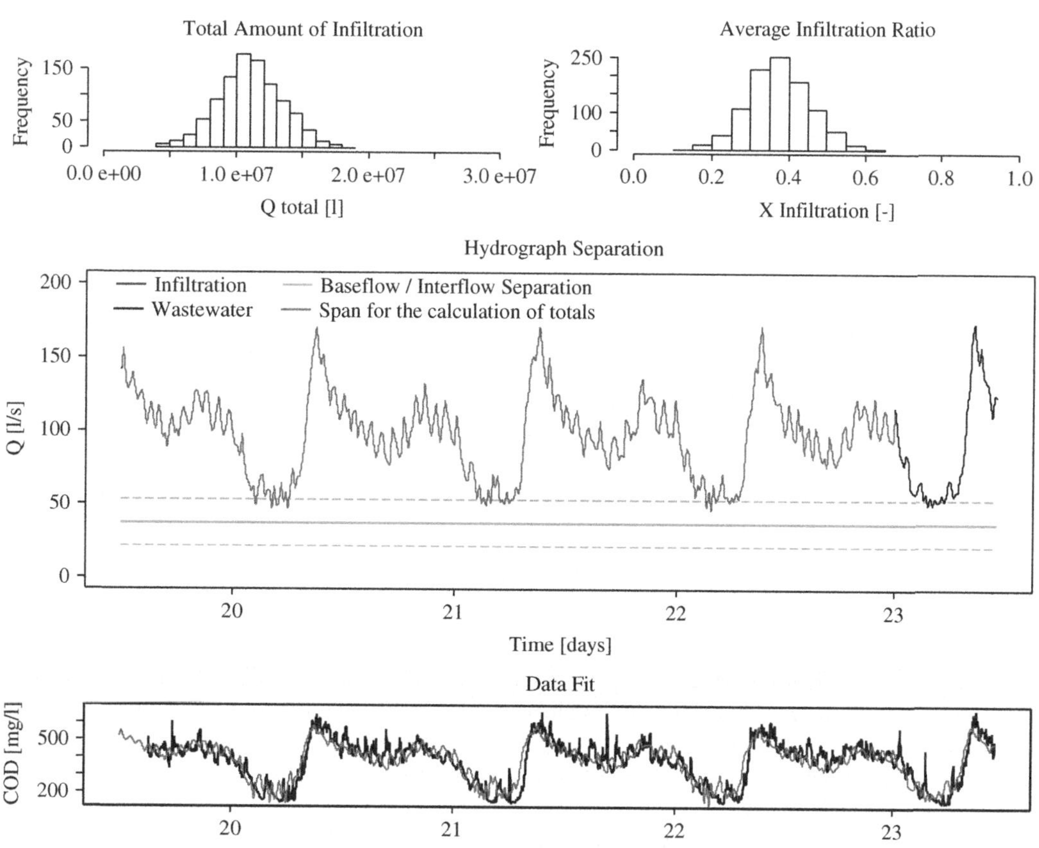

Figure 12: Outputs for the shortened data set.

Table 6: Accumulated results (for 95% confidence limit)

Time Span	Σ Total infiltration	Infiltration ratio
19.09 to 23.09	Min: 4.2×10^3 m^3	Min: 14.3%
	Median: 10.9×10^3 m^3	Median: 37.1%
	Mean: 11.1×10^3 m^3	Mean: 37.4%
	Max: 18.5×10^3 m^3	Max: 62.5%

The infiltration ratio relates the volumes of infiltration to the total amount of wastewater discharged within the considered time span. The histograms of Figure 12 illustrate the frequency distributions for both total infiltration and infiltration ratios for the time span. The results are given for a 95% confidence level which solely account for the effect of random errors that are superimposed upon the measured time series. The validity of the parameter estimates might also be influenced by systematic errors affecting the accuracy of the measured data series and which have not been considered here.

The hydrograph separation in Figure 12 predicts an infiltration rate of approximately 36 l/s under non-rainfall conditions. This figure is based on the combined modelled COD and flow time series as determined from the mixing equation. Inspection of the observed diurnal flow data (Figure 13) as recorded at the sewage treatment works, indicates that the minimum night-time dry weather flow is about 48 l/s. Based on the US EPA estimation that some 85% of such flow is directly contributed by infiltration waters, this would yield a corrected figure of 40.8 l/s for the infiltration rate, which is close to the modelled value. The 36 l/s predicted infiltration value, if averaged over a 4 day period, yields a total infiltration volume of 10.9×10^3 m^3 which compares favourably to the median and mean values determined from the Monte Carlo simulation (see Table 6).

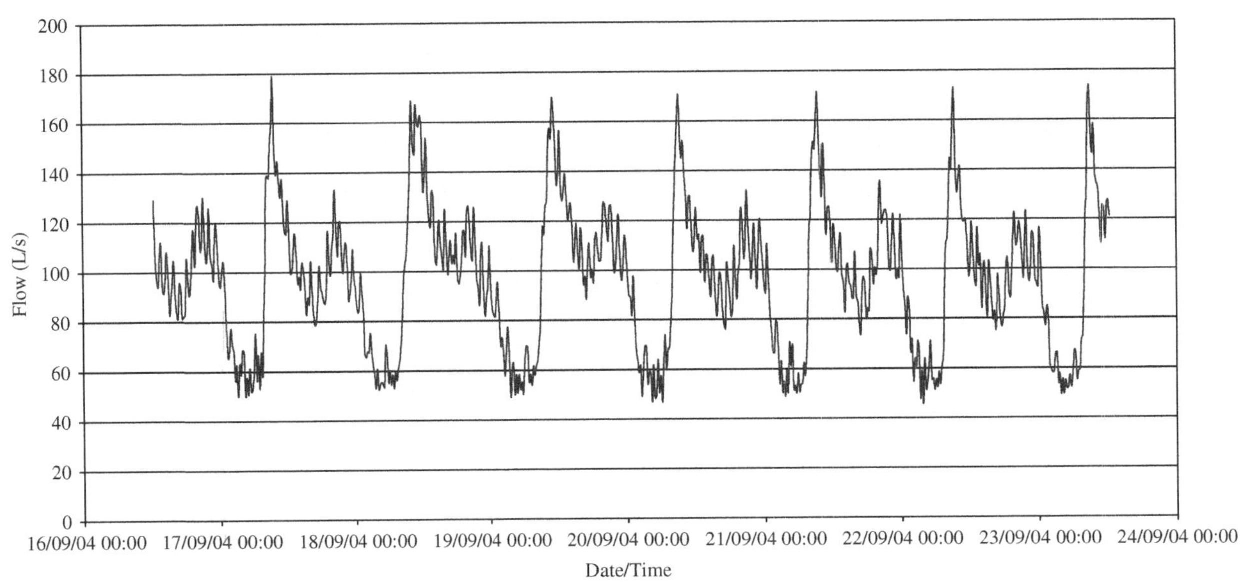

Figure 13: Flow time series for Berkhamsted STW 16–24 September 2004.

Dry weather flow as empirically derived from Formula A (DWF = PG+I+E), would give estimates for the total 25,000 population load from the Berkhamsted catchment varying between 43.4 l/s to 63.7 l/s based on 150 l/hd.day and 200 l/hd.day water consumption (G) respectively, assuming no trade effluent (E) contributions.

Assuming the observed flow time series as recorded in Figure 13 represents discharges over a dry weather (i.e non-rain) period, and that there are therefore negligible inflow or interflow infiltration contributions, this time series can also be used to derive an average "dry weather flow" rate for the period of 87.04 l/s. If this DWF value is used in Formula A, for a water consumption (G) value of 150 l/d.day, the infiltration rate would be 43.64 l/s whilst at 200 l/hd.day, the infiltration rate would be 29.2 l/s. Based on a water consumption (G), the methodology appears to be reasonably robust and enables a quantification of (non-polluted) infiltrating groundwater to be made based on the measured time series of wastewater flow and pollutant equivalent concentrations and without recourse to any additional external information. The procedure demonstrates the potential utility of the submersible spectrometer probe for accurate determination of COD-equivalent. The ability of the methodology to also consider natural storage and interflow phenomena significantly broadens the practical application of the method. Further tests during periods of elevated groundwater level particularly following rain events would enable recessional and storage components to be identified and quantified, thus supporting a better understanding of the dynamics of infiltration processes in urban sewer networks and should in turn support more effective remediation strategies. The most critical benchmark for the more general application of the methodology would be the attainable minimisation of overall uncertainty in the infiltration estimates. In such circumstances it would also be necessary to monitor the wastewater flows.

The data analysed in this report was collected in September when groundwater levels in the Berkhamsted catchment were low. It would be interesting to test the developed methodology in the winter months when elevated groundwater levels exist. Performing this at the inlet to the Berkhamsted STW would allow a direct comparison to the low groundwater situation described in this chapter. The ability of the methodology to identify the interflow component following a substantial rainfall event also needs to be tested.

ACKNOWLEDGEMENTS

The authors wish to acknowledge the cooperation of Thames Water Utilities plc in providing access to the sewer and for supervising safety procedures during the field work. We are also indebted to Peter Lister and Chris Willgress for their full and efficient technical assistance both before and during field work.

REFERENCES

Adams, E.E., Stolzenbach, K.D., Lee, J.J., Caroli, J. and Funk, D. (1998) Deposition of contaminated sediments in Boston Harbor studied using fluorescent dye and particle tracers. *Estuarine Coastal and Shelf Science*, **46**(3), 371–382.

Blanford, W.J., Brusseau, M.L., Yeh, T.C.J., Gerba, C.P. and Harvey, R. (2005) Influence of water chemistry and travel distance on bacteriophage PRD-1 transport in a sandy aquifer. *Water Res.*, **39**(11), 2345–2357.

Dierberg, F.F. and DeBusk, T.A. (2005) An evaluation of two tracers in surface-flow wetlands: Rhodamine WT and lithium. *Wetlands*, **25**(1), 8–25.

Edwardson, K.J., Bowden, W.B., Dahm, C. and Morrice, J. (2003) The hydraulic characteristics and geochemistry of hyporheic and parafluvial zones in Arctic tundra streams, north slope, Alaska. *Adv. Water Resour.*, **26**(9), 907–923.

Flury, M. and Wai, N.N. (2003) Dyes as tracers for vadose zone hydrology. *Review of Geophysics*, **41**(1), Article No. 1002.

Gray, J.L. and Sedlak, D.L. (2005) The fate of estrogenic hormones in an engineered treatment wetland with dense macrophytes. *Water Environ. Res.*, 2005, **77**(1), 24–31.

Giulianelli, M., Mazza, M., Prigiobbe, V. and Russo, F. (2003) Assessing exfiltration in an urban sewer by slug dosing of a chemical tracer. 1–12 in: Marsalek, J., Sztruhar, D., Giulkianelli, M and Urbonas, B (Edits.), *Proceedings of NATO Workshop on Urban Runoff; Enhancing the Urban Environment.* Rome, Italy, 5–8 November, 2003.

Gruenfeld, M. (2000) *Exfiltration in Sewer Systems: A Draft Report to the US Environmental Protection Agency*. National Risk Management Research Laboratory, Radison, New Jersey, USA, 2000.

Harman, J., Robertson, W.D., Cherry, J.A. and Zanini, L. (1996) Impacts on a sand aquifer from an old septic system: Nitrate and phosphate. *Groundwater*, 1996, **34**(6), 1105–1114.

Hess, K.M., Davis, J.A., Kent, D.B. and Coston, J.A. (2002) Multispecies reactive tracer test in an aquifer with spatially variable chemical conditions, Cape Cod, Massachusetts: Dispersive transport of bromide and nickel. *Water Resour. Res.*, **38**(8), Article No. 1161.

Keefe, S.H., Barber, L.B., Runkel, R.L., Ryan, J.N., McKnight, D.M. and Wass, R.D. (2004) Conservative and reactive solute transport in constructed wetlands. *Water Resour. Res.*, 2004, **40**(1), Article No. W01201.

Lin, A.Y.C., Debroux, J.F., Cunningham, J.A. and Reinhard, M. (2003) Comparison of rhodamine WT and bromide in the determination of hydraulic characteristics of constructed wetlands. *Ecol. Eng.*, 2003, **20**(1), 75–88.

Matamoros, V., Garcia, J. and Bayona, J.M. (2005) Behavior of selected pharmaceuticals in subsurface flow constructed wetland: A pilot-scale study. *Environ. Sci. Technol.*, **39**(14), 5449–5454.

Pang, L.P., Close, M., Greenfield, H. and Stanton, G. Adsorption and transport of cadmium and rhodamine WT in pumice and sand columns. (2004) *New Zealand Journal of Marine and Freshwater Research*, **38**(2), 367–378.

Revitt, D.M., Ellis, J.B. and Paterakis, N. (2006) Comparison of tracer techniques for monitoring sewer losses. *Journal of Environmental Monitoring*, **8**(5), 564–571.

Rieckermann, J. (2005) *Quantification of exfiltration from sewers with tracers.* Unpublished PhD thesis, Swiss Federal Institute of Technology, Zurich, Switzerland.

Rieckermann, J. and Bares, V. (2005) *Standard Operation Procedures (SOP): Quantification of Exfiltration from Sewers with Artifical Tracers* (QUEST). EAWAG, Zurich, Switzerland., 48pp.

Rieckermann, J. and Gujer, W. (2002) Quantifying exfiltration from leaky sewers with artificial tracers. *Proceedings of the International Conference "SOM 2002 Sewer Operation and Maintenance"*, Bradford, UK, 26–28 November, 2002, pp 8.

Rieckermann, J., Borsuk, M., Reichert, P. and Gujer, W. (2005) A novel tracer method for quantifying sewer exfiltration. *Water Resour. Res.*, **41**(5), Art. No. W05013.

Rutsch, M., Rieckermann, J. and Krebs, P. (2005) Quantification of sewer leakage–a review. In: Mikkelsen, P. (Edit), *Proceedings of 10th International Conference on Urban Drainage(ICUD),* Copenhagen, Denmark, 21–26 August 2005, IWA Publishing, London, UK. ISBN 978 1843395 744.

Seguret, F. and Racault, Y. (1998) Hydrodynamic behaviour of a full scale-scale submerged biofilter and its possible influence on performances. *Water Sci. Technol.*, **38**(8–9), 249–256.

Vasudevan, D., Fimmen, R.L. and Francisco, A.B. (2001) Tracer-grade rhodamine WT: Structure of constituent isomers and their sorption behaviour. *Environ. Sci. Technol.*, **35**(20), 4089–4096.

Yang, Y., Lerner, D.N., Barrett, M.H. and Tellam, J.H. Quantification of groundwater recharge in the city of Nottingham, UK. (1999). *Environ. Geol.*, **38**(3), 183–198.

Chapter 10

Modelling sub-catchment approaches to leakage rates

Tomas Metelka, Karol Pryl, Torsten Franz, Zdenik Pliska, Peter Krebs and Mandy Rutsch

ASSOCIATED MODELS AND TOOLS

Measurements of exfiltration and infiltration rates in sewer pipes and house connections provide data and results which need to be integrated and adequately displayed in order to give end-users a clear representation and understanding of the phenomena affecting their sewer systems. Moreover, field data obtained by application of the tracer methods developed in APUSS Work Area 1 (see Chapters 2, 4 and 5), need to be developed in conjunction with appropriate models, for both calibration purposes and comparison of modelled simulation results to field measurements. The suite of complementary models and software tools that were developed for this purpose within the APUSS Work Area 3 (WA3) are briefly presented in this chapter and comprise:

- a similarity approach
- a leakage approach
- the AquaBase software package and
- a large scale application example.

THE SIMILARITY APPROACH

The approach

Exfiltration and infiltration are basically driven by water head and leakage of the pipe according to Rauch and Stegner (1994), although as indicated in Chapter 6, other parameters such as matrix potential are significant variables. Nevertheless, detailed knowledge concerning cause-effect relationships between independent pipe characteristics and dependent infiltration/exfiltration rates is still fairly limited. There is a significant lack of data not only due to the large number of relevant processes and influencing factors (Davies *et al.*, 2001) but also due to the insufficient and generalised data situation for large sewer systems that characterise most national situations. Thus, both deterministic (e.g. Dupasquier, 1999) and empirical models (e.g. Gustafsson *et al.*, 1991) are difficult to apply and depend strongly on extensive field measurements. Due to the high personnel and financial expenditure associated with such investigations, an optimisation of measurement campaigns in terms of both cost efficiency and information yield is therefore essential. To this end, data-driven models within APUSS were developed using the **similarity approach** based on the precept that "similar pipe characteristics will lead to similar infiltration/exfiltration rates", i.e. similarities in and between groups of

© 2010 IWA Publishing. *Assessing infiltration and exfiltration on the Performance of Urban Sewer Systems (APUSS).* Edited by Bryan Ellis and Jean-Luc Bertrand-Krajewski. ISBN: 9781843391494. Published by IWA Publishing, London, UK.

sewer pipe reaches can be used to improve the information yield and to facilitate the transfer of available information on a sub-catchment scale.

In order to verify this approach, statistical methods were applied to comprehensive data sets for the city of Dresden and the cities of Bottrop and Gladbeck, all in Germany. The data sets contained in total, 27 sub-catchments together with their network topology and a full characterisation of the individual sewer pipe reaches as identified by their basic attributes. For every sub-catchment, the infiltration rate was estimated by measurements, with the corresponding groundwater table being known for most sub-catchments. The characteristics of the sub-catchments were compared using analysis-of-variance methods (Müller, 1991). The application of these methods can evaluate whether or not several grouped samples, i.e. sewer reach groups, belong to one basic population. Significant dissimilarities were detected between sub-catchments and reach populations having different infiltration rates. Due to the wide range of observed infiltration rates (min ÷ max = 0.18 ÷ 2.95 L/[m^2.h] = 1:16), it seemed reasonably probable that the observed dissimilarities between the reach populations can be linked in some basic way to the infiltration rates.

The relationship between the reach attributes and the infiltration rates was analysed using multidimensional scaling, MDS (Borg and Groenen, 1997). This method efficiently transforms a high dimensional arrangement of objects, such that a low-dimensional and interpretable configuration with the optimum approximation of the observed pattern is reached. Thus, any number of independent parameters can be represented by one or two figures. A relation between the measured infiltration rates and the MDS results for the Dresden data set, without reference to groundwater information, could not be found. However, a relationship between sewer attributes and the general urban district type (e.g. inner central district, broader urban centre, suburb etc.) was identified. Given this link between urban type and network development, it can be concluded that in principle a larger network consists of identifiable homogenous areas.

The MDS results for the Dresden data set with groundwater information are shown in Figure 1. The parameter set was reduced to one dimension, (*dim. 1*). A good correlation between *dim. 1* and the reach-surface-specific infiltration rate (q_f) was found. A comparison with the time weighted head shows that this correlation, i.e. the specific infiltration rate, is not dominated by the groundwater head. The analysis of a combination of the Dresden and the Bottrop/Gladbeck data sets led to similar results.

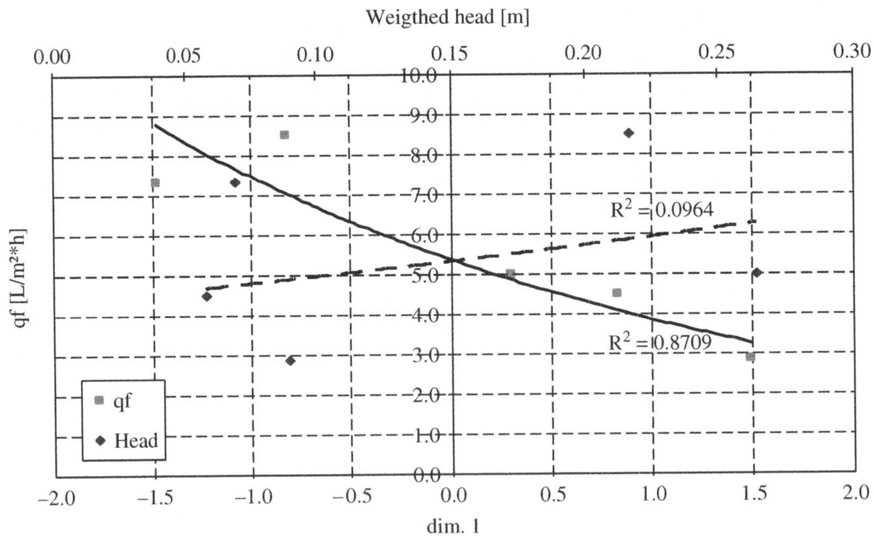

Figure 1: MDS result of Dresden data set with groundwater information and water head vs. reach-surface-specific infiltration rate q_f.

Because of the reasonable correlation determined between *dim. 1*, which serves as a surrogate for the reach attributes, and the infiltration rate, it can be concluded:

(i) that there is a recognisable relationship between the independent parameters and the infiltration rate and

(ii) that the parameters - or some sub-set of them - are sufficient to describe infiltration.

Based on these assumptions, it can be reasonably concluded that the similarity approach can be regarded as being an applicable methodology.

Application 1: Method to identify optimum location of measurements

An optimally defined measurement gauge can be located at the outlet of a more or less homogeneous sub-catchment. Thus, the reach-specific error of the measured infiltration rate is minimised and the information content gained is high. The developed method consists of two steps: i) describing the homogeneity of the upstream sub-catchment with a similarity figure (φ) for every potential placement and ii) determining the most homogenous sub-catchments using an optimisation algorithm.

The calculation of the similarity figure was adapted from cluster analysis (Kaufman and Rousseeuw, 1990). The input consists of a data set containing all relevant sewer reaches with their given attributes. Due to the scale invariance of distance measures, the parameters must be standardised and can be weighted by means of their relevance for infiltration. The distance or dissimilarity between two reaches, represented by their parameter vectors is calculated by the Euclidean and the Mahalanobis distance relation (Fahrmeir *et al.*, 1996). The distances are aggregated on sub-catchment level to determine one characteristic value, with the mean distance between all reaches and the mean distance of all reaches to the groups centroid. These aggregations are based on the idea of a more or less equal distribution of extraneous water in sewers. To determine the optimal placement, an optimisation algorithm is run. With an iterative procedure, the catchment is divided until a given number of sub-catchments, i.e. measurement gauges, is reached. The conditions for separating manholes are:

- the similarity figure φ is minimal: a minimal φ stands for a maximal possible homogeneity.
- sum of the size measures of relevant reaches > critical value: the size measure can be the sewer length, the reach surface, the connected area, connected inhabitants etc. This constraint results from measurement requirements such as minimal flow or length.
- the manhole is not within a mesh: an infiltration rate measured in a mesh cannot be allocated to reaches.

The method was tested on six sub-catchments. Reach-related infiltration rates were modelled with a conceived infiltration model based on real independent parameters and a constant groundwater table on an arbitrary level. For every sub-catchment, 50 random distributions of five gauges were generated and compared with the optimised positioning. The comparison showed an information improvement and error reduction respectively, of mean 14% to a maximum of 35%. The main reason for these results, which might seem to be rather mediocre, is the fact that the procedure is not a classification of independent objects. The network topology, especially the number and size of meshes, as well as the spatial distribution of attributes as a result of the sewer building history, have a strong influence on the results. With a better knowledge about infiltration and exfiltration processes, the optimisation results might be much higher. Furthermore, the method was tested in three smaller catchments having relatively detailed information on infiltration. The application of the method resulted in a good coherence between the optimised arrangement of gauges and detected changes of measured infiltration rates.

The proposed optimisation method is not based on typical input/output functions. It compares and classifies states. Therefore, the results must be considered within the boundary conditions of the given data set, and as such will always have a significant uncertainty. Information about groundwater conditions has an overwhelming relevance for their quality. Advantageously, a definite parameter set is not necessary and the method can be adapted to nearly every data situation.

Application 2: Integrated model to assign and merge infiltration rates of sub-catchments

With models which allow the transfer of I/E measurements from measured sub-catchments to other unmonitored watersheds, it is possible to aggregate the available information on a catchment scale. Two models based on statistical methods have been developed within the APUSS project.

The first method is based on the analysis of variance and treats sub-catchments, represented by means and variances, as one coherent unit. It determines parameter similarities between sub-catchments with unknown I/E class and sub-catchments with known I/E class. In cases of high similarities, the I/E class is transferred. The parameters are then analysed with post-hoc tests, e.g. the Fisher Least Significant Difference test (Winer *et al.*, 1991), and multiple comparison tests (Siegel and Castellan, 1988) respectively. The number of not rejected tests (i.e. where two sub-catchments are determined as belonging to one population for the considered parameter) is summarised. Pairs of sub-catchments with a high sum can be assumed as being similar and should have a similar I/E class.

Figure 2 shows a matrix of the similarities between paired sub-catchments with modelled infiltration classes. As five parameters were used for the infiltration model, five is therefore the maximum number of similarities. Due to their distribution, the minimum number of similarities to be investigated was set to three, or 60%. From the 15% of pairs identified as being similar, only 50% were correctly assigned (shown by crosses in Figure 2), i.e. they were similar with regard to all 5 parameters.

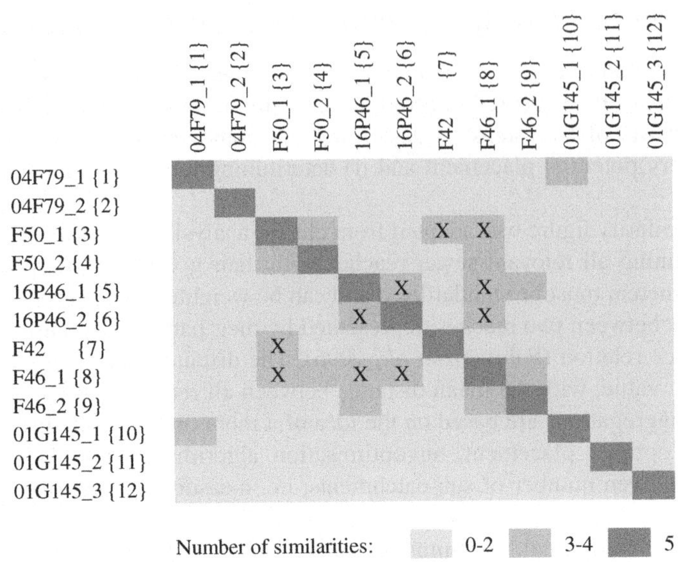

Figure 2: Frequency of similarities between sub-catchments; modelled infiltration classes.

The second transfer-method is based on discriminating analysis (Fahrmeir *et al.*, 1996) and considers single sewer pipe reaches. The method assigns the reaches of a sub-catchment with unknown I/E class to sub-catchments with known I/E class based on classification probabilities. The resulting I/E class of the investigated sub-catchment is calculated as a weighted mean.

The results of the above mentioned example shown in Figure 2 lead to an assignment rate of 50% assuming a suitable assignment error of 0.66 class width (equivalent to one third of the range). Assuming a softer error of 1.0 class width, 75% of the sub-catchments were correctly assigned. Furthermore, the method was applied to the Dresden data set with groundwater information. Assuming a suitable assignment error of 0.66 class width, 1 of the 5 subcatchments was correctly assigned. Assuming a softer error of 1.0 class width, 4 of 5 sub-catchments were correctly assigned.

With the transfer methods, it is possible to improve the information about the I/E status of complete sewer networks. With a suitable error term being defined as being acceptable, the transfer of infiltration classes was successful achieved for approximately half of the sub-catchments. Keeping in mind that only few data were available for these systems. a positive verification of the methods is indicated but nevertheless cannot be accepted as being completely satisfactory given the uncertainties involved.

THE LEAKAGE APPROACH

Background

The leakage approach can be traced back to the Darcy equation. It has been widely used for modelling hydrological interactions between aquifer and surface water and is related to the wetted area of the river bed, the difference between groundwater and surface water level, and a specific leakage factor representing the ability of the reach to infiltrate groundwater or exfiltrate river water. As an integrative parameter the leakage factor describes various attributes of the sub-soil layer of the river bed and thereby the potential of exchange between the compartments of ground- and surface water. The approach is applicable both for infiltration and exfiltration.

According to Gustafsson *et al.* (1991) and Karpf and Krebs (2004a) the leakage approach can be modified for the simulation of groundwater infiltration into sewer systems:

$$Q_{infiltration} = k_L A_S (h_G - h_S) \tag{1}$$

With a requirement that: $h_G > h_S$
and where:

$Q_{Infiltration}$	infiltration of groundwater (m^3/s)
A_S	groundwater-influenced pipe surface (m^2)
h_S	water level in sewer pipes (m)
h_G	groundwater level (m)
k_L	leakage factor (s^{-1}).

An extended model, which takes into consideration regional differences and house connections was proposed for further development and implementation:

$$Q_{in} = (h_{GWL} - h_w)P_{w1}LK_lK_r + q_{0in} + \overline{q_{inHC}}N_{HC} \tag{2}$$

With a requirement that: $h_{GWL} > h_w$
and where:

Q_{in}	infiltration flow (m^3/d)
h_w	water level in the pipe (m)
h_{GWL}	groundwater level around the pipe (m)
P_{w1}	external wet perimeter (m)
L	length of the pipe (m)
K_l	local coefficient (d−1)
K_r	regional coefficient (−)
$q_{0\ in}$	infiltration flow from other infiltration sources (m^3/d)
$q_{in\ HC}$	mean infiltration flow for a single house connection (m^3/d)
N_{HC}	number of house connections to the pipe.

The achievable spatial-temporal resolution of the approach is relatively high but depends on the data situation. It is, at least theoretically possible to determine infiltration at single pipe level and at daily scale but only very limited information about processes can be expected. The quality of modelling and prediction of infiltration rates is high for large catchments (Karpf and Krebs, 2004b).

Application

Data needs

The data needs for the application are listed in Table 1. For the calibration of the model coefficients, time series of groundwater levels and infiltration rates Q_{in} at the end of the catchment, i.e. at the wastewater treatment plant (WWTP), must be available. To cover various groundwater conditions, a long time period and a high temporal resolution of data is necessary. Monthly values over a period of several years are recommended.

Table 1: Data needs for the leakage model approach.

Type	Data	Source
Structural data	L	The structural data are contained in the sewer data base. Profile shape, diameter and slope are necessary to calculate PW1.
	N_{HC}	
	Profile shape	
	Diameter	
	Slope	
Water levels	h_w	The water level in the sewer pipe can be modelled. However, it is also feasible to estimate it by measurements of water levels in the system at known conditions.
	h_{GWL}	The estimation of groundwater level at each pipe is based on the interpolation of groundwater measurements.
Coefficients	K_l	The coefficients have to be calibrated.
	K_r	
Specific infiltration rates	$Q_{0\ in}$ $q_{in\ HC}$	The constants must be determined by separate investigations (e.g. by balancing). At least $q_{in\ HC}$ seems to be not essential for a successful application.

The infiltration rates can be balanced by calculating the difference of wastewater flow and the average consumption of drinking water. Due to uncertainties associated with balanced drainage rates (Karpf and Krebs, 2003), the variation of infiltration rates may be smoothed.

Calibration

The regional coefficient K_r is determined for a group of pipes according to field measurements representing regional peculiarities. Unless external inputs from measurements are available, it should be set to a default value equal to 1.

The calibration of the local coefficient K_l is based on the following equation:

$$K_{l,T} = \frac{Q_{in,T}}{\sum_{i=1}^{n}\left[\left(h_{GWL,i,T} - h_{w,i,T}\right)\cdot A_{i,T}\right]} \qquad (3)$$

With a requirement: for all reaches that $h_{GWL,i,T} > h_{w,i,T}$
and where:

$K_{l,T}$ integral leakage factor at time T (d^{-1})
$Q_{in,T}$ balanced infiltration in the catchment at time T, without Q_{inHC} and Q_{0in} (m^3/d)
$h_{w,i,T}$ water level in the sewer pipe i at time T (m)
$h_{GWL,i,T}$ groundwater level at the sewer pipe i at time T (m)
$A_{i,T}$ groundwater-influenced pipe surface of pipe i at time T (m^2).

The calculated leakage factor represents an integral parameter for all groundwater-influenced pipes at a certain time T. In order to refer individual leakage factors to pipes, the calculation has to be carried out for a number of time spots. Thereby, the leakage factor of each groundwater-influenced pipe can be estimated to a weighted average of all calibration cases:

$$K_{l,i} = \frac{\sum_{T_i} K_{l,T}}{n_i} \qquad (4)$$

With a requirement that: equidistant time steps
and where:

$K_{l,i}$ calibrated leakage factor for pipe i
$K_{l,T}$ integral leakage factor at time T
T_i time spot when $h_{GWL,i,T} > h_{w,i,T}$
n_i number of time spots when $h_{GWL,i,T} > h_{w,i,T}$.

A simplified example for the calibration procedure is shown in Figure 3. For every time T (March and October), one leakage factor was calculated. The weighting for every pipe was done by averaging these $K_{l,T}$-values. Thus, the calibrated leakage factor of pipe 1, which is not groundwater-influenced in March, is equal to $K_{l,October}$; the factor of pipe 2 is equal to the average of $K_{l,October}$ and $K_{l,March}$.

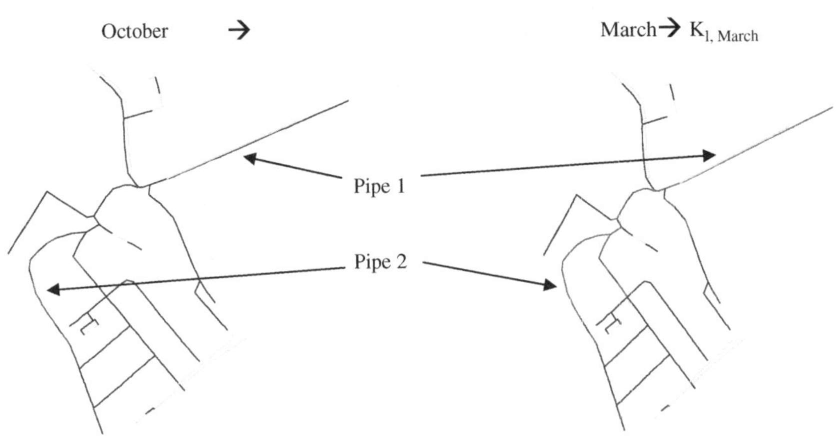

Figure 3: Example for model calibration.

THE AQUABASE PACKAGE

The development of a computerised data based management tool offered additional value to the APUSS project through the provision of relevant data directory and visual display menus of value to end users. The development of a conceptual model and an integrated approach resulted in software allowing the estimation of I/E in specific sewers together with integration over a catchment scale. The database software holds all necessary data for a particular catchment and is able to apply mainstream APUSS project results to a sewer system described by attributed and stored data.

It was decided to make use of the existing software AquaBase tool with an inner structure corresponding to all demands on sewer system data description and with the possibility of additional development in order to implement new features as defined in the APUSS project (Figure 4).

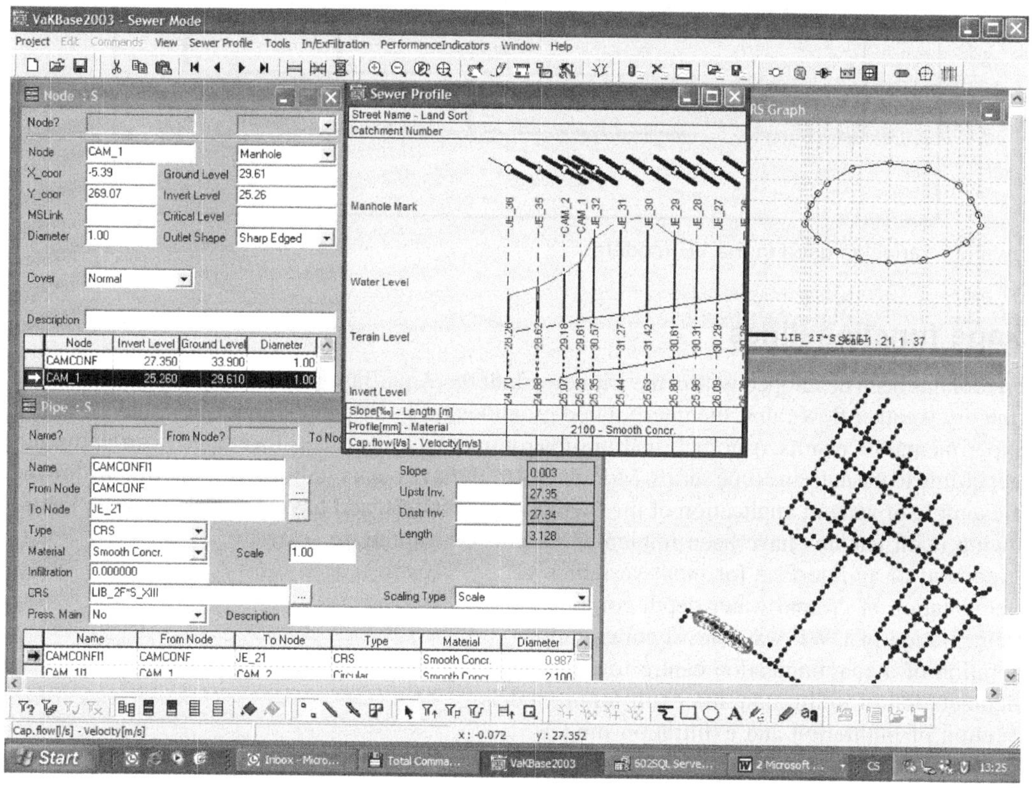

Figure 4: The aquaBase environment.

The AquaBase program represents a very useful tool for "easy-to-use" asset data management by water utility companies. The AquaBase functions in principle as an urban drainage database with GIS functionalities, a user-friendly interface and a number of specialised functions related to sewerage system data management. The tool has been developed as a client-server architecture based on the WinBase database engine. There are no particular hardware requirements related to the use of this application.

AquaBase has been developed over the course of several years and the current version profits from numerous practical comments and suggestions from water utility specialists. The tool is very flexible and gives the user insight into the asset (or model) data that is being manipulated. Specialised queries can detect random as well as systematic errors in the data and propose ways of correcting the incorrect data. A system of flagging and selections allows the user to work with subsets of the whole data set and to focus on areas of priority. Very fast graphics accompany standard GIS use.

AquaBase was chosen as a user-friendly software tool, enabling the incorporation of field data as well as developing internal tools and routines specifically requested for the APUSS project. The AquaBase data structure has been extended to consider new data inputs, variables and coefficients used for the calculation of infiltration and exfiltration rates. The system was modified to support the graphical presentation (using colour thematic maps) of the input and result data associated with pipe networks and catchment polygons.

The models implemented

Exfiltration is measured at the sewer reach scale, whilst infiltration is measured at subcatchment scale. Accordingly, conceptual models respectively at sewer reach and subcatchment scales for both exfiltration and infiltration have been proposed. The involved key state variables are water levels in the sewer pipes h_W, surrounding groundwater levels h_{GWL} and I/E wetted perimeters P_W (Figure 5). The model parameters at reach and catchment scales reflect the structural state of the sewer pipes and need to be calibrated by means of field experiments and supporting data.

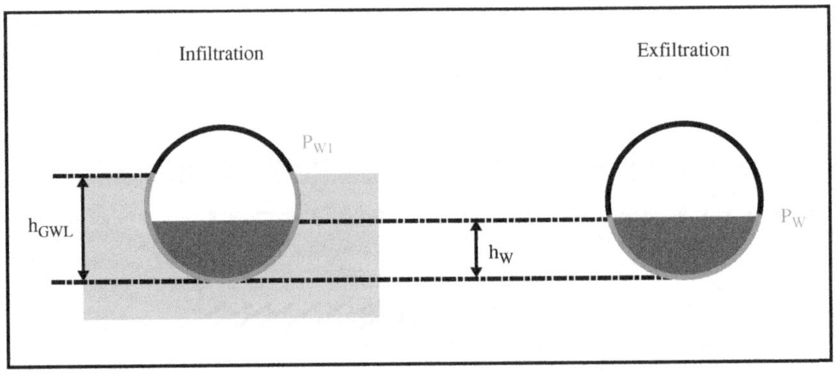

Figure 5: Key state variables used in the I/E models.

The package functionalities

An integrated tool has been developed within the framework of the AquaBase software which displays a map of the sewer system and the dry weather flow rates, the interpolated groundwater levels, and which also includes data and time series containing experimental I/E results. Automatic calibration routines have also been developed in order to calibrate the model parameters according to the data and time series. Models and tools have been established and tested within the framework of a large scale example of general application of the methods and models, and having an end-user application perspective.

The following main features have been implemented into the AquaBase software:

- development of an interface for input variables
- implementation of ground water depth computation
- implementation of DWF water level computation
- elaboration of a communication centre for separate tools (I/E data processing)
- implementation of infiltration and exfiltration models
- calibration of infiltration and exfiltration models.
- performance indicators calculation, analyses and links to other PI platforms (Figure 6)
- uncertainty analysis of the obtained results (Figure 6).

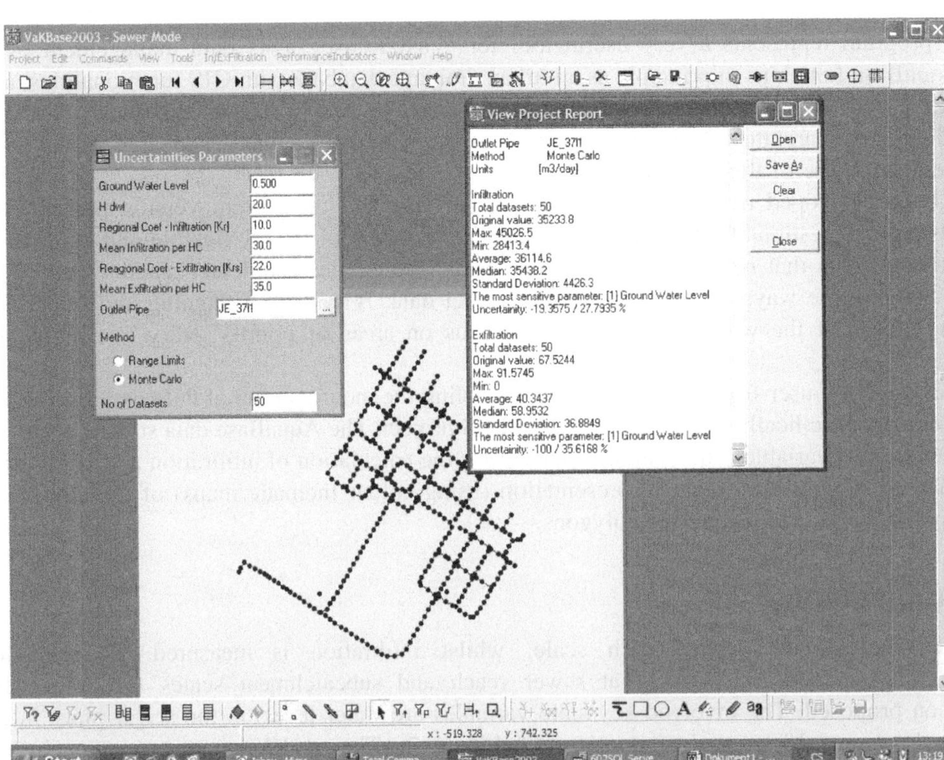

Figure 6: The uncertainty analyses interface.

A number of multi-parametric formulas describing the infiltration and exfiltration process in sewer pipes have been implemented in the AquaBase software, giving the user the possibility to calculate rough estimates based on sensitivity analysis for distinct parameters. I/E can then be calibrated based on the new tracer methods of measurement to achieve a better accuracy in the final estimate. In order to cope with the uncertainty of specific phenomena, an external module in AquaBase was designed and implemented, based on the Monte Carlo method. This module develops a large spectrum of potential results which are then statistically post-processed and the uncertainty, within the selected range of confidence interval defined. This functionality brings an extra value to the AquaBase and the I/E calculation, and propagates also to the values of performance indicators (PI), which are generated at the end of the calculation. PIs can be then used by practitioners for the planning of rehabilitation strategies and investment in urban drainage systems.

The AquaBase with the I/E extension has been developed as a clear product of the APUSS project. As such it can be freely used by both scientists as well as practitioners within the urban drainage domain. It is expected that this tool will be useful for water utility companies in terms of I/E assessment, WWTP functions as well as sewer system rehabilitation. The AquaBase software is delivered by an installation CD, and is accessible through the APUSS website (*www.insa-lyon.fr/:Laboratoires/URGC-HU/apuss*).

LARGE SCALE TEST APPLICATION

A large scale area application for final testing of the APUSS methodology and the Aquabase system was undertaken, with the Prague sewer system catchment being selected to perform this task.

Data collection

The main data source for the application was the Sewer System Master Plan for Prague, processed and elaborated in 1999–2001 and which provided an appropriate and readily accessible source for the APUSS methodological testing. Despite this accessible data source, further necessary data collection posed an "uphill struggle" in terms of required human resources. In order to fully describe the selected system, the following necessary data had to be collected:

- sewer system topology data,
- full catchment data,
- ground water level maps,
- monitoring campaign data.

The **Sewer system topology** model covered only the main trunk sewer system but the available basic sewer system data consisted of more than 7000 nodes with a pipe length of approximately 190 km.

Each node is described by a set of information:

- X and Y coordinates of the node centre,
- bottom (invert) level,
- level of terrain,
- type of node (manhole, outlet, object (e.g CSO, pump).

Each pipe is described by a set of data:

- upstream (from) and downstream (to) the node,
- material (to recognise roughness),
- shape (circle, egg . . .),
- level of downstream connection.

Catchment data: the overall drained area exceeding 200,000 hectares and was divided into almost 1000 catchments which are described in detail.

Each catchment is described by a set of information:

- node to which the catchment is connected,
- catchment area,
- hydraulic length,
- slope,
- number of connected inhabitants,
- percentage of impervious area.

Ground water level maps were digitized in detail. This resulted in more than 10 Gbytes of data contained in over 120 files, i.e. approximately 200,000 points (Figure 7). Such an amount of XYZ-point data was too much to handle so therefore, the data was diluted to grid 100 × 100 m cells by a program specifically developed to perform this task. This resulted in approximately 50,000 points, which was possible to work with.

Each ground water level point is defined by:

- node of artificial type,
- water level value in the node time series table.

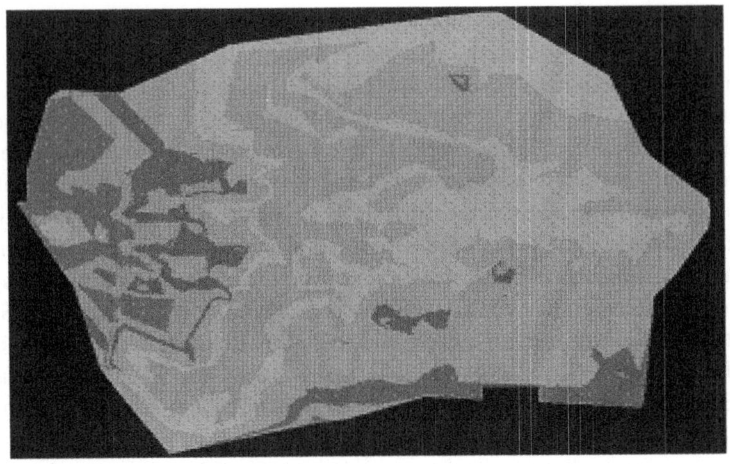

Figure 7: Elaborated and evaluated map of groundwater level.

Monitoring campaign data. Within the scope of the Prague Urban Drainage Master Plan, the monitoring campaign was divided into three phases over a period of time with over 100 monitoring profiles. For the APUSS implementation, 45 calibration profiles monitoring discharge were selected (see left part of Figure 8) and for each calibration profile, several discharge time series were measured. All the discharge time series have been elaborated in order to find a relevant minimum discharge value.

Each calibration profile is defined by:

- a pipe with a monitoring profile location,
- I/E calibration discharge value.

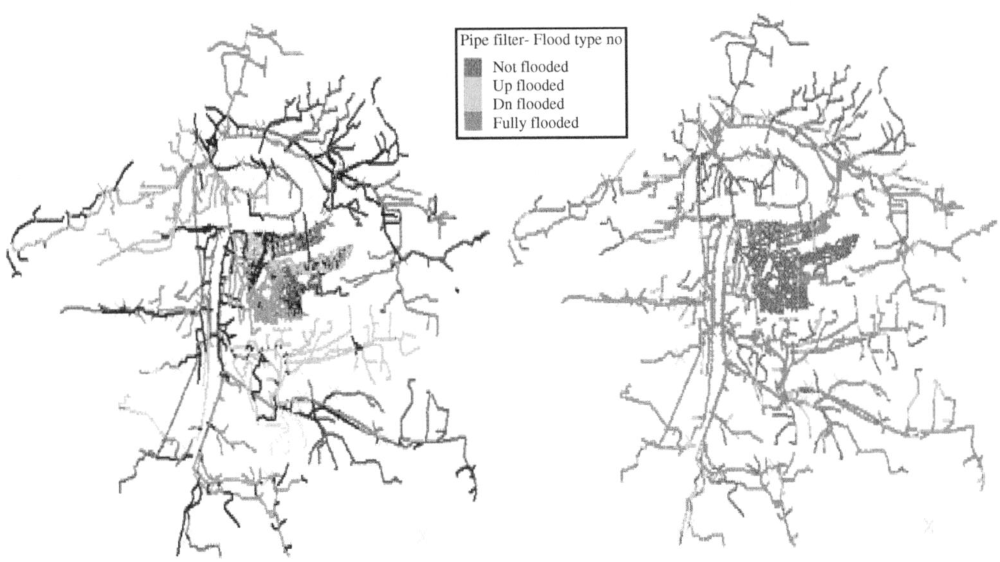

Figure 8: Thematic maps of sewer system. (left = I/E groups definition, right = flood by ground water level).

I/E Computation

Processing of the APUSS features implemented in the AquaBase software has been carried out according to elaborated training notes with all tasks performed according to a pre-released written operational procedure. The very large data sets however, called for performance and logical implementation changes in the AquaBase software to enable secure and accurate calculations.

The changes were essentially the result of:

- ground water level computation. The extreme number of water level points made the software unable to process the computations. A maximum number of 50,000 was found sufficient to describe water levels within the Prague area and the software demonstrated satisfactory performance in the subsequent calculation process.

– I/E group definition. From the 45 calibration profiles, an equivalent number of 45 I/E groups had to be created and this was probably the main problem in data elaboration. The sewer system is linked and it was thus necessary to find the proper location of group borders in order not to get the system looped. The auto-calibration process of infiltration/exfiltration calculation was not found to be stable or fast enough, but following the implementation of changes in the logical structure of data the calculation performance increased to acceptable levels.

Data collection and elaboration for the purposes of I/E modelling on a large catchment scale is undoubtedly time consuming. Moreover, calibration of I/E is not an easy or fast process despite developed enhanced functions and robust hardware. Users have to be on the one hand advanced enough to understand the computational background and meaning of the results values (from an engineering point of view) and, on the other hand, users also have to be advanced enough to know how, where and what to change in software settings to obtain expected results of a satisfactory quality and robustness.

REFERENCES

Borg, I and Groenen, P. (1997) *Modern Multidimensional Scaling*. New York (USA): Springer Verlag. ISBN 0-387-94845-7.

Davies, J.P., Clarke, B.A., Whiter, J.T. and Cunningham, R.J. (2001) Factors influencing the structural deterioration and collapse of rigid sewer pipes. *Urban Water*, **3**, 73–89.

Dupasquier, B. (1999) *Modélisation hydrologique et hydraulique des infiltrations d'eaux parasites dans les réseaux séparatifs d'eaux usées.* (Hydrological and hydraulic modelling of infiltration in separated sewers). PhD thesis: ENGREF, Paris. (in French).

Fahrmeir, L., Hamerle, A. and Tutz, G. (Editors). (1996) *Multivariate statistische Verfahren.* (Multivariate statistical methods.) Walter de Gruyter, Berlin, Germany. ISBN 3-11-013806-9.

Gustafsson, L.G., Lindenberg, S. and Olsson, R. (1991) Modelling of the indirect runoff components in urban areas. *Proc. Int. Conf. on Urban Drainage and New Technologies*, Dubrovnik, Yugoslavia, 127–133.

Karpf, C. and Krebs, P. (2003) Definition und Bilanzierung von Fremdwasser. (Definition and Balancing of Parasiting Water.) *Umweltpraxis*, **10**, 35–38.

Karpf, C. and Krebs, P. (2004a) Sewers as drainage systems – Quantification of groundwater infiltration. *Proceedings of Novatech 2004*, Lyon, France, 6–10 June, Vol. 2, 969–975.

Karpf, C. and Krebs, P. (2004b) Application of the leakage model to assess exfiltration. *Proceedings of the 6th International Conference on Urban Drainage Modelling*, Dresden, Germany, 15–17 Sept., 217–224. Also in Koegst, T. and Krebs, P. (eds. 2005): Urban Drainage Modelling. IWA Publishing, London, UK. ISBN 978 1843394976.

Kaufman, L. and Rousseeuw, P.J. (1990) *Finding Groups in Data – An Introduction to Cluster Analysis*. John Wiley, New York, US. ISBN 0-471-87876-6.

Müller, P.H. (ed.). (1991) *Wahrscheinlichkeitsrechnung und mathematische Statistik*. (Probability calculation and mathematical statistics). Akademieverlag, Berlin, Germany. ISBN 3-05-500608-9.

Rauch, W. and Stegner, T. (1994). The colmation of leaks in sewer systems during dry weather flow. *Water Sci. & Tech.*, **30**(1), 205–210.

Siegel, S. and Castellan, N. (1988) *Nonparametric Statistics for the Behavioural Sciences*. McGraw-Hill, New York, US. 2nd edition. ISBN 0-071-00326-6.

Winer, B., Michels, K. and Brown, D. (1991) *Statistical Principles in Experimental Design*. McGraw-Hill. New York, US. ISBN 0-070-70982-3.

Chapter 11

Generic multi-criteria methodology to compare investment/rehabilitation strategies: principles and case study

Jean-Luc Bertrand-Krajewski, Sylvie Barraud, Pascal Le Gauffre and Emilie Baer

INTRODUCTION

Infiltration and/or exfiltration (I/E) in sewer systems have technical, environmental and economic impacts. Sewer operators have to make decisions and choices between several possible investment / rehabilitation strategies aiming to solve problems related to I/E. Most frequently, the comparison of the possible strategies is based on a traditional cost-benefit approach, which consists of minimising the costs of investment and operation for an expected given benefit. Two of the main limitations in such an approach are the difficulty:

- of evaluating all criteria in monetary units (e.g. what is the cost in Euros of any kg of chemical oxygen demand (COD) discharged into a river by a wastewater treatment plant (WWTP) overloaded by infiltration?);
- of taking into account some criteria that cannot be easily calculated, such as for example, environmental aspects.

The choice of any investment strategy dealing with I/E phenomena should obviously include environmental, technical and economic aspects. Within the APUSS project, a specific workpackage (WP9) was devoted to a methodology aiming to help practitioners in the evaluation and the comparison of investment/rehabilitation strategies. The two main objectives here were:

- to propose a general methodology based on an integrated and multi-criteria approach to compare and rank strategies to remedy I/E problems;
- to illustrate the application of this methodology to a semi-virtual case study.

This chapter presents successively:

- the methodology
- the description of the sewer system used for the case study
- the evaluation of the impacts of infiltration
- the investment/rehabilitation strategies
- the multi-criteria comparison of the strategies and some results of the case study.

It should be emphasised that the most important part of this chapter is the methodology itself, which can be applied by any operator with their own information and with necessary adaptations. The semi-virtual case study is only given to illustrate how the methodology can be applied to a specific case, with site specific simplifications and hypotheses. Consequently, the conclusions drawn from the case study should be neither generalised nor extrapolated.

Edited by Bryan Ellis and Jean-Luc Bertrand-Krajewski. ISBN: 9781843391494. Published by IWA Publishing, London, UK.

DESCRIPTION OF THE METHODOLOGY

The methodology is represented in Figure 1, with the eight main steps discussed in the following paragraphs. It will then be illustrated by means of a case study.

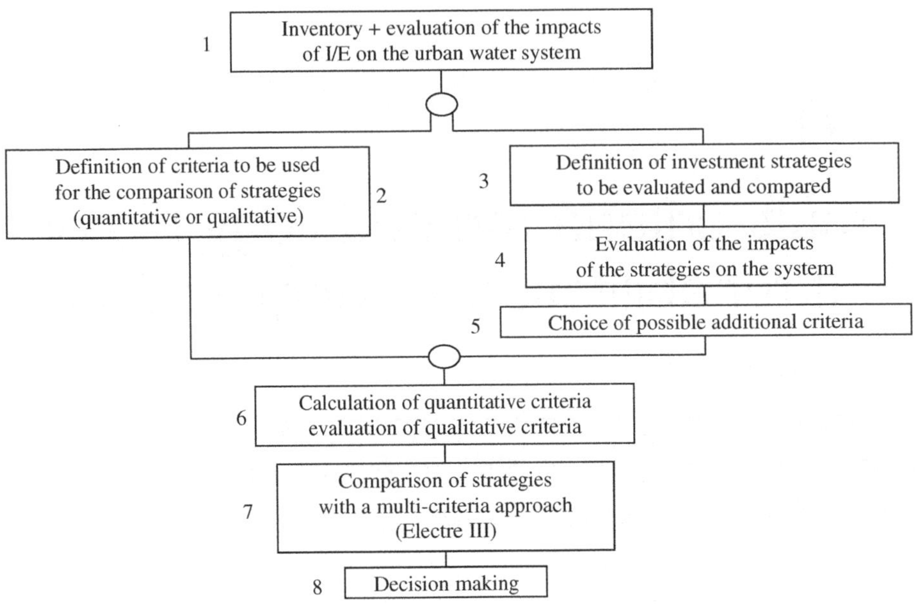

Figure 1: Methodology for definition, comparison and choice of investment/rehabilitation strategies regarding infiltration/exfiltration based on a multi-criteria analysis.

Step 1: inventory and evaluation of I/E impacts on the urban water system

Infiltration and exfiltration may have numerous impacts on the urban drainage and water system. The inventory and the evaluation of these impacts are fundamental in order to obtain an exhaustive view of the different aspects and to avoid neglecting any of them. Based on a literature review carried out in the APUSS project (Neitzke, 2002), the impacts of infiltration and exfiltration are summarized respectively in Figure 2 and Figure 3.

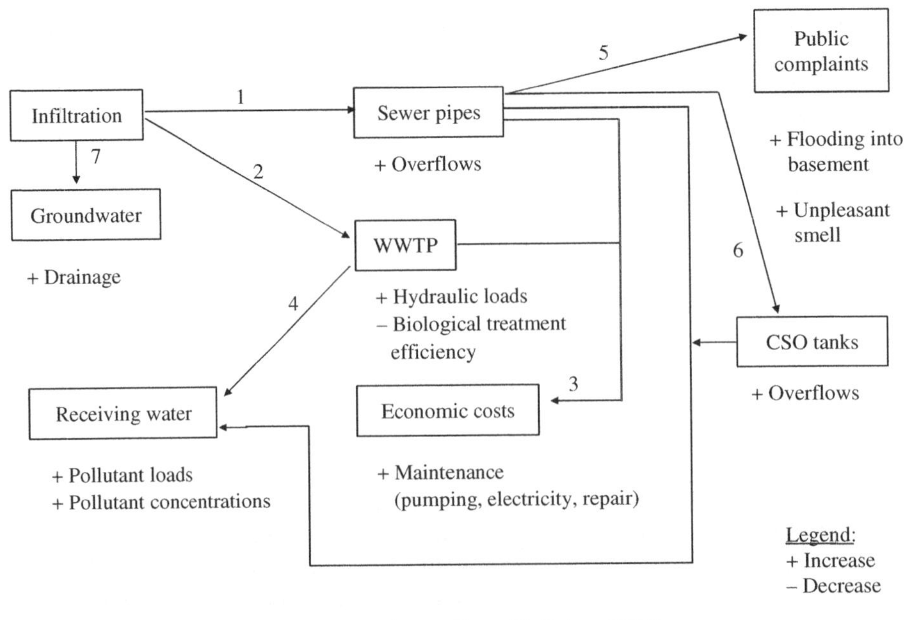

Figure 2: Impacts of infiltration on urban drainage systems (from Neitzke, 2002).

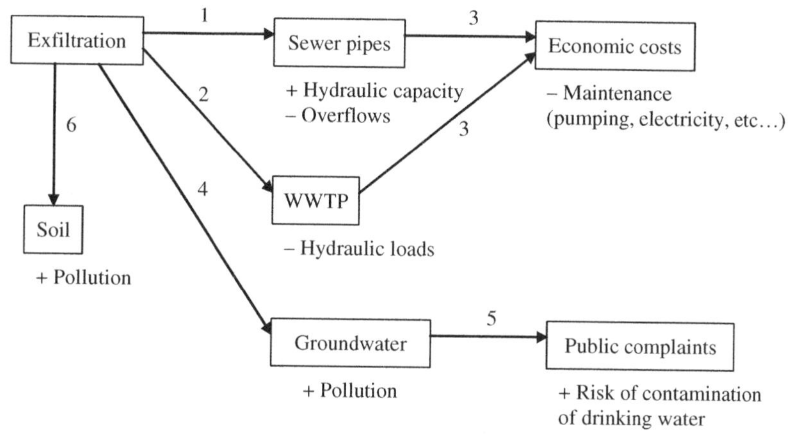

Figure 3: Impacts of exfiltration on urban drainage systems (from Neitzke, 2002).

Infiltration increases the hydraulic loads in sewer pipes, which lead to more frequent and premature combined sewer overflows CSOs (1), which can be partly avoided by storage of excess water in CSO tanks (6). Emptying CSO tanks increases pumping times and energy consumption, leading to higher energy costs (3). CSOs and CSO tanks both increase the pollutant loads discharged into the receiving waters (4). Overflows may also lead to flooding in house basements, unpleasant smells and public complaints (5).

Infiltration also impacts the WWTP operation. An increased hydraulic load leads to higher pumping costs (3). The biological treatment efficiency decreases because of the dilution of influent concentrations (2). Due to permanent infiltration, the capacity of the WWTP is also reduced during rainfall events, which results in both more frequent discharges of untreated water and higher risk of sludge losses from the secondary clarifier. As a consequence, the pollutant loads discharged into the receiving waters increase (4). Drainage of groundwater into sewers by infiltration may also lead to the lowering of groundwater levels, with potential geotechnical risks for building and urban infrastructures (7).

Exfiltration decreases the hydraulic loads in the sewer pipes (1) and the hydraulic loads entering the WWTP (2). This leads to economic savings because less electricity is needed for pumping and for operation of the WWTP (3). However, the main risk linked to exfiltration is the pollution of soil (6) and groundwater (4), which may lead to the contamination of drinking water resources (5).

For any specific catchment, all the above impacts have to be checked (are they all pertinent in this particular catchment?) and evaluated (flows, volumes, transfer of pollutants, etc.) by means of appropriate measurements and modelling. The quality of this evaluation is a key element for the quality of the following steps in the proposed methodology. All necessary efforts should be devoted to obtain and elaborate the requested knowledge in respect of the I/E phenomena and their impacts.

Step 2: definition of criteria

Impacts identified in step 1 should be expressed through criteria that can be used for their evaluation and for the comparison of investment strategies. In the multi-criteria approach Electre III (see Roy, 1978, 1996; Vincke, 1992 for detailed information on this approach) has been used in step 7, although it is not possible to express all criteria with a common unit (such as for example, Euros in traditional cost-benefit approaches). The natural units used to evaluate the impacts can be retained, for example, pollutant loads expressed in kg, costs in Euros, overflow volumes in m^3, etc.

Some impacts may be difficult to evaluate, such as groundwater drainage, flooding in basements, groundwater pollution risk, etc., because of the lack of knowledge regarding the phenomena involved in the corresponding processes or the lack of data and information to calculate the criteria, etc. This is why, when criteria cannot be calculated with numerical values, they can be expressed in Electre III by evaluation grades based on operator's expertise and estimation.

Some impacts could also be considered as not been really important for a given case, and could be intentionally neglected after exhaustive review and evaluation has been completed. This choice, to be made by the operator should always be site-specific. In the following case-study, only pollutant loads discharged into receiving water and economic costs have been considered. Individual sewer pipe overflows, flooding in basements and groundwater drainage have not been modelled as they are not considered to be relevant in this particular case.

Step 3: definition of investment/rehabilitation strategies

A review of all possible investment/rehabilitation strategies should be made. Regarding infiltration, the strategies may concern for example, the sewer pipes (rehabilitation), and/or the WWTP (e.g. larger treatment capacities) and/or CSO storage tanks (larger tank capacities). Regarding exfiltration, only rehabilitation of sewer pipes may solve the negative impacts on soil and groundwater generated by exfiltration. After review, each possible strategy should be evaluated using the best available level of detail.

Regarding the selection of rehabilitation technologies (RTs), many reviews exist in journals, conferences, textbooks and on the web. In France, the French National research project RERAU (RERAU, 2004) is a very useful source of information. At European level, the FP5 project CARE-S (part of the CityNet cluster) has made an inventory of 73 techniques (Villanueva *et al.*, 2004). Rehabilitation technologies can be distributed in three categories: i) repair, ii) renovation and iii) replacement, depending on the importance of the planned rehabilitation. Within the CARE-S project, CLABSA s.a. and SINTEF s.a. have developed both a database for RTs (Villanueva *et al.*, 2004) and a cost tool (Freni *et al.*, 2005). The database allows the user to select appropriate RTs based on the observed defects, on the pipe characteristics (diameter, material, accessibility, etc.) and on the technical function expected from the RT (structural improvement, sealing, etc...). O'Brien and Gere Engineers (O'Brien and Gere Companies, 2001) have compared different RTs (mostly trenchless technologies) based on their impacts, i.e. costs, pumping requirements, surface disruption, lateral interruption and upsizing possibilities.

The two main impacts are costs and traffic disruption. Surface disruption appears frequently as the main reason for the non-applicability of a given RT because the disturbance of traffic is considered as a very critical negative impact. Usually, taking into account the environment and the characteristics of the sewers, the number of RTs potentially applicable to a given catchment is rather limited. Costs may vary a lot from one technology to another. Investment costs also depend on the context (traffic, accessibility) and on the characteristics of the damaged pipes such as diameter, depth, etc. Consequently, it is difficult and sometimes even not pertinent to give an average cost of the different RTs. Any evaluation should be site specific and based on the local circumstances.

Step 4: impacts of investment/rehabilitation strategies on the urban water system

After review and selection in step 3, each possible strategy should be defined in detail and its impacts on the urban water system should be evaluated by means of modelling, scenarios, etc.

Step 5: possible additional criteria

In addition to the usual criteria selected in step 2, it might be interesting to have a long-term view and to integrate the expected life time (life cycle) of rehabilitation in the comparison of the investment strategies. Indeed, expected life times vary dramatically among the various RTs that may be used; for example, Frehmann (2004) gives orders of magnitude of the expected life time of sewer pipes depending on the RT used (Table 1).

Table 1: Amortisation: life cycle of sewer pipes (from Frehmann, 2004).

	Threshold value	Cover depth H = 4 m	Cover depth H = 10 m	
Replacement	80 to 100 years	45 years to 61 years to 88 years	65 years to 103 years to 141 years	minimum expected value maximum
Rehabilitation	50 to 60 years	47 years to 65 years	19 years to 31 years	minimum maximum
Interest calculation is not considered.				

Delleur *et al.* (1998) have proposed the RT expected life times be used to calculate the annual unit construction costs AUCC as follows:

$$AUCC_j = \left[\frac{r(1+r)^{n_j}}{(1+r)^{n_j}-1}\right]UCC_j$$

where r is the interest rate, n_j is the expected life time of the rehabilitation technology j and UCC_j the unit construction cost of the rehabilitation technology j.

The higher the expected life time of the RT is, the lower the AUCC value will be. An investment strategy i could have a higher UCC than a strategy k, but taking into account its longer life time, its AUCC could be lower. This expected life time is difficult to estimate because it varies according to the context (traffic, depth, etc.). A study carried out by Orditz (2004) on the durability of rehabilitations in sewer systems showed that very few studies have been produced on this topic. Moreover, most rehabilitation technologies have only been applied over limited time periods (up to a maximum 25 years) to provide long-term knowledge on their durability.

Contrary to investment strategies in WWTPs or in CSO tanks, rehabilitation of sewer systems may also increase the asset serviceability of the sewer system. The concept of asset serviceability has been defined by Ewan Associates and Mac Donald (2001) as "*the ability of an asset to deliver a defined service to customers and safeguard the environment*". Increasing the serviceability implies minimizing the dysfunctions of the considered system. For sewer systems, a list of all possible dysfunctions has been established in the European project CARE-S according to the European Standards EN 752-2 (1996) and EN 13508-1 (2003):

- ongoing degradation from abrasion
- blockage
- risk of collapse
- ongoing corrosion
- excessive spillage
- exfiltration (seepage loss)
- flooding
- decrease in hydraulic capacity
- infiltration
- ongoing degradation from root intrusion
- sand silting
- destabilisation of the ground-pipe system.

The sewer serviceability may be an important criterion because it expresses that sewers rehabilitated for infiltration purposes could contribute to the improvement of pipes which should have been rehabilitated for other dysfunction purposes such as risk of collapse, ongoing corrosion, etc. The increase of the asset serviceability following sewer rehabilitation could be expressed by quantitative criteria such as reduction in the number of flooding events. However, for most dysfunctions, it may be difficult to estimate the impact of the sewer rehabilitation. Consequently, only qualitative notes may be used which are essentially based on the operator's knowledge and expertise.

Other additional criteria could be defined by the operator depending on their objectives, on available knowledge, data and models. The advantage of the proposed methodology lies in its adaptabitility to the each specific case. All information, impacts and phenomena that should be taken into account can be introduced and evaluated either quantitatively or qualitatively by means of grades.

Step 6: calculation and evaluation of criteria

Once defined in steps 2 and 5, all quantitative criteria have to be calculated (by simple calculations or more elaborated simulation models) and all qualitative criteria have to be estimated (by expertise, empirical knowledge, etc.).

Step 7: multi-criteria analysis with Electre III

The analysis and the ranking of the strategies should be carried out by means of a multi-criteria approach accounting for criteria expressed in their own natural units, which avoids the problems due to averaging of or compensating for criteria. Among the several existing multi-criteria methods, the Electre III method has been chosen (Roy, 1978; Vincke, 1992), which also accounts for uncertainties in the evaluation of the criteria. It is thus particularly appropriate for studies on environmental aspects, because of the high uncertainties in phenomena, data and modelling results (see for example, Bertrand-Krajewski *et al.*, 2002).

For readers who are not familiar with Electre III, a short introduction is provided here aiming to facilitate the understanding of the following case study. Of course, in the case of a real application by an operator, they would need to be trained adequately before using Electre III, at least to the minimum level allowing a meaningfull discussion with a subcontractor applying it.

Electre III is a ranking method (Roy, 1978; Vincke, 1992) using "pseudo criteria" that explicitly take into account an indifference threshold qj and a strict preference threshold spj for each criterion j. The indifference threshold is particularly interesting, as it can be used to integrate the specific uncertainties in each criterion; Electre III is then less sensitive to variations of the input data and parameters than other methods such as Electre II (Roy and Bertier, 1971; Roy, 1996). Electre III requires the definition of a weight *wj* for each criterion *j*. The weights are chosen by the decision maker, according to their own hierarchy and preferences. Then, actions or strategies are compared in ordered pairs (*Si*, *Sk*). Each pair of actions is characterised by an outranking degree *S*(*Si*, *Sk*) expressing the credibility of the statement "*Si* outranks *Sk*". The outranking degree is calculated according to two types of index; a concordance index and discordance indices.

The following variables are thus defined:

wj weight of criterion *j*
pij performance of a strategy *i* according to the performance indicator or criterion *j*
qj(*pij*) indifference threshold which is either a constant or expressed as a percentage of *pij*
sp(*pij*) strict preference threshold which is either a constant or expressed as a percentage of *pij*

The concordance index is expressed as follows (see Figure 4):

$$C(S_i, S_k) = \frac{\sum_{j \in J} w_j c_j(S_i, S_k)}{\sum_j w_j}$$

where

$$c_j(S_i, S_k) = \begin{cases} 1 \text{ if } p_{ij} + q_j(p_{ij}) \geqslant p_{kj} \\ 0 \text{ if } p_{ij} + sp_j(p_{ij}) \leqslant p_{kj} \\ \quad \text{linear in between} \end{cases}$$

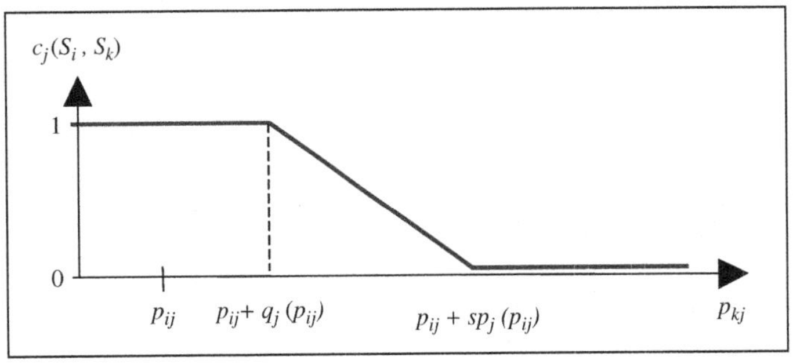

Figure 4: Calculation of a partial concordance index *cj* (*Si*, *Sk*) for a criterion *j*.

The definition of the discordance index for each criterion *j* requires the introduction of a veto threshold *vj*(*pij*) refusing the outranking of *Si* by *Sk* when *Si* is not preferred to *Sk* on criterion *j* and when the difference between the two performances is greater than *vj*(*pij*). The discordance index for each criterion *j* is expressed as follows (see Figure 5):

$$D_j(S_i, S_k) = \begin{cases} 0 \text{ if } p_{kj} \leqslant p_{ij} + sp_j(p_{ij}) \\ 1 \text{ if } p_{kj} \geqslant p_{ij} + v_j(p_{ij}) \\ \quad \text{linear in between} \end{cases}$$

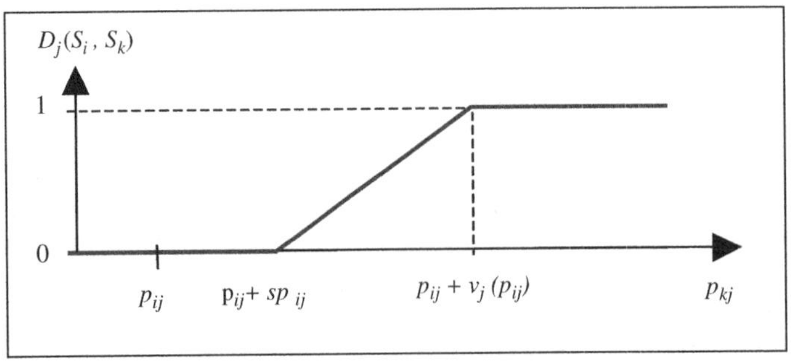

Figure 5: Calculation of a partial discordance index D_j (S_i, S_k) for a criterion *j*.

The degree of outranking $S(S_i, S_k)$ is then defined by:

$$S(S_i,\ S_k) = \begin{cases} C(S_i,\ S_k) \text{ if } \forall j\ D_j(S_i,\ S_k) \leq C(S_i,\ S_k) \\ C(S_i,\ S_k) \cdot \prod_{j \in J} \frac{1-D_j(S_i,\ S_k)}{1-C(S_i,\ S_k)} \quad \text{in other cases} \end{cases}$$

The valued outranking relationships are then used in order to build a simple outranking graph. An outranking arc between two actions S_i and S_k is defined if $S(S_i, S_k)$ is close to the maximum λ of all outranking degrees and if the difference between $S(S_i, S_k)$ and $S(S_k, S_i)$ is significant. For that purpose, a threshold $s(\lambda)$ has to be determined.

There is an arc in the graph between S_i and S_k if $S(S_i, S_k) \geq \lambda - s(\lambda)$ and if $|S(S_i, S_k) - S(S_k, S_i)| \leq s(\lambda)$. For each action S_i a qualification $Q(S_i)$ is calculated according to the difference between the number of actions which are outranked by S_i and the number of actions which outranks S_i. The solutions that present the best qualification receive a rank equal to 1. These actions are then removed from the initial set of actions and the same procedure is started again in the remaining set (new λ, new graph, new rank, etc).

A similar procedure is carried out but from the worst actions (worst qualification) to the best ones. These two ways of ranking are called distillations. The final ranking is obtained through the average of the ranks of the two intermediate distillations.

Multi-criteria analysis helps the decision maker to rank strategies, and consequently to make decisions. However, it should be clearly emphasised that the weights and the thresholds involved in the use of the method remain partly subjective (i.e. they depend on the decision maker's priorities, perception, etc.) and could be of great influence in the final results. Consequently, the proposed solutions have to be reviewed critically before making the final decision in step 8.

Step 8: decision making

This is the last step, based on the ranking provided by Electre III. At this level, various alternatives or scenarios can still be proposed and evaluated, and a robustness analysis has to be carried out to check that the ranking is not too sensitive to small variations of the values of criteria, weights, thresholds, etc.

DESCRIPTION OF THE ECULLY SEMI-VIRTUAL URBAN WATER SYSTEM

The MCA methodology established here has been tested, for demonstration purposes, on a semi-virtual case study based on the Ecully catchment in France. Ecully is a 245 hectare urban catchment in the western suburbs of Lyon. It is one of the experimental catchments of the OTHU project (http://www.othu.org). The urban habitat is mostly residential, with a population of approximately 7700 inhabitants (census 1999). The imperviousness of the catchment is evaluated as being 46%. The drainage system is a combined system with a 0.027 m/m mean slope. There is no WWTP at the outlet of the actual catchment, but a trunk sewer lead the effluents to one of the two largest WWTPs in Lyon. For the purpose of this case study, a virtual WWTP has been introduced, designed and simulated by means of the ASIM software developed at EAWAG (Gujer and Larsen, 1995). In order to limit the impact of CSOs, a virtual CSO storage tank has also been introduced in the case study. All components are schematized in Figure 6, which illustrates some basic elements of information for the study.

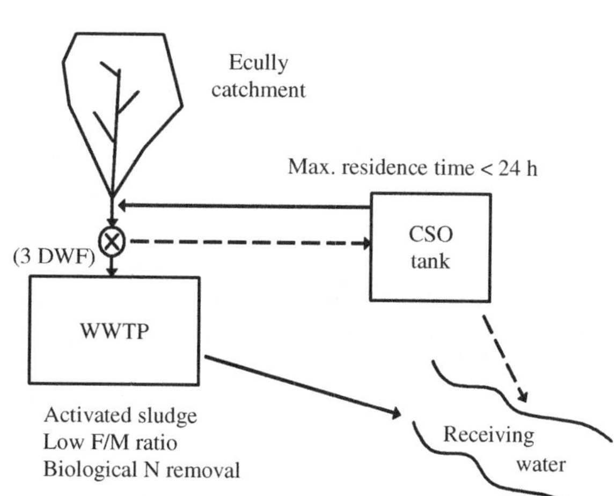

Figure 6: Components of the Ecully semi-virtual case study.

The case study may appear as simple. However, its complete description and its modelling to illustrate the application of the methodology proposed above, require a lot of information. When adequate or detailed information is not available, simplifications and hypotheses have been made, which do not affect the application of the methodology itself but which may lead to some possible bias in the numerical values and in the conclusions.

Available data from the OTHU monitoring station and from measurement campaigns are :

- 2 years of flow data at the outlet of the catchment, with a 2 minutes time step.
- 15 years of rainfall data from the nearest raingauge located in Champagne-au-Mont-d'Or, with a 6 minutes time step.
- 8 measurement campaigns carried out during dry weather, with hourly data on suspended solids, COD and TKN concentrations.
- 15 measurement campaigns carried out during rainfall weather (Musso, 1997), with data on suspended solids, COD and BOD concentrations.

Both the rainfall–runoff process over the catchment and the flow propagation in the sewer network have been simulated by initial and continuous rainfall loss functions and by means of a single linear reservoir, with a 6 minutes time step. The model has been calibrated with 10 rainfall events and validated with 5 other events. Initial (in mm) and continuous (expressed as a fraction of the rainfall intensity) losses are significant; respectively 1 mm and 0.6 of rainfall intensity, which means that many impervious surfaces are not connected to the sewer system. For our demonstration case study, the continuous losses have been finally set to 0.6 in order to work with a more common value.

The typical mean foul sewage flow Q_{fs} (i.e. without any infiltration) has been calculated based on observed dry weather days and infiltration measurement campaigns carried out in 2003. Q_{fs} corresponds to a volume of 1901 m^3/day. The virtual WWTP design and its functioning have been simulated by means of the ASIM software developed by EAWAG (Gujer and Larsen, 1995). It is an activated sludge plant with biological nitrogen removal (nitrification-denitrification). The design has been calculated for a maximum inflow *Qmax* equal to 3 times the mean dry weather flow DWF, for an influent design standard temperature of 13°C and for a maximum nitrogen concentration in the effluent equal to 10 mg N/L (i.e. 83% removal) which leads to a COD removal rate equal to 95%. The nitrate recirculation rate is limited to a maximum of 4 times the dry weather flow as recommended by French design standards. Annual influent temperature was assumed to follow a sinusoidal function over time, which was calibrated from the mean daily effluent temperatures of 80 different days. Mean annual effluent temperature was found to be equal to 17.4°C. The amplitude of the sinusoidal function is 10.4°C (minimum temperature of 12.2°C in February and maximum temperature of 22.6°C in August). As the observed mean temperature is higher than the design temperature, higher nitrogen removal rates may be expected than design removal rates.

According to measurement campaigns, foul sewage and surface runoff mean concentrations of COD and TKN are respectively equal to 519 and 224 mg O_2/L, and 59 and 19 mg N/L. In case of overflow, the excess flow is temporarily stored in the CSO tank. If the CSO tank is full, the effluents are discharged directly into the receiving water. The water stored in the tank is then pumped towards the WWTP, its storage duration being less than 24 hours, in order to avoid septicity and its detrimental consequence on the WWTP. If the storage duration is longer than 24 hours, the water is discharged into the receiving water. These rules have been defined to simplify the case study. For real systems, other and more complex rules may be used.

During dry weather days, simulations are carried out with a 1 hour time step and then recalculated for a daily time step. During wet weather days (i.e. days during which at least one rainfall event occurs), simulations are carried out with a 6 minutes time step and results are then recalculated for hourly and daily time steps.

The infiltration ratio IR is defined as the percentage of infiltrated water compared to the dry weather foul sewage. In the absence of a rainfall event, IR = 50% corresponds to a daily volume of infiltrated water equal to 50% of the daily foul sewage volume. Seven simulation series (scenarios) have been carried out over a 15 years long period (to account for annual variability) including both dry and wet weather conditions, with infiltration ratios of 0, 25, 50, 75, 100, 150 and 200%. The reference simulation series (named scenario A) is calculated with IR = 0%, with a well designed WWTP and without CSO tank. The results of each simulation series are i) economic criteria: operation costs (Euro/year) and ii) environmental criteria: annual COD and total nitrogen (NGL) loads discharged into the receiving water by the WWTP (treated effluent) and by the CSO structure (untreated effluent; kg/year).

It was assumed that i) infiltration of clear water does not bring any pollutant into the sewer system and does not change the influent temperature which follows a sinusoidal function over the year, and ii) infiltration occurs only between January and July, which corresponds to the high groundwater level period.

MODELLING RESULTS FOR SCENARIO A

Annual rainfall variability

The annual rainfall height corresponding to all rainfall events in the 15 years data series is given in Figure 7. The mean annual height is equal to 500 mm. This quite low value is due to the fact that the 15 years series of rainfall records used for the simulations has been elaborated by the Greater Lyon Water Department by removing many rainfall events which were considered as not being strong enough to generate overflows in the sewer system. In other words, the series has not been elaborated to simulate all rainfall events but to simulate only the events which may have clear hydraulic impacts on the sewer system. Consequently, this bias in the rainfall data series leads to an underestimation of the total volumes and overflow during wet weather periods, compared to the ideal situation where all rainfall events would have been kept in the data base.

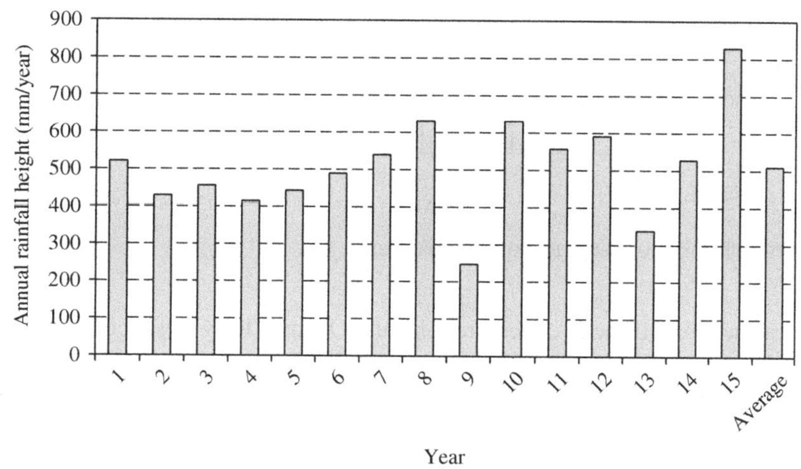

Figure 7: Annual rainfall heights (mm) used for simulations.

Foul sewage and total wastewater flows and pollutant loads

The typical foul sewage flow Q_{fs} (i.e. without any infiltration) has been calculated as the average of observed dry weather days in 2003. Its variation during a 24 hours period is given in Figure 8: it corresponds to a volume of 1901 m^3/day. This daily foul sewage flow pattern has been kept constant for all 15 years of simulation. From this curve, hourly flow rate values have been calculated for dry weather day simulations of the WWTP.

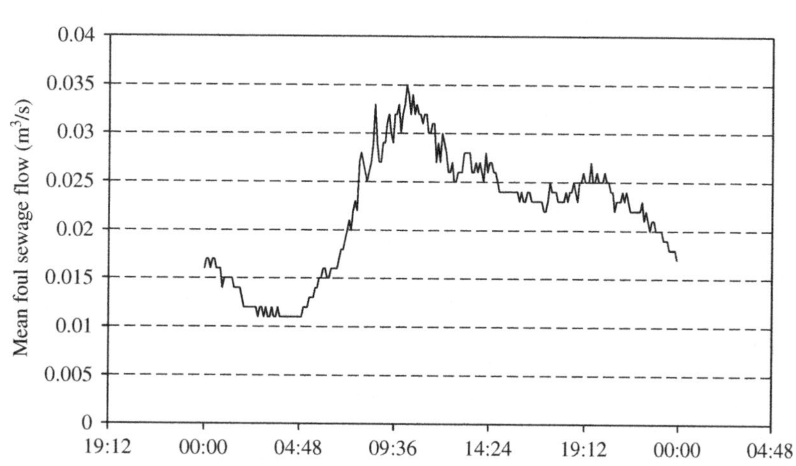

Figure 8: Mean daily foul sewage flow pattern (m^3/s).

The total flow Q_T is the sum of the foul sewage flow Q_{fs} and of the runoff flow Q_{run} calculated from the rainfall data by means of the linear reservoir. The distribution of the volumes at the outlet of the catchment is shown in Figure 9. Annual volumes at the outlet of the catchment vary from one year to another one depending on the annual rainfall height. Runoff volumes contribute from 19% to 45% of the total annual volume.

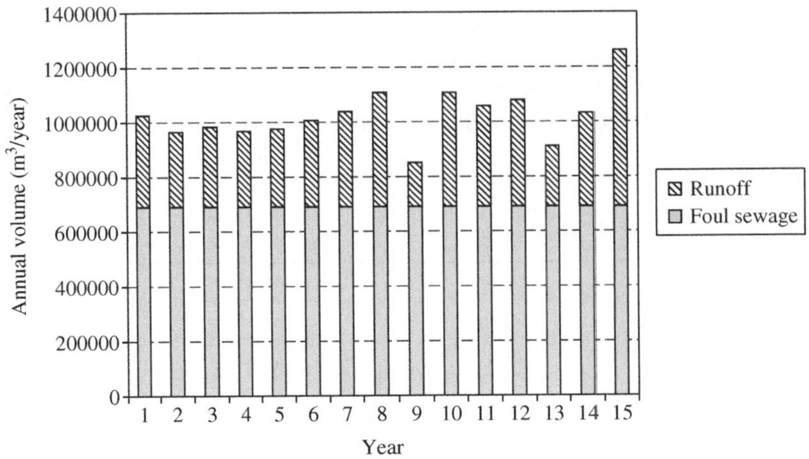

Figure 9: Distribution of runoff and foul sewage annual volumes.

The foul sewage and surface runoff mean concentrations of COD and TKN (respectively 519 and 224 mg O_2/L, and 59 and 19 mg N/L) have been kept constant for all simulations. This hypothesis is of course not realistic, as it is known and observed that pollutant concentrations widely vary from rainfall event to rainfall event and also within events. However, these variations were difficult to simulate adequately during this case study as it is very difficult to calibrate a rainfall event pollutant model with limited data sets (see for example, Mourad *et al.*, 2004). It has been decided to keep the case study simple and to use constant values to avoid the introduction of many additional hypotheses that would have been hard to justify or validate. This may affect the final results, but not their orders of magnitude.

Total pollutant loads have been determined by summing foul sewage loads and surface runoff loads. Due to the low concentrations during wet weather, the foul sewage contributes annually 74% to 91% of the total COD loads and 79% to 93% of the total nitrogen loads.

CSO tank

During dry weather, no overflow occurs because the total flow is lower than the maximum WWTP design flow ($Qmax = 3 \times$ DWF). As a consequence, on an annual basis, 64% to 87% of the total volume is treated by the WWTP, as shown in Figure 10. The annual volume treated by the WWTP is almost constant. The same pattern is observed for pollutant loads. Consequently, the operation of the WWTP and the impact of the infiltration on the WWTP have been simulated only for one year (year n° 1), assumed to be representative of the complete 15 years period.

However, the annual total non-treated pollutant load varies significantly: the non-treated COD load discharged in the year 15 is four times higher than the load discharged in year 9 and two times higher than the average year. In order to take into account this significant variability, three years have been analysed more specifically in the following paragraphs:

- year 1 as an average rainfall year (521 mm/year)
- year 9 as the minimum rainfall year (246 mm/year)
- year 15 as the maximum rainfall year (828 mm/year).

Wastewater treatment plant (WWTP)

As indicated above, annual volume and pollutant loads treated by the WWTP are almost constant for all 15 years. Consequently, the WWTP operation and the impacts of infiltration have been simulated only for year n° 1. The analysis of all days of the year reveals that the efficiency slightly decreases during wet weather days. The annual distribution and pathways of pollutant loads are shown respectively in Figure 11 for COD and in Figure 12 for nitrogen.

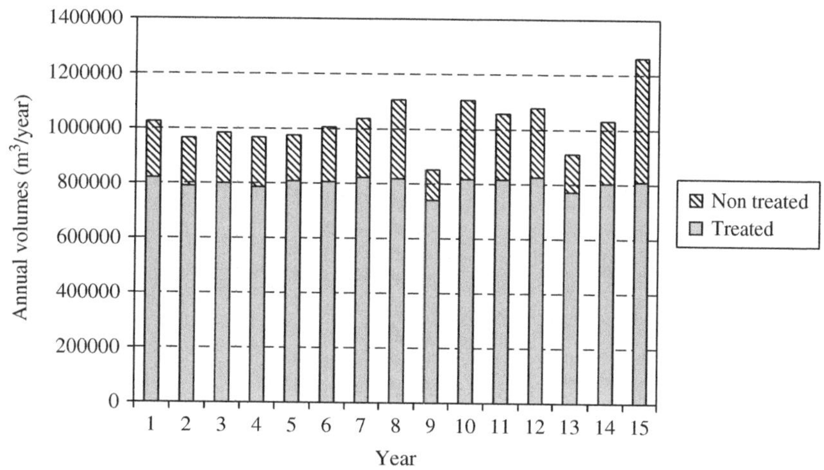

Figure 10: Distribution of annual treated and non-treated volumes.

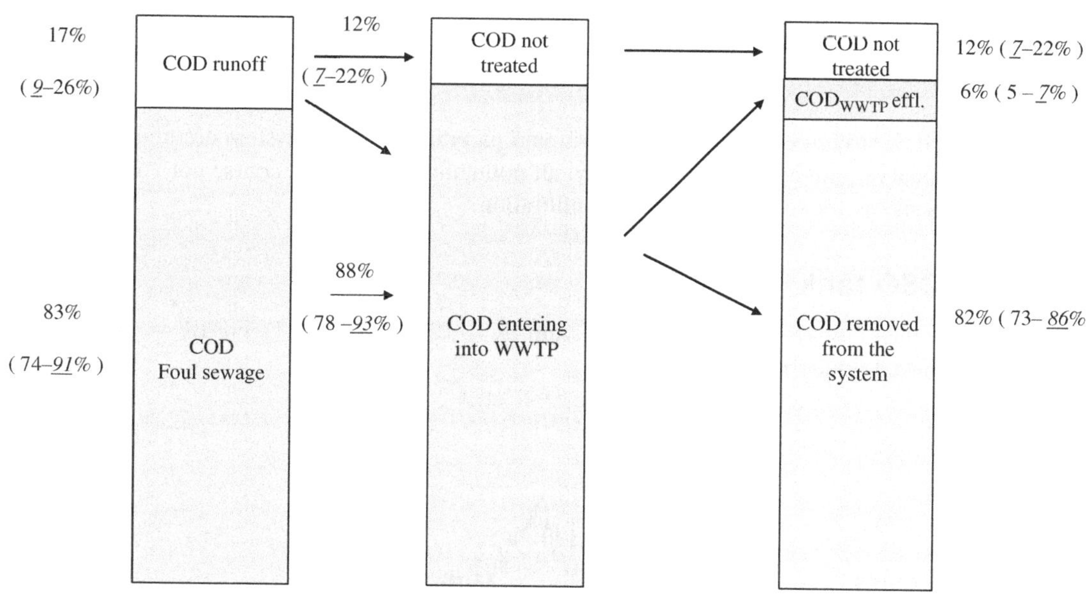

Figure 11: Distribution and pathways of the annual COD load through the WWTP. "COD_{WWTP} effl." means the COD load discharged by the WWTP into the river after treatment. In bold: year 1 (average); in italic: year 9 (minimum); in normal characters: year 15 (maximum).

Regarding COD loads, 14% to 27% of the loads produced over the catchment are discharged into the receiving water. On average, two thirds are directly discharged by overflows without any treatment and one third is discharged by the WWTP.

Regarding nitrogen loads, 24% to 34% are discharged into the river. These higher percentages compared to COD are due to the lower efficiency of the WWTP regarding nitrogen removal. On average, one third is directly discharged by the overflows without any treatment and two thirds are discharged by the WWTP.

IMPACTS OF INFILTRATION ON POLLUTANT LOADS AND ENERGY DEMAND

The infiltration flow $Qinf$ is expressed as the infiltration ratio IR which corresponds to the percentage of the foul sewage flow Qfs : IR $= 100\ Qinf/Qfs$. The total flow QT at the WWTP inlet is equal to $Qrun + (1+\text{IR}/100)\ Qfs$. Six series of simulations were carried out with IR values equal to 25%, 50%, 75%, 100%, 150% and 200%.

In this case study, the catchment and the sewer system are simulated as a whole lumped system. All overflows are supposed to occur only at the outlet of the catchment, which is a rather realistic hypothesis for the Ecully catchment

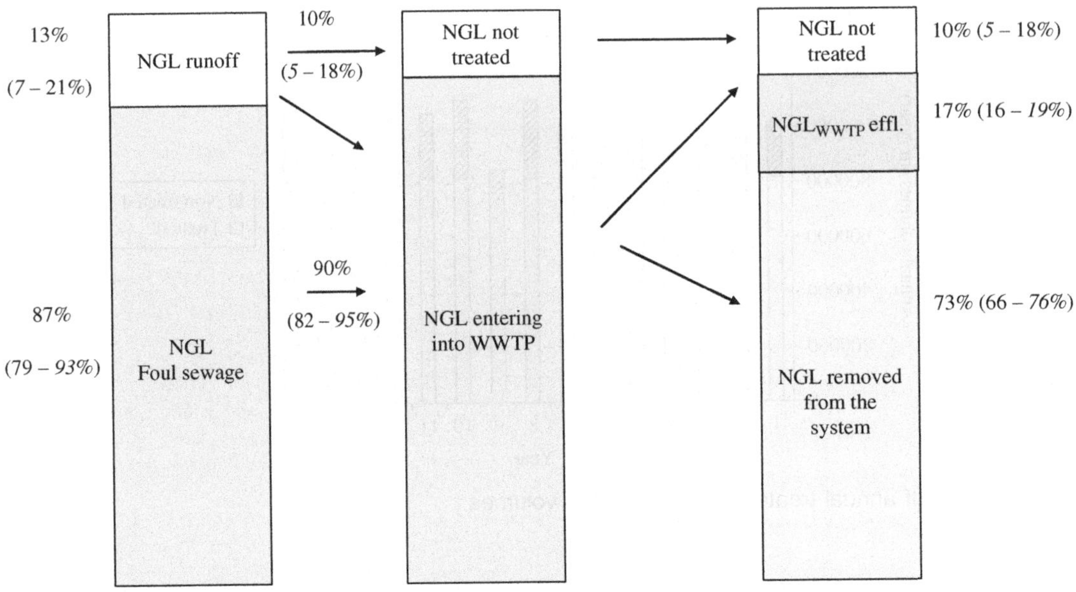

Figure 12: Distribution and pathways of the annual NGL load through the WWTP."NGL$_{WWTP}$ effl." means the NGL load discharged by the WWTP into the river after treatment. In bold: year 1 (average); in italic: year 9 (minimum); in normal characters: year 15 (maximum).

where the 5 small overflow structures located in the upstream parts of the sewer system are almost never in function, except for very exceptional rainfall events. Consequently, all pollutant loads (but of course not volumes) at the outlet of the catchment are the same as for scenario A without infiltration.

Impacts on the CSO tank

During dry weather, the total flow Q_T is lower than Q_{max} until IR = 100% (Figure 13). For IR = 150% and 200%, Q_T is higher than Q_{max} for several hours per day.

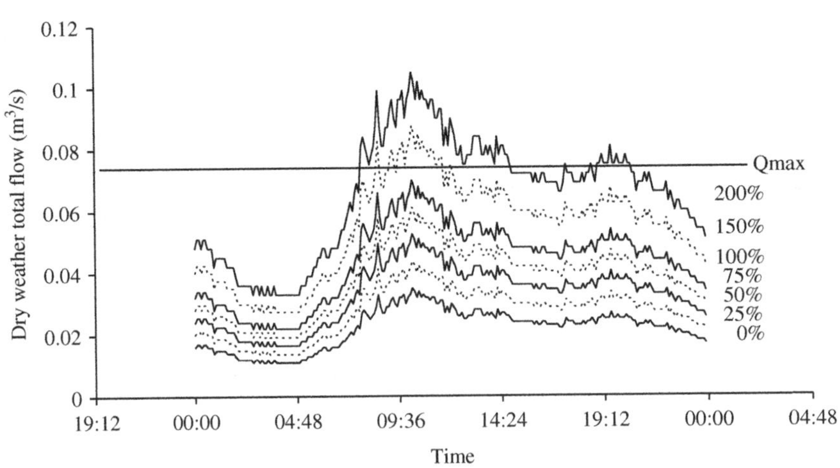

Figure 13: Dry weather flow patterns for IR = 0, 25, 50, 75, 100, 150 and 200%.

In case of IR = 150%, the volume overflowed into the CSO tank during dry weather is redirected to the WWTP during lower inflow periods; all water is treated. But in the case of IR = 200%, the CSO tank cannot be fully emptied and after 24 hours, according to the simulation rules, the remaining (but here negligible) volume (typically 2 m^3/day) is discharged directly into the receiving water.

During wet weather, the time required to fully empty the tank is higher because both volumes transferred in sewer pipes and volumes overflowed into the tank are higher. The volume of non-treated wastewater discharged into the receiving water is consequently increasing when IR is increasing. On an annual basis, the total non-treated COD load is multiplied by a factor ranging from 1.08 (year 15) to 1.33 (year 1) in the case IR = 200%, as shown in Figure 14. As expected, infiltration has less impact on the total non-treated pollutant load discharged into the receiving water when the rainfall height is more important (year 15). For all the different years, the non-treated loads increase linearly until IR = 100%, but then exponentially from IR = 150%. The same trend is observed for the total non-treated nitrogen loads.

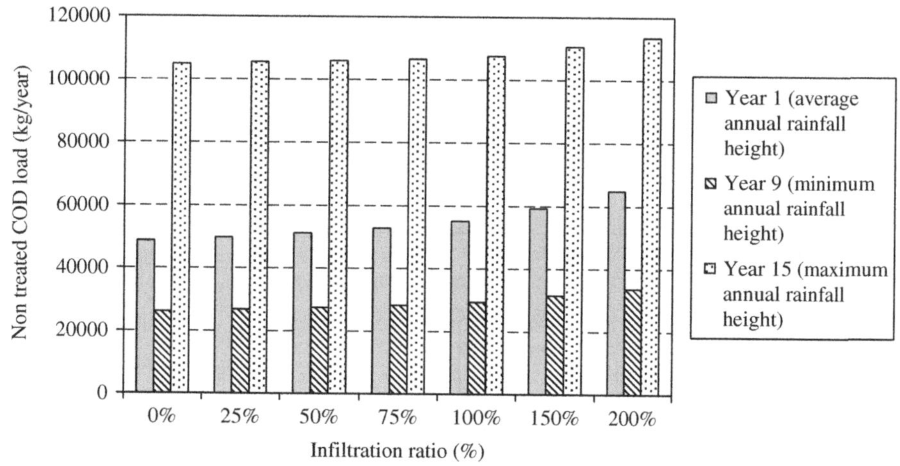

Figure 14: Non-treated COD Load (kg/year) discharged Into the receiving water vs. the infiltration ratio IR for different years.

Impacts on the WWTP

Pollutant Loads

As it is assumed that infiltrated water does not bring any pollutant load, only inflow volumes are modified by infiltration, which corresponds to dilution, as shown in Figure 15.

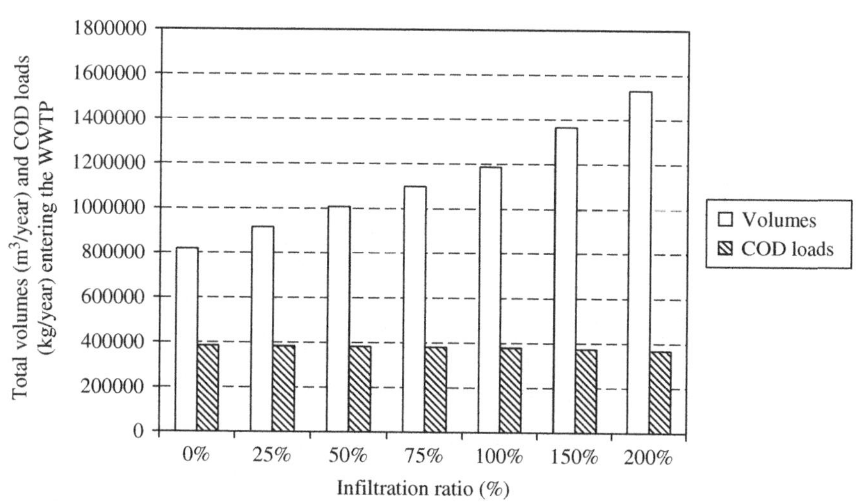

Figure 15: Total volumes and total COD loads entering the WWTP vs. the infiltration ratio.

This dilution of the influent and the decrease of residence time in biological tanks significantly affect the efficiency of the WWTP, especially regarding nitrogen, as shown in Table 2. Consequently, the annual total pollutant loads (sum of non treated loads and loads discharged by the WWTP) discharged into the receiving water increase with infiltration (see Figure 16 and Figure 17).

As the WWTP has been designed for high nitrogen removal, the COD removal remains high and is not significantly affected by increasing infiltration ratios. The COD load discharged by the WWTP remains constant. However, the total COD load discharged into receiving water increases significantly due to the increase of the non-treated loads (Figure 16).

Table 2: WWTP Removal efficiency for COD and nitrogen vs. the infiltration ratio.

IR (%)	Efficiency for COD (%)	Efficiency for N (%)
0%	94%	83%
25%	94%	81%
50%	94%	80%
75%	94%	78%
100%	94%	76%
150%	94%	75%
200%	94%	72%

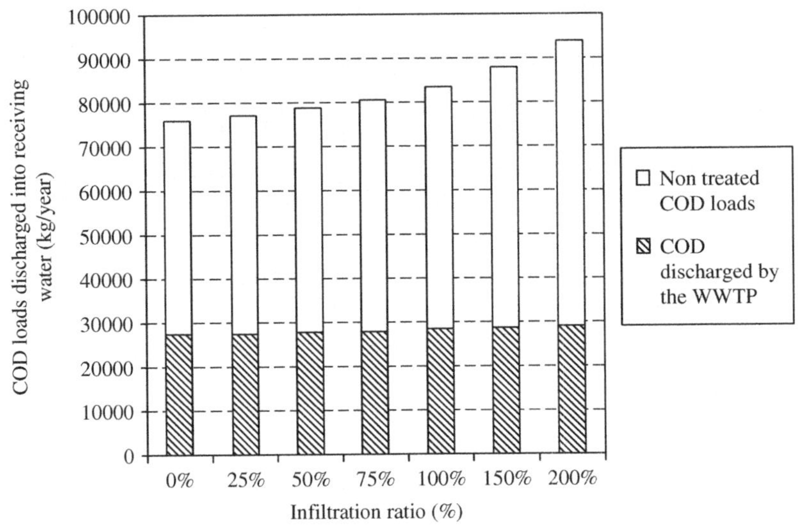

Figure 16: COD loads discharged into the receiving water vs. the infiltration ratio (year 1).

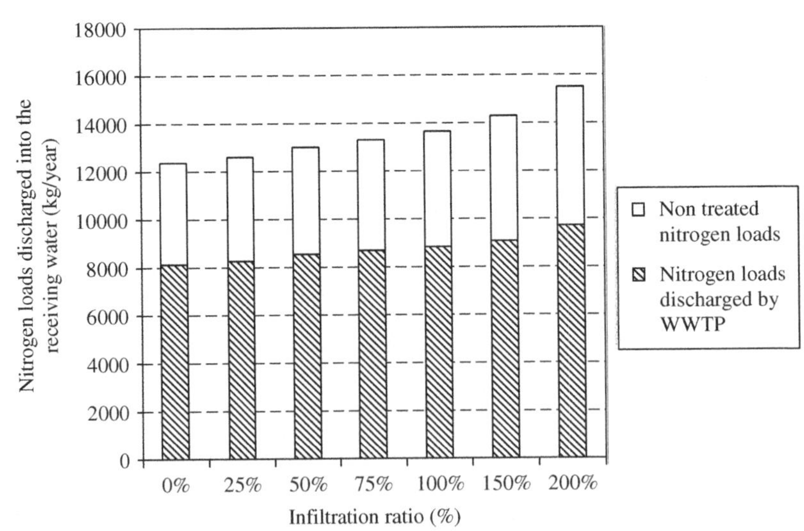

Figure 17: Nitrogen loads discharged into the receiving water vs. the infiltration ratio (year 1).

Regarding nitrogen, there is a more pronounced decrease of the WWTP efficiency, but a less significant direct discharge of non-treated nitrogen into the receiving water (Figure 17). The corresponding total annual pollutant loads values are given in Table 3 for the three selected years.

Table 3: Increase of annual pollutant loads discharged into the receiving water vs. the infiltration ratio

YEAR 1				
IR	**Total COD load emitted in the receiving water (kg COD/year)**	**Total NGL load emitted in the receiving water (kg N/year)**	**Total COD load with infiltration/ total COD load without infiltration (scenario A)**	**Total NGL load with infiltration/total NGL load without infiltration (scenario A)**
0%	76005	12385	1.00	1.00
25%	77129	12617	1.01	1.02
50%	78879	13012	1.04	1.05
75%	80626	13315	1.06	1.08
100%	83503	13662	1.10	1.10
150%	87949	14298	1.16	1.15
200%	93866	15498	1.23	1.25

YEAR 9				
IR	**Total COD load emitted in the receiving water (kg COD/year)**	**Total NGL load emitted in the receiving water (kg N/year)**	**Total COD load with infiltration/ total COD load without infiltration (scenario A)**	**Total NGL load with infiltration/total NGL load without infiltration (scenario A)**
0%	53537	10416	1.00	1.00
25%	54194	10605	1.01	1.02
50%	55257	10938	1.03	1.05
75%	56146	11163	1.05	1.07
100%	57723	11392	1.08	1.09
150%	60134	11838	1.12	1.14
200%	62643	12699	1.17	1.22

YEAR 15				
IR	**Total COD load emitted in the receiving water (kg COD/year)**	**Total NGL load emitted in the receiving water (kg N/year)**	**Total COD load with infiltration/ total COD load without infiltration (scenario A)**	**Total NGL load with infiltration/total NGL load without infiltration (scenario A)**
0%	132346	17217	1.00	1.00
25%	133100	17417	1.01	1.01
50%	133858	17723	1.01	1.03
75%	134493	17926	1.02	1.04
100%	136078	18157	1.03	1.05
150%	139318	18676	1.05	1.08
200%	142611	19612	1.08	1.14

Based on values given in Table 3 and by accounting for both measurement and model uncertainties, one may consider, for the multi-criteria analysis to be carried out later on, that differences in total annual pollutant loads are not significant until the difference reaches 5%. In other words, for IR values lower than 75% for year 1 for example, there is no meaningful change in total annual nitrogen and COD loads. Similarly, it can also be assumed that differences in total annual pollutant loads higher than 10% have a detrimental impact on the environment.

The above threshold values, which will affect the final ranking by Electre III, have been chosen partly arbitrarily for this semi-virtual case study. Depending on the context, they should be modified in a real case, where uncertainties in data and modelling results should be better assessed and impacts on receiving water should also be further analysed. For example, 2000 kg/year of total nitrogen may have a negative impact on a small river whereas it can be negligible for a large river with high flows. Nevertheless, such a choice will always remain partly subjective and should be discussed between engineers, operators and decision makers.

It also appears that impacts of infiltration vary from one year to another one depending on the rainfall variability. For example, the total annual COD loads discharged into the receiving water with IR = 200% compared to the scenario A (without infiltration) is higher for the year 15 than for the year 1. Consequently, IR values to be exceeded, to consider meaningful changes in total annual pollutant loads vary depending on the studied year, which will affect the final ranking by Electre III. Taking into account this variability, it may be interesting for stakeholders to compare the results for several years before taking their decision regarding investment strategies.

Operation costs

Based on design requirements and pump manufacturers specifications, the pumping station at the WWTP inlet has been designed with a total head of 6 m, an efficiency of 55% and a power of 13.5 kW. As the annual volume entering the WWTP is multiplied by 1.9 for IR = 200% (infiltration is considered only from January to July, during the high groundwater level period), the annual energy consumption of the pumping station is also multiplied by 1.9 (Figure 18).

In the WWTP, it has been assumed that aeration is made by means of 2 fine bubbler diffuser devices, with a daily aeration time of 12 hours per day. Regarding the mixing of activated sludge, there is one mixer in the anoxic tank functioning 24 hours per day and one mixer in the aerobic tank functioning 12 hours per day. The return sludge and the nitrate recirculation pumps are assumed to be in function 12 hours per day. With increasing values of IR, the sludge recirculation is also increasing (Figure 19 and Table 4) as it is correlated to the influent volume. Aeration and mixing durations do not increase as pollutant loads, volumes of the tanks and the F/M ratio are constant.

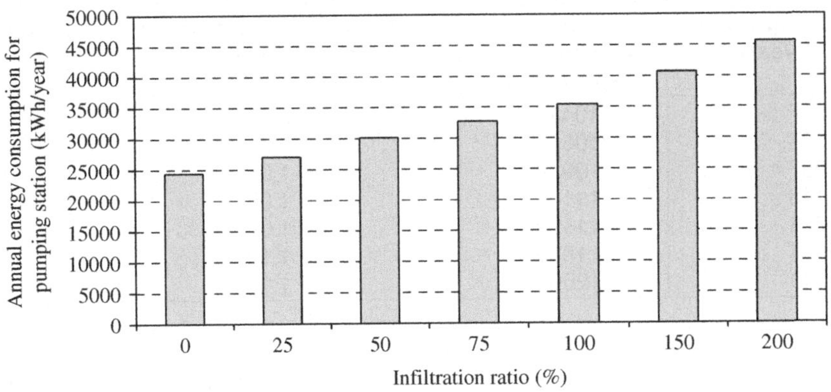

Figure 18: Annual energy consumption of the pumping station vs. the infiltration ratio.

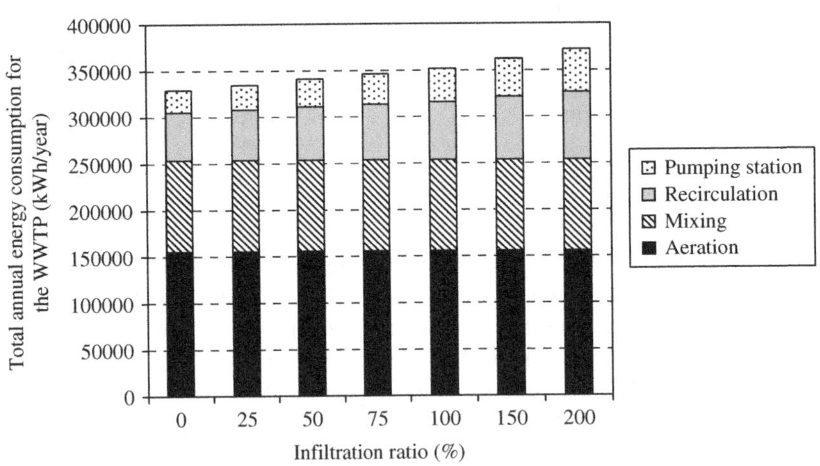

Figure 19: Total annual energy consumption in the WWTP vs. the infiltration ratio.

Energy costs have been calculated with a constant price of 0.06 Euros/kWh. The difference in the total annual energy consumption becomes significant from IR = 150%. (ratio 1.10). The additional costs for the WWTP operation reach 13% for IR = 200%. As the main conclusion for this case study, for IR = 25% and 50%, impacts of infiltration are low on both the pollutant loads emitted by the WWTP and the energy consumption for all the years 1, 9 and 15. No rehabilitation or investment is really needed. From IR = 75% for years 1 and 9 and from IR = 150% for year 15, negative impacts on pollutant loads and on operation costs may lead the operator to consider a rehabilitation of the drainage system.

Table 4: Total relative annual energy consumption (Reference value = 1 for Scenario A) and total annual costs in Euros vs. the infiltration ratio.

IR	Energy for pumping station	Energy for aeration	Energy for mixing	Energy for recirculation	Total energy	Total costs (Euros/year)	Additional costs due to infiltration (Euros/year)
0%	1.00	1.00	1.00	1.00	1.00	19770	0
25%	1.11	1.00	1.00	1.05	1.02	20063	323
50%	1.23	1.00	1.00	1.11	1.04	20466	696
75%	1.34	1.00	1.00	1.16	1.05	20781	1011
100%	1.45	1.00	1.00	1.21	1.07	21100	1330
150%	1.67	1.00	1.00	1.32	1.10	21737	1967
200%	1.87	1.00	1.00	1.41	1.13	22336	2566

CHOICE OF INVESTMENT/REHABILITATION STRATEGIES

Investment strategies should be defined by taking into account the specific context of the catchment. For example, no replacement will be carried out if excavation is not possible due to very critical traffic problems. In the considered semi-virtual case-study, no major problem limits the application of any strategy. Three main investment strategies have been considered:

- investment in larger treatment plant capacities to account for infiltration, in order to have the same pollutant removal efficiency as without infiltration,
- investment in larger CSO tank volume,
- investment in rehabilitation of the sewer system.

Investment in larger treatment plant capacities

Impacts on pollutant loads

The objective is to reach an efficiency of 95% regarding COD removal and 83% regarding nitrogen removal. The new design made with the ASIM software is based on the dry weather flow and pollutant loads by taking into account infiltration. Consequently, the capacity of the treatment plant increases and larger volumes and pollutant loads are accepted into the WWTP. This increase in WWTP capacity also reduces the volume of overflows emitted by the CSO tank (Figure 20).

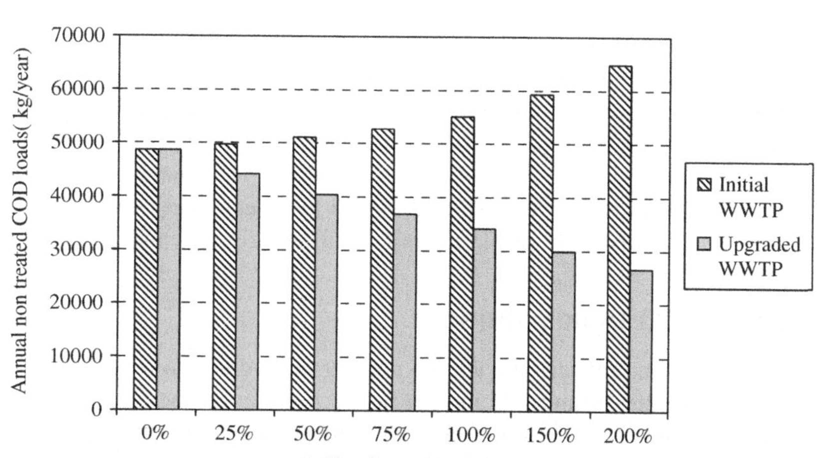

Figure 20: Annual non-treated COD load with initial and upgraded WWTP vs. the infiltration ratio (year 1).

Thanks to the upgraded WWTP, total COD loads emitted from both the WWTP and the CSO tank into receiving water are lower compared to scenario A without infiltration and with the initial WWTP (Figure 21). Similar trends are observed for the total nitrogen loads.

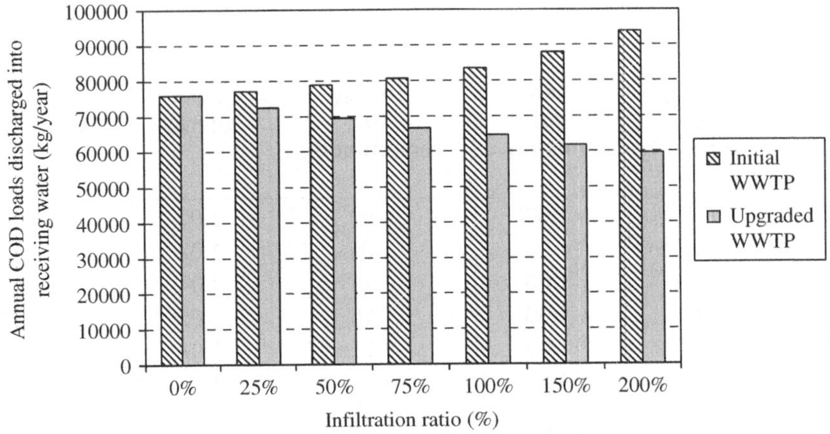

Figure 21: Annual total COD loads emitted Into the receiving water with initial and upgraded WWTP vs. the infiltration ratio (year 1).

Impacts on operation costs

The WWTP efficiency regarding nitrogen could be increased by at least two ways:

– increase of the residence time in the biological tank for a better nitrification by increasing the size of the tanks;
– increase of the recirculation for a better denitrification by keeping the recirculation flow below 4 times the average dry weather flow.

The effects of infiltration can be compensated until IR = 100% by increasing only the recirculation. This is why the size of the biological tank remains equal to 1 in Table 5. From IR = 150%, the maximum limit for recirculation flow is reached, and the size of the biological tank size has to be increased in order to fully account for infiltration. The results of the simulations are given in Table 5. The secondary clarifier is designed based on hydraulic constraints and its size increases as a function of the increasing IR. For IR higher than 150%, the impacts on the WWTP are very significant and obviously show that increasing the WWTP capacity is not the most pertinent solution.

Table 5: Increase of the relative sizes of biological tank and clarifier to account for infiltration.

IR (%)	0%	25%	50%	75%	100%	150%	200%
Biological tank volume	1.00	1.00	1.00	1.00	1.00	1.25	3.50
Secondary clarifier volume	1.00	1.14	1.41	1.69	1.98	2.08	2.36

Regarding investment costs, only the biological tank and the secondary clarifier are taken into account because they are the main components affected by infiltration. For cost calculations, we have assumed that the mean cost of 1 m^3 of biological tank or secondary clarifier was approximately 600 Euros, including land, building and equipments. This cost is based on Swiss data and could be lower in other European countries.

Regarding operation costs, the results are presented in Figure 22. The energy consumption for pumping and recirculation is slightly increasing because the WWTP is designed to accept larger volumes. The energy consumption for aeration and mixing also increases because of the increase of the sludge age and of the size of the tanks.

Investment in a larger CSO tank volume

Investing in a larger CSO tank volume reduces the annual number and volume of overflows. Various volumes of the CSO tank have been simulated for 20, 50 and 100 m^3 per active hectare (the active area is the fraction of the total catchment area which really contributes to the runoff entering into the sewer system). However, in terms of pollutant loads, the increase of the CSO tank size is effective only for low values of IR (Figure 23). These results are mainly explained by the maximum storage duration of 24 hours in the CSO tank. When infiltration increases, the time necessary for emptying the CSO tank also increases. Consequently, on the one hand, investment in a higher CSO tank capacity allows the storage of larger volumes, i.e. limits the number of overflows. On the other hand, for high IR values, effluents cannot be redirected to the WWTP because of high flows in the sewer system. An increasing fraction of the stored volume is thus discharged into the receiving water after 24 hours of storage. The results for the years 9 and 15 follow the same pattern as the results for year 1 showed in Figure 23.

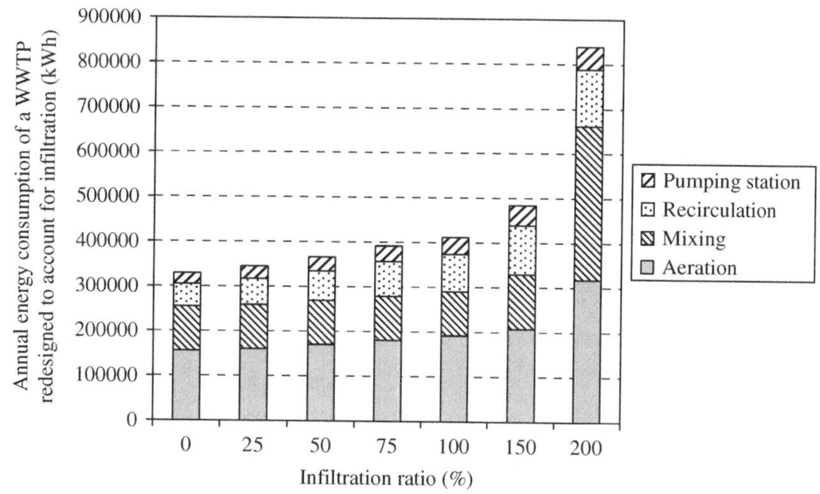

Figure 22: Annual energy consumption of a WWTP re-designed to account for infiltration.

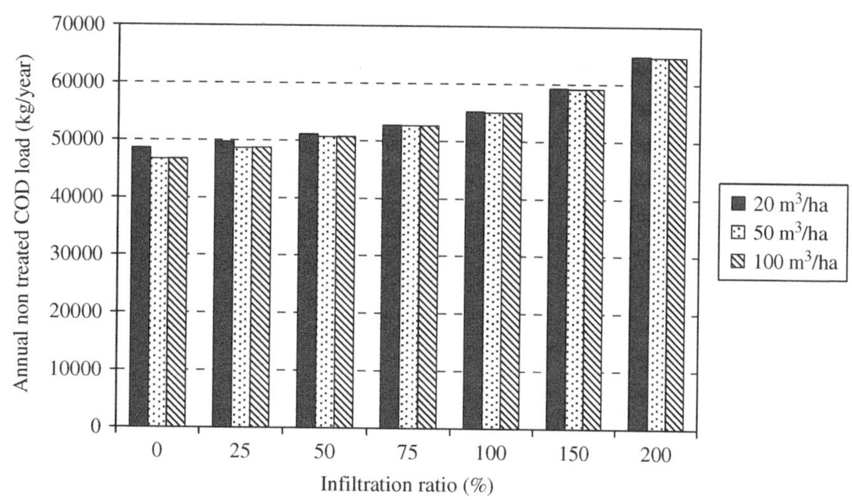

Figure 23: Annual non-treated COD loads vs. the infiltration ratio and the CSO tank volume (year 1).

The total pollutant COD loads (from both CSO tank and WWTP) emitted into the receiving water do not significantly decrease with increasing CSO tank volumes (Figure 24). In this particular semi-virtual case study, it appears that investment in a larger CSO tank volume is not very promising; this investment strategy will not be considered further.

Rehabilitation of the sewer system

Four rehabilitation technologies (RT) have been selected for this case study:

- repair, which consists in local actions such as injections, masonry, etc.
- sliplining : insertion of a liner, i.e. a new pipe into the existing damaged pipe (renovation).
- Cured-In-Place-Pipe (CIPP) : a flexible prefabricated tube impregnated with a thermosetting resin is inserted in the existing pipe and either winched or inverted into place with water pressure. Injected steam or hot water cures the resin and shapes the tube into the form of the existing pipe. Application of heat hardens the resin after a few hours, forming a jointless pipe-within-a-pipe (renovation).
- conventional trench (replacement).

Table 6 summarizes the advantages and the disadvantages of each selected RT.

In the present case study, all selected RTs are supposed to be applicable. The structural state of sewer pipes in the Ecully catchment is not precisely known, due to non-systematic CCTV inspections. Among 15 km of sewer pipes, only 1 km is man-entry. In non man-entry pipes, only a few CCTV examinations have been carried out, mainly for reactive

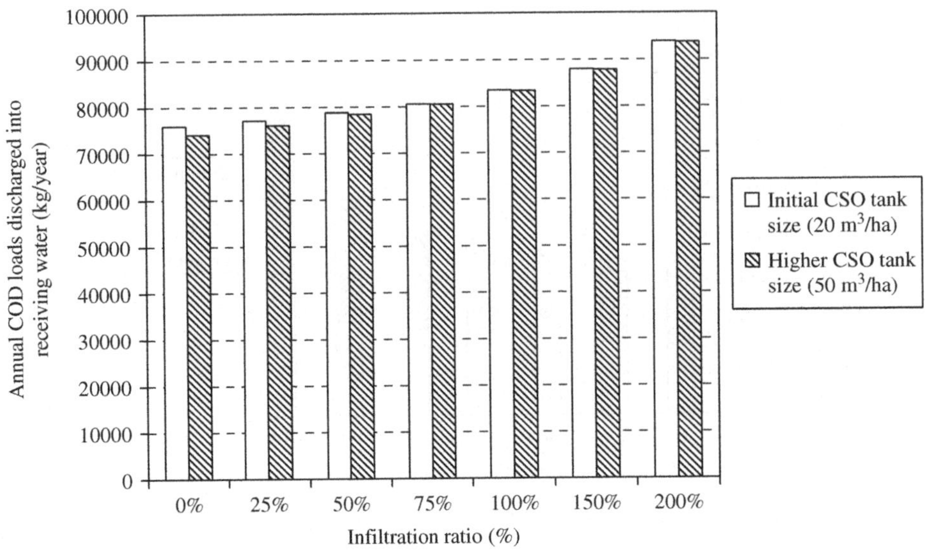

Figure 24: Annual total COD load emitted Into the receiving water vs. IR and CSO tank volume.

Table 6: Advantages and disadvantages of some rehabilitation technologies.

Rehabilitation technology	Advantages	Disadvantages
Repair (injections, etc.)	– no excavation – not expensive	– no restoration of the site (correction of structural problems) – remaining of other defects
Sliplining (renovation)	– no excavation – restoration of the site	– capacity losses – construction inspection by CCTV
CIPP (renovation)	– no excavation – jointless pipe	– capacity losses – no entire restoration of the site – construction inspection by CCTV
Conventional trench (replacement)	– restoration of the site – construction inspection from the surface	– excavation

actions. Only pipes which are assumed to be critical (according to the knowledge of sewermen) are inspected. Moreover, the exact location of infiltration is not really known. Facing this lack of data, the following hypotheses have been made for the case study:

- infiltration occurs only in the pipes located in the river aquifer,
- infiltration is diffuse and occurs along all pipes located in the aquifer.

With the above hypotheses, 3 km of pipes are concerned (2 km of non man-entry pipes and 1 km of man-entry pipes), which represents a rehabilitation of 20% of the complete sewer system. If these 3 km are assumed to be rehabilitated at once, this rate of 20% is very high compared to the average rehabilitation rate in French cities which is around 2% of the total sewer length per year.

Investment costs are difficult to estimate due to their great variations depending on the context of the catchment. However, the following average prices have been applied to the present case study :

- repair costs are given by French specialists based on experience. Costs are estimated to be 75 Euros per linear meter and 20% additional costs are added for planning, manpower and supervision.
- sliplining costs and conventional trench costs are estimated using data from Frehmann (2004), which depend on diameter and depth of the pipes.
- CIPP costs have been given by C. Montero from the CARE-S project (extract from the CARES-S cost tool, Freni *et al.*, 2005). They depend on the diameter of the pipe and include re-opening of service connections and renovation of the manhole bottom. 20% of additional costs are added for planning and construction supervision.

For the semi-virtual case study, total investment costs for rehabilitation are summarized in Table 7. Regarding the two renovation technologies, CIPP appears less expensive than sliplining. Consequently only repair, CIPP and conventional trench have been selected for further analysis.

Table 7: Investment costs (in Euros) for the selected rehabilitation technologies.

	Minimum	Mean	Maximum
Injections (repair)	–	1434551	–
Sliplining (renovation)	5161572	6474694	7787815
CIPP (renovation)	–	5443871	–
Conventional trench (replacement)	5257106	6806425	12413309

Another important hypothesis is that by rehabilitating 100% of the sewer pipes subject to infiltration problems, 100% of the infiltration problem will be solved. This optimistic hypothesis should be considered with extreme caution because:

- the elimination of infiltration could not be complete due to bad installation or incomplete repair,
- groundwater may find other ways to enter the sewer systems (HCs, manholes, etc.).

However, we have used this simple hypothesis for convenience. Indeed, pollutant loads emitted into the receiving water and operation costs after sewer rehabilitation will be the same as in scenario A without infiltration.

As explained previously in step 5, sewer system rehabilitation linked to infiltration may contribute to increase the global serviceability of the sewer system. If data and models are available, it can be expressed in a decrease in flooding, decrease of silting, etc. In our case, this criterion has not been considered because many additional hypotheses would have been needed.

Mixed investment: renovation of 50% of the pipes and investment in upgrading the WWTP

In the mixed investment strategy, only 50% of the sewer pipes submitted to infiltration are renovated and the WWTP is re-designed in order to account for the remaining infiltration. Renovation technology has been chosen because it is an intermediate solution; it is cheaper than replacement and it improves the serviceability better than repair. One assumes that the renovation of 50% of the pipes is eliminating 50% of infiltration, and that investment costs for sewer pipes are 50% less expensive than the rehabilitation of 100% of the pipes (equal distribution of costs for all pipes). This investment appears as an intermediate solution regarding pollutant loads and investment costs.

COMPARISON OF THE STRATEGIES

Impacts of the investment strategies on each selected criterion

As observed in the above paragraphs, in the proposed case study, infiltration significantly affects the drainage system from IR = 75% for years 1 and 9 and from IR = 150% for year 15. For IR = 25% and 50%, impacts are considered as negligible and no investment strategy has been defined.

Figure 25 to Figure 28 present the impacts of investment strategies on each selected criterion for the year 1. The conclusions described for year 1 can also be applied to the results of years 9 and 15 (not shown here): only the amplitude of the results is modified.

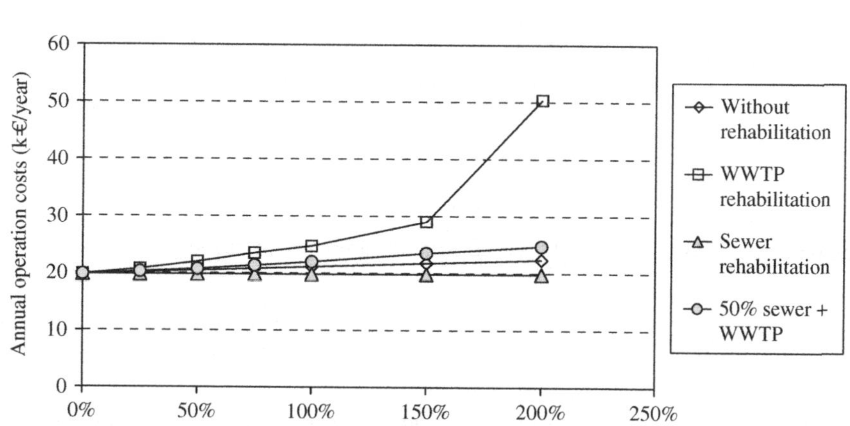

Figure 25: Annual total COD load emitted into the receiving water vs. the infiltration ratio and the investment strategy.

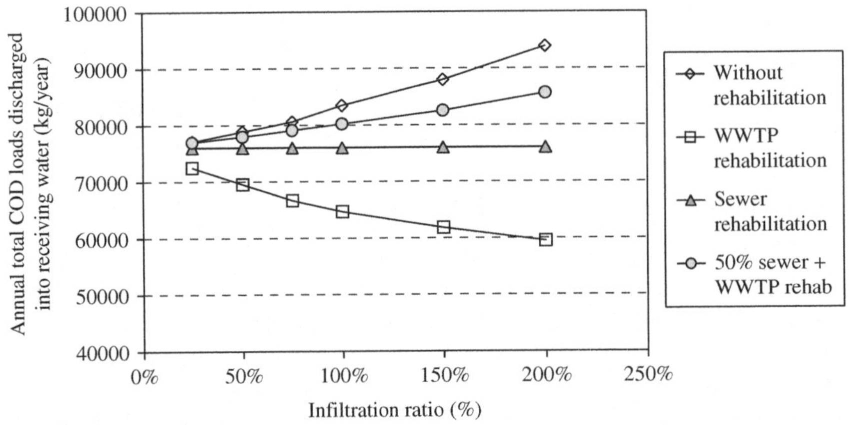

Figure 26: Annual total nitrogen loads emitted into the receiving water vs. the infiltration ratio and the investment strategy.

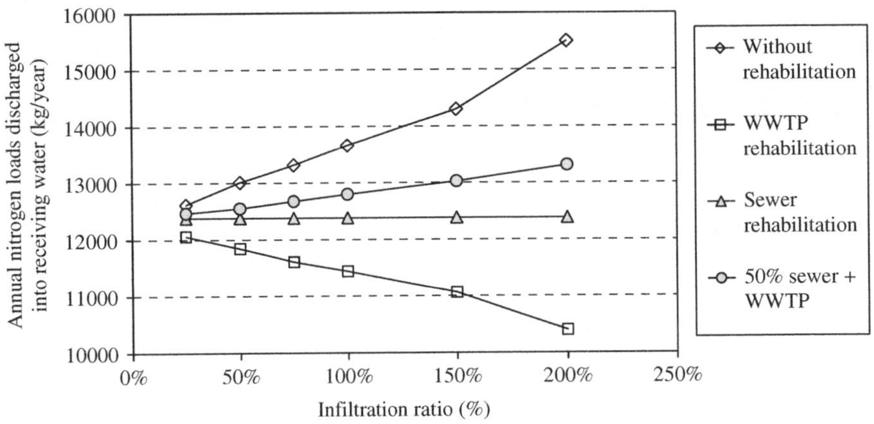

Figure 27: Annual operation costs vs. the infiltration ratio and the investment strategy.

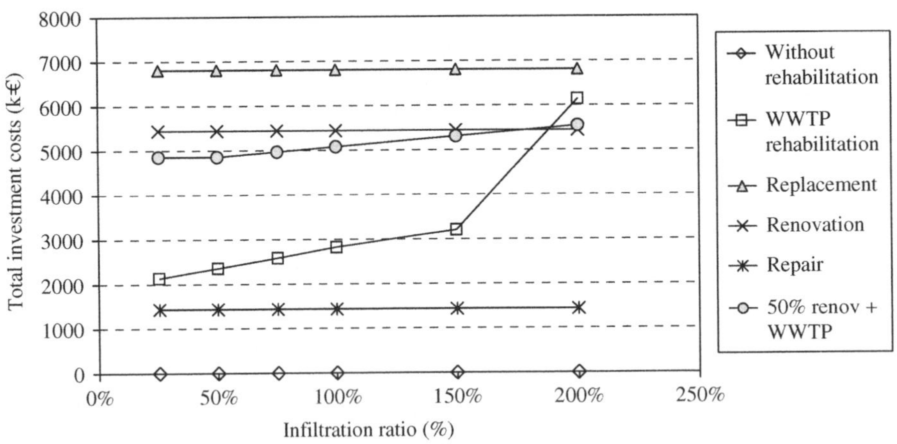

Figure 28: Total investment costs vs. the infiltration ratio and the investment strategy.

The operation, financial and total annual costs have been calculated for one year only because the annual volume and pollutant loads treated by the WWTP are constant from one year to another one as previously shown. Consequently, operation, financial and total annual costs are equivalent for all years.

From IR = 75%, investment in higher WWTP capacities represents the best solution regarding pollutant loads emitted into the receiving water (Figure 25 and Figure 26). The COD loads discharged using WWTP investment strategy are 17% to 37% lower than with sewer rehabilitation. Regarding total nitrogen loads, they are 13% to 33% lower than with sewer rehabilitation. Taking the same uncertainty thresholds as previously quoted (5% and 10%), investment in WWTP appears better than sewer rehabilitation for COD and total nitrogen loads. As expected, "no investment" strategy appears as the worst solution.

Regarding operation costs, sewer rehabilitation is the cheapest solution whereas WWTP rehabilitation is the most expensive one, due to the increase of energy consumption in the pumping station and for the treatment (Figure 27).

Regarding total investment costs, repair is the cheapest solution among the investment strategies (Figure 28). WWTP investment costs are cheaper than replacement, renovation or mixed investment until IR = 150%; replacement being the worst solution. From IR = 200%, investment in WWTP becomes more expensive than replacement, renovation or mixed investment solutions due to the increase of the tank volumes. Replacement remains the most expensive solution for all IR values.

If the total investment costs (or unit construction costs UCC) are transformed into annual unit construction costs (AUCC) based on the life cycle of the rehabilitation technology as proposed in step 5, the annual total costs can be calculated as the sum of the AUCC and of the annual operation costs (Figure 29). The annual total costs represent the amount to be paid annually by the operator. For the calculation of AUCC, the interest rate has been taken to be equal to 5%. The expected lives of the selected rehabilitation technologies have been set to 80 years for the replacement, 50 years for the repair and renovation and 30 years for the WWTP based on the life cycle previously proposed.

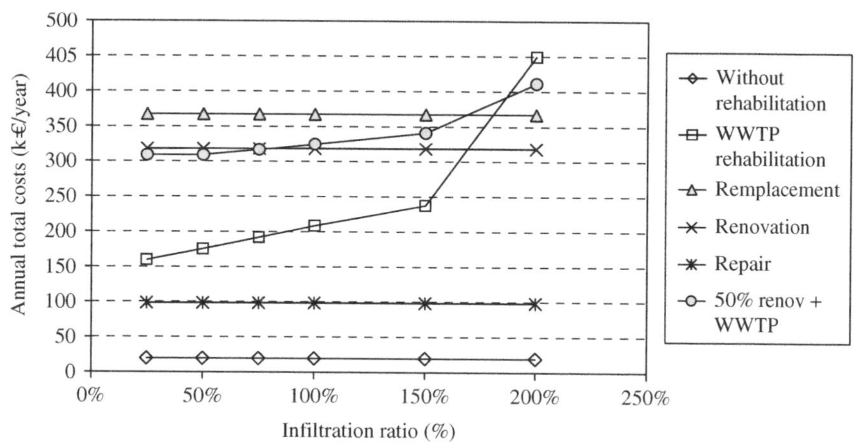

Figure 29: Annual total costs vs. the infiltration ratio and the investment strategy.

Repair remains the less expensive solution. However, the mixed investment strategy becomes more expensive than renovation from IR = 100% and replacement from IR = 200%. The WWTP investment strategy becomes the worst solution for IR = 200%.

Multi-criteria analysis of each IR scenario

Each scenario for IR = 75%, 100%, 150% and 200% has been analysed with Electre III. The final Pareto graph for IR = 150% is given in Figure 30. (Pareto graphs for the other scenarios are not shown here). Figure 30 shows that "no investment" is the worst solution regarding COD loads and nitrogen loads criteria. However, it is better than WWTP rehabilitation and the mixed investment regarding operation costs and is the cheapest solution regarding investment costs and annual total costs.

WWTP rehabilitation is the best solution regarding COD and nitrogen loads but the worst one regarding operation costs. However, it is better than sewer replacement, sewer renovation and mixed investment strategies for investment and total annual costs.

Sewer replacement, renovation and repair are intermediate solutions regarding pollutant loads and the best solutions regarding operation costs. However, sewer replacement is the most expensive solution regarding investment costs and sewer renovation requires higher investment costs than WWTP rehabilitation. Taking into account the life times of the rehabilitations, sewer renovation outranks the mixed investment for annual total. Sewer repair appears as the less expensive solution among the investment strategies. Finally, the mixed investment is an intermediate strategy for all the criteria.

To summarize, none of the selected strategies appears as the best for all criteria. Electre III is then used to compare and rank the different solutions.

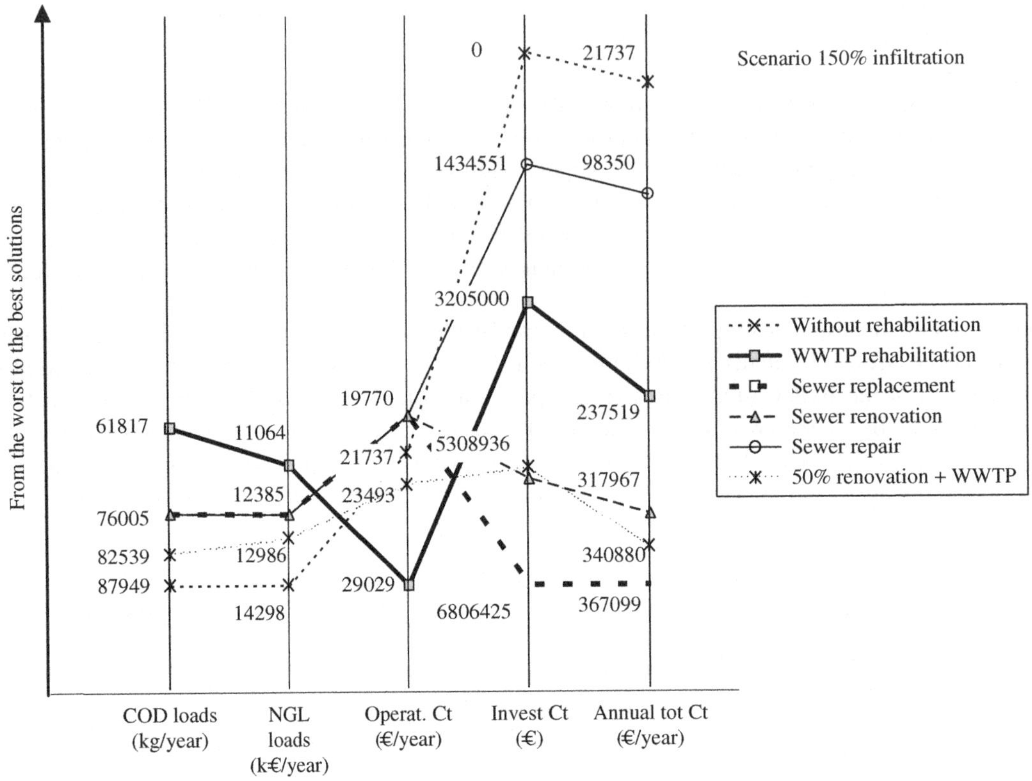

Figure 30: Comparison of investment strategies according to the different criteria (Pareto graph). (COD and NGL loads: total COD and nitrogen loads discharged into receiving water, Operat. Ct: operation costs, Invest Ct : investment costs, annual tot Ct : annual total costs).

Weights, indifference and strict preference thresholds should be defined for each criterion. Indifference threshold corresponds to the uncertainty. If, for a given criterion, the difference between two strategies is lower than the indifference threshold, the values are considered to be equivalent for this criterion. If the difference is higher than the preference threshold, one of the strategies is strongly preferred to the other one for this criterion. If the difference threshold is between indifference and preference thresholds, one strategy is slightly preferred to the other one for this given criterion.

The weights have been defined in order to equitably compare environmental and economic points of view, although the choice depends on each end user and on each case.

As previously proposed, indifference and strict preference thresholds have been respectively set to 5% and 10% for pollutant loads.

Regarding costs, it is assumed here that they are established without uncertainty (which is not true in reality). Nevertheless, the decision maker may consider, for example, that a difference of 2000 Euros per year is not significant to distinguish one strategy from another for operation costs, but that a difference of 5000 Euros per year is very discriminating. An indifference threshold has then been set to 2000 Euros per year and the preference threshold to 5000 Euros per year. Similarly, for investment costs, indifference and preference thresholds have been set respectively to 50 000 and 100 000 Euros, and to 10 000 and 20 000 Euros per year for total annual costs.

The veto threshold is an eliminatory threshold, which means that if the difference between two strategies is higher than this threshold, one of the strategies is eliminated. In the case study, none of the values is considered as eliminatory. However, the veto threshold could be defined by the decision maker, e.g. as a limit for a budget or as a threshold pollutant concentration or load which shall not be exceeded.

All above values are summarised in Table 8. An analysis of the robustness and of the sensitivity of the different thresholds and weights should be carried out in order to check if the calculated ranking is stable or not. The results however (not shown here), appear quite stable for the case study. The application of the Electre III method leads to the results given in Table 9.

For IR = 75%, the "no investment" strategy appears as the best solution for all years. It is outranked by the WWTP investment strategy but it is only slightly outranked by sewer rehabilitation regarding pollutant loads, and it is the cheapest solution regarding operation, investment and annual total costs.

Table 8: Set of electre III parameters without accounting for sewer serviceability.

Point of view	Environmental criteria		Economic criteria		
Criterion	Total COD load (kg/year)	Total nitrogen load (kg/year)	Operaion costs (€/year)	Investment costs (€)	Annual total costs (€/year)
Weight	3	3	2	2	2
Indifference threshold	5%	5%	2000	50000	10000
Preference threshold	10%	10%	5000	100000	20000
Veto threshold	–	–	–	–	–

Table 9: Final Ranking of the Investment Strategies After Simulations Using Electre III. ("Without" means "no rehabilitation" and "50% R" means the mixed investment strategy).

	IR scenario	Ranking results
75%	Year 1	Without → WWTP → Repair → Renov → Replac
	Year 9	Without → WWTP → Repair → Renov → Replac
	Year 15	Without → WWTP → Renov → Replac Repair
100%	Year 1	Repair → WWTP → Without → Renov → Replac → 50% R
	Year 9	Repair → Without → WWTP → Replac Renov
	Year 15	Without → WWTP → Renov → Replac Repair
150%	Year 1	Repair → WWTP → Without → 50% R → Replac Renov
	Year 9	Repair → Without → Replac WWTP Renov
	Year 15	Without → WWTP → Renov → Replac Repair
200%	Year 1	Repair → WWTP → Without → 50% R Renov Replac
	Year 9	Repair → Without → Replac WWTP Renov
	Year 15	Repair → WWTP → Without → Replac Renov

The WWTP investment strategy is ranked in second place because it strongly outranks the "no investment" strategy and it slightly outranks sewer rehabilitation regarding pollutant loads. It also strongly outranks the renovation and replacement strategies for investment and annual total costs and it is slightly outranked by sewer rehabilitation in terms of operation costs.

For year 15, the repair strategy is also ranked in second place because it is equivalent to the "no investment" strategy regarding pollutant loads and it is a cheap investment.

For IR = 100%, the repair strategy becomes the best solution because it strongly outranks the "no investment" strategy regarding pollutant loads for year 1 and strongly outranks the WWTP investment strategy regarding operation, investment and total annual costs. The WWTP investment strategy remains in second place because it strongly outranks the "no investment" strategy regarding pollutant loads.

For year 9, the repair strategy also becomes the best solution but the "no investment" and the renovation strategies are in second place before the WWTP investment strategy. The repair strategy slightly outranks the "no investment" strategy for pollutant loads; it is equivalent to the "no investment" strategy for operation costs and it is strongly outranked by the "no investment" strategy regarding investment and annual total costs. The "no investment" strategy is equivalent to the WWTP investment strategy for pollutant loads and it strongly outranks the WWTP investment strategy regarding investment and total annual costs.

For year 15, the "no investment" strategy remains the best solution as it is equivalent to the sewer rehabilitation regarding pollutant loads.

For IR = 150%, renovation is at the second place for year 1 as it strongly outranks the "no investment" strategy regarding pollutant loads. The WWTP investment strategy becomes equivalent to "no investment" and renovation strategies because it strongly outranks the "no investment" strategy regarding pollutant loads.

For IR = 200%, repair becomes the best solution for year 15 as it strongly outranks the "no investment" strategy regarding COD loads and slightly outranks the "no investment" strategy for NGL loads.

The replacement strategy appears as the worst solution for all years due to its high investment and annual total costs.

As a conclusion, it appears that the ranking of investment strategies varies from one year to another one depending on the rainfall height, which may increase or reduce the impacts of infiltration, i.e. the annual pollutant loads discharged into the receiving water. During the decision making process, the decision maker should be reminded of this variability and make decisions based on several years simulations.

DISCUSSION OF THE RESULTS: INFLUENCE OF HYPOTHESES

Evaluation of pollutant loads

Runoff pollutant concentrations were taken as being constant for all rainfall events. In reality, these concentrations may vary significantly from one event to another and even within one single event. Moreover, the treatment in the CSO tank (settling) and the WWTP secondary clarifier have not been explicitly modelled. If they would have been modelled, conclusions regarding pollutant loads may have been different (e.g less pollutants discharged from the CSO tank, a little bit more from the WWTP). This is one of the limits of this simplified case study.

The hypothesis that the rehabilitation of 100% of the pipes submitted to infiltration leads to the elimination of 100% of the infiltration could be further discussed. Groundwater will likely find other ways to enter the sewer system. For the same reason, the hypothesis that the rehabilitation of 50% of the pipes will lead to a 50% decrease of the total infiltration is even more questionable. There is not much literature on this topic. Usually, the water tightness is checked immediately after the sewer rehabilitation has been completed, but no long term observations are usually made. Heijs and Moroney (2003) showed that integrated repair programmes including public lines and private properties (such as house connections) lead to good results but that partial rehabilitation programmes (for example 80% of public pipes) can lead to high investments without significant final improvements regarding infiltration. In that case, pollutant loads emitted into the receiving water remain at a high level even after rehabilitation. For this semi-virtual case study, this means that the benefits expected from the rehabilitation have been very likely overestimated. Consequently, only integrated programmes (100% of rehabilitation of the all infiltrating pipes) appear pertinent regarding infiltration problems. An investment in WWTP upgrading would probably be preferred to a partial sewer rehabilitation strategy.

Evaluation of costs

The sewer rehabilitation rate applied in our case-study (20% of the total length of the drainage system) is high compared to the normal mean rate of 2% applied in French cities. By keeping the objective of 100% of rehabilitation of infiltrating pipes, investment costs could likely be reduced by precise determination of the locations where infiltration mainly occurs. The first step consists of identifying the pipes which are influenced by groundwater, i.e. pipes that are located within groundwater. Then the APUSS tracer methods, which give results with lower uncertainties than traditional flow measurement methods, could be applied to decide which pipes have to be firstly rehabilitated. This could drastically reduce investment costs of sewer rehabilitation, especially for renovation and replacement techniques which can then appear less expensive than investment in WWTP upgrading. But this would not solve the above mentioned problem of partial rehabilitation programmes.

Up to now, the choice among possible sewer rehabilitation technologies has frequently been based on total investment costs and on the applicability of the technologies. However, their durability may have a high influence on this choice if the operator is interested in long-term investments. Unfortunately, most of the techniques are too recent to allow a serious estimation of their expected life times and rough hypotheses on their durability have to be made.

CONCLUSION

A generic multi-criteria methodology is proposed to assess environmental, technical and economic impacts of infiltration (or exfiltration) on urban water systems, and to evaluate the consequences of possible investment/

rehabilitation strategies. Electre III, a method with a high adaptability to diverse contexts and objectives, has been used to compare and rank the strategies.

The Ecully semi-virtual case study illustrates the application of the methodology. It shows that the methodology is actually applicable, but also that validated and site specific information is necessary in order to carry out a valuable analysis. In most real cases, neither all impacts of infiltration and exfiltration nor the consequences of rehabilitation techniques can be properly and exhaustively evaluated because of a lack of data, of models and of knowledge on the involved phenomena. Consequently, many hypotheses will still be introduced in the analysis, with unknown bias in the application of the methodology. Results should then always be reviewed carefully.

The final decision making process should also take into account the context of the study and not forget other aspects that may not have been modelled explicitly. Additionally, the methodology, and especially the multi-criteria analysis with Electre III, facilitates the dialogue between all stakeholders and the comparison of their respective approaches and preferences. The dialogue and the decision making process itself are more important in such a collaborative work, than the final numbers and values that would have been obtained by a single person aiming to consider all points of view.

ACKNOWLEDGEMENTS

Special thanks to Christian Abegglen (EAWAG), Carlos Montero (CLABSA) and Torsten Frehmann (Emschergenossenschaft).

REFERENCES

Bertrand-Krajewski, J.-L., Barraud, S and Bardin, J.-P. (2002). Uncertainties, performance indicators and decision aid applied to stormwater facilities. *Urban Water*, 4(2), 163–179. ISSN 0273-1223.

Delleur, J.W., Reyna, S and Vanegas J. (1998). Sewerage rehabilitation. 414–450 in J. Marsalek et al. (eds.) : *Hydro-informatics Tools for Planning, Design, Operation and Rehabilitation of Sewer Systems*. NATO ASI Series 2. Environment. Vol.44. Kluwer Academic Publishers, Dordrecht, The Netherlands. ISBN 0792350979.

EN 752-2 (1996). *European Standard EN 752-2 : Drain and Sewer Systems Outside Buildings–Part 2: Performance Requirements*. AFNOR. Paris, France. 18 p.

EN 13508-1 (2003). *European Standard EN 13508-1 : Conditions of Drain and Sewer Systems Outside Buildings- Article 1: General Requirements*. November 2003. CEN. Brussels, Belgium.

Ewan Associates, Mott Mac Donald Ltd. (2001). *Development of Enhanced Serviceability Indicators for Sewerage Assets*. Reports ref.01-G003/02/Ofwat/od/dg/002/63452. OFWAT. Birmingham, UK. Available at OFWAT website: http://www.ofwat.gov.uk.

Frehmann, T. (2004). *Cost-benefit Analysis*. Report D11.2. APUSS project. Available on the APUSS website at http://www.insa-lyon.fr/Laboratoires/URGC-HU/apuss

Freni, G., Hafskjold, L.S and Montero, C. (2005). *Rehabilitation Technologies Costs Tool*. In unpublished CARE-S project Report D12 "Rehabilition Technologies Survey". SINTEF. Trondheim, Norway.

Gujer, W and Larsen, T.A. (1995). The implementation of biokinetics and conservation principles in ASIM. *Water Science and Technology*, 31(2), 257–266. Available at the ASIM website: http://www.asim.eawag.ch

Heijs, J and Moroney, P. (2003). *Is Sewer Rehabilitation Really Effective?* Unpublished draft report from Water Services, North Shore City Council, Auckland, New Zealand.

Mourad, M., Bertrand-Krajewski, J.-L and Chebbo, G. (2004). Calibration and validation of multiple regression models for stormwater quality prediction: effect of number and characteristics of available data. 67–74 in *Proceedings of the 4th International Conference on Sewer Processes and Networks (4th SPN)*, 22–24 Novemeber 2004. Funchal, Madeira.

Musso, E. (1997). *Evaluation des Rejets de Temps de Pluie de Grands Basins Versants Urbains: Contribution Méthodologique à la Comparaison des Outils et à L'estimation des Incertitudes*. PhD thesis, January 1997. Dept of Civil Engineering, INSA of Lyon, Villeurbanne, France. 363 p.

Neitzke, S. (2002). *Evaluation Expérimentale de L'infiltration et de L'exfiltration dans les Réseaux D'assainissement. Etude Bibliographique et Identification des Sites Expérimentaux*. Master thesis. July 2002. Dept of Civil Engineering, INSA of Lyon, Villeurbanne, France. 105 p. + appendix (in French).

O'Brien and Gere Companies. (2001). *Trenchless Technologies: An Economical Alternative for Sewer Rehabilitation*. Available at O'Brien and Gere Companies web site : *http://www.obg.com/*

Orditz, D. (2004). *Contribution à L'étude de la Durabilité des Réhabilitations de Réseaux D'assainissement*. Report RERAU. CSTB. Paris, France. 52 p.

RERAU. (2004). *Projet National Réhabilitation des Réseaux d'Assainissement Urbains : Restructuration des Collecteurs Visitables*. Guide Technique, Volume 2. Tec et Doc, Paris, France. 250 p.

Roy, B. (1978). Electre III : un algorithme de classements fondé sur une représentation floue des préférences en présence de critères multiples. *Cahier du CERO*, 20(1), 3–24.

Roy, B. (1996). *Multicriteria Methodology for Decision Aiding*. Kluwer Academic Publishers, Dordrecht, The Netherlands. 292 p. ISBN 978 07923 4166 6.

Roy, B and Bertier, P. (1971). *La Méthode ELECTRE II*. Working Paper 142. SEMA-Metra International. Paris, France. 32 p. (Also presented at 6th Int. Conf. of Operational Research, 21–25 April 1972. Dubmin, Ireland.

Villanueva, A., Montero, C., Raclavsky, J. *et al.* (2004). *Rehabilitation technologies database*. In unpublished CARE-S project Report D12 "Rehabilition Technologies Survey". SINTEF. Trondheim, Norway.

Vincke, P. (1992). *Muticriteria Decision-Aid*. John Wiley & Sons, Chichester, England. 154 p. ISBN 978 0471 93184 3.

ACKNOWLEDGEMENTS

REFERENCES

Chapter 12

Economic and operational aspects

Adriana Cardoso and Torsten Frehmann

INTRODUCTION

Municipalities and/or sewer operators have to decide investment strategies to rehabilitate and upgrade both the efficiency and quality of their sewer systems. Frequently, such decisions are based on limited information and do not consider simultaneously the sewer pipes, the wastewater treatment plant and the receiving environment, especially groundwater. In order to provide some basic working elements to help operators in making such decisions, Work Area 4 (WA4) within the APUSS project included three topics:

(i) the definition of a set of infiltration/exfiltration specific performance indicators, PIs,
(ii) the proposition of a generic multi-criteria methodology to compare investment strategies which is elaborated separately in Chapter 10 and
(iii) the analysis of cost structures of sewer systems and costings for the new tracer methods proposed in the APUSS methodology (as outlined in Chapter 2).

Detailed reports and publications are available on the APUSS website (www.insa-lyon.fr/Laboratoires/URGC-HU/apuss).

I/E PERFORMANCE INDICATORS

Objectives

A set of specific performance indicators (PIs) has been developed to assess the impact of infiltration/exfiltration (I/E) on sewer systems and applied to three case studies. PIs are considered to be a means of aggregating information on system characteristics and experimental data gathered from monitoring or modelling, and translated into performance values. PIs have been calculated from fundamental sewer system characteristics and I/E rates measured with the tracer methods developed in Work Area 1 (see Chapter 2). Once calculated, PIs can be classified in relation to "good" or "bad" performance, which allows a standardised and objective comparison of the performance of different sewer systems and constitutes a means to technically support the establishment of priorities for rehabilitation and/or construction investments, taking into account I/E impacts (Cardoso *et al.*, 2002; 2005). Links have also been established between the AquaBase software platform (Chapter 10) and the PI calculation software.

 Assessing infiltration and exfiltration on the Performance of Urban Sewer Systems (APUSS). Edited by Bryan Ellis and Jean-Luc Bertrand-Krajewski. ISBN: 9781843391494. Published by IWA Publishing, London, UK.

Methodology

The principal requirements considered for the performance assessment methodology are:

(i) flexibility, in order to accommodate with ease the different sensitivities, interpretations or objectives of the analysis;
(ii) a certain degree of standardisation in order to facilitate a multi-disciplinary approach, where the various aspects to be considered may be brought down to the same quantified basis; and
(iii) a quantitative, numerical base.

The envisaged tool should be translatable computationally in order to afford intensive use, either from within or as post-processor to, the current modelling techniques or monitoring data (Cardoso *et al.*, 1999).

The methodology consists of the selection and development of three components for each aspect of performance analysed (Cardoso and Coelho, 2004) and include:

(i) the numerical value of a sewer network property or state variable, which is expressive of the particular I/E aspect being scrutinized, i.e. the Performance Indicator (PI) definition;
(ii) a classification of the PI values, scoring them in relation to good or bad performance;
(iii) an operator, which allows the performance values at element level to be aggregated across the system, or parts of it, to obtain an overall system performance figure.

The performance assessment scale depends on the scale of available data. In cases where data are available at element level, there are

(i) a global value that is achieved through a particular operator in order to represent the performance assessment of the network, e.g. a simple or weighted average, and conversely
(ii) a population of elementary values, which leads itself to a basic statistical treatment, giving information on the PI value dispersion on the system through, for example, the use of percentile bands. In cases where data are only available at catchment level, only a global PI value for that catchment is achieved.

Performance indicators for infiltration and exfiltration

Seven PIs for infiltration were established for the assessment of impacts of infiltration in sewer systems (Cardoso and Coelho, 2004) and are presented in Table 1. Five PIs for exfiltration were established for performance assessment of impacts of exfiltration in sewer systems (Cardoso and Coelho, 2004) and are presented in Table 2.

These PI were applied to four experimental catchments that have been studied in the APUSS project: two in Italy, Rome (Infernetto and Torraccia) and two in France, near Lyon (Yzeron and Ecully); further details on the catchments are given in Chapters 7 and 8. Torraccia is about 85 ha, with 55 ha being residential in character and 30 ha being rural. The sewer network consists of an egg-shaped combined system (1×1.2 m, 1.5×1.8 m and 1.5×2.1 m) laid about 4 to 9 m below ground level. It is 14 years old and the pipe material consists of concrete. Infernetto is about 5.5 km2 and it is characterized by residential houses with private gardens. The groundwater level changes between 6 to 7 m under the ground in the north-east part and 1.5 m under the ground in the south-west. The system is separated and sewer cross sections are circular (0.3 m, 0.6 m, 0.8 m and 1 m) laid about 2 to 6 m below ground level. It is about 19 years old and the pipe material consists of concrete (size 1.0 m), ceramic (sizes 0.6 m and 0.8 m) and PVC (size 0.3 m). Infernetto and Torraccia have been chosen because they have very different geological, hydrogeological and sewer characteristics and can be considered to be representative of almost all urban areas within Rome. Infiltration and exfiltration rates have been assessed by the application of the APUSS tracer methods described in Chapter 2.

The Yzeron catchment in Lyon, France has an area of 150 km2 and has been subject to increasing urbanization and imperviousness over the last 20 years. Its drainage system is combined, with a total length of 294 km and sewer sizes ranging from 200 mm circular pipes to 2 m egg-shaped pipes. In 1992, a diagnostic study revealed high infiltration rates linked to a degraded state of the sewer system. The Ecully catchment is smaller, with an area of 245 ha and is mostly residential with an imperviousness evaluated as 46% (de Bénédittis, 2004). Its drainage system is combined, with a total length of 15 km, a mean slope of 0.027 m/m, and sewer sizes ranging from 200 mm circular pipes to 1.8 m egg-shaped pipes. Infiltration and exfiltration rates have been measured using two of the developed APUSS methods.

Results of performance indicators for infiltration and exfiltration

Some results obtained for the application of infiltration and exfiltration PIs to the case studies described, are presented in Figures 1 to 4. In these figures the performance of the different systems can be compared on the same graph (Cardoso *et al.*, 2004; 2005). In the case of the Yzeron catchment, infiltration PIs were calculated for 8 sub-catchments and also for the entire catchment as a simple mean. There was no data available on exfiltration. Flow data used for the calculation was the average daily infiltration or exfiltration flow and the average daily dry weather flow as obtained from the APUSS experimental measuring campaigns.

Table 1: Infiltration performance indicators.

Definition	Unit	Designation	Concept/Explanation
$\frac{Q_{inf}}{Q_{full}}$	(%)	PI_{I1} Infiltration full capacity utilization	*Proportion of the sewer full section flow capacity used by the infiltration flow. This indicator supplies information on the hydraulic performance of the analysed sewer(s), but gives no indication about the net infiltration flow. White et al. (1997) refer that, for design purposes, in UK 10% of sewer capacity should be allocated to infiltration in separate domestic systems according to the Water Authorities Association (WRc, 2006).*
$\frac{Q_{inf}}{Q_{avdwf}}$	(%)	PI_{I2} Infiltration proportion of dry weather flow	*Infiltration flow expressed as a percentage of the daily mean dry weather flow. A drawback of this indicator is its dependency on the values of the dry weather flow. This indicator gives an estimate of the weight of infiltration flow on treatment costs. It may be expressed either as percentage of volume or cost. Stevens (1998) presents a study in New York where about 50% of the total flow reaching WWTP was infiltration flow; Belhadj et al. (1995) report values of 42% of dry weather flow.*
$\frac{Q_{inf}}{Q_{avdwf}-Q_{inf}}$	(%)	PI_{I3} Infiltration proportion of domestic waste-water	*Infiltration flow expressed as a percentage of the daily mean domestic wastewater. A drawback of this indicator is its dependency on the values of the dry weather flow. The German guidelines ATV-A118E (ATV, 1999) considers an infiltration/ inflow flow equal to 100% of domestic flow, for design purposes of separate domestic systems.*
$\frac{Q_{inf}}{\text{sewer longitudinal wall area}}$	(m^3/day/ (cm.km))	PI_{I4} Infiltration flow per unit sewer wall area	*Infiltration flow per unit area of sewer wall. This indicator is related to the sewer wall area that is potentially subject to infiltration. It can be the wall area below groundwater level. In this case it was considered the total sewer wall area. Heaney et al. (2001) refers to that in USA this value can vary from 0.05 to 1.39 m^3/day/(cm.km). Tchobanoglous et al. (2003) refer that these values can range from 0.1 to 10 m^3/day/(cm.km).*
$\frac{Q_{inf}}{\text{sewer length}}$	(m^3/s/km)	PI_{I5} Infiltration flow per unit sewer length	*Mean infiltration flow per unit length of sewer. This indicator is related to the sewer length that is potentially subject to infiltration. This indicator is relevant in systems where infiltration takes place predominantly along the sewers. NRC (2003) refers 150 m^3/day/km as being excessive.*
$\frac{Q_{inf}}{\text{n° manholes}}$	(m^3/s)	PI_{I6} Infiltration flow per manhole	*Mean infiltration flow per manhole unit. Manholes are an important source of infiltration. This indicator gives an idea of the influence of the number of manholes on total infiltration.*
$\frac{Q_{inf}}{\text{n° service connections}}$	(m^3/s)	PI_{I7} Infiltration flow per service connection	*Mean infiltration flow per service connection. Service connections are an important source of infiltration. This indicator gives an idea of the influence of the number of service connections on total infiltration. Ellis (2001)refers that in UK house connections contribution represents 30% to 40% infiltration in sewers.*

From the results it is possible to conclude that, for the global Ecully catchment, the contribution of infiltration is very high, representing about 40% of the dry weather flow that reaches the WWTP (PI_{I2}). This fact has no impact with regard to the hydraulic capacity of the system as PI_{I1} presents a value of 0.3%. This means that the main impacts are expressed in terms of operational costs and pollutants that will be transferred to the WWTP or discharged into the receiving water.

For the global Yzeron catchment, the contribution of infiltration represents 55% of the dry weather flow that will reach the WWTP (PI_{I2}). There are impacts with regard to the hydraulic capacity of the system as PI_{I1} represents a value of 13%, but this is not very significant. This means that there are large impacts in terms of operational costs, as the wastewater volume to be transported and treated in the system due to infiltration flow, is twice the dry weather flow. This reflects the fact that the Yzeron sewers are in a bad structural state as detected some years ago by the Greater Lyon Sewer Department, and rehabilitation projects are currently under preparation.

Table 2: Exfiltration performance indicators.

Definition	Unit	Designation	Concept/Explanation
$\frac{Q_{exf}}{Q_{avdwf}}$	(%)	PI_{E1} Exfiltration proportion of dry weather flow	*Exfiltration flow expressed as a percentage of the daily mean dry weather flow. A drawback of this indicator is its dependency on the values of the dry weather flow. Amick and Burgess (2000) present values of experimental studies with values varying between 11.9% and 49%. Ellis et al. (2002) refer that probably this value is not greater than 10%.*
$\frac{Q_{exf}}{\text{sewer length}}$	(m^3/s/km)	PI_{E2} Exfiltration flow per unit sewer length	*Mean exfiltration flow per unit length of sewer. This indicator will give relevant results in systems where exfiltration takes place predominantly along the sewers. Literature values are between 0.000027 and 0.001 m3/day/m (Amick and Burgess, 2000).*
$\frac{Q_{exf}}{\text{sewer longitudinal wall area}}$	(m^3/day/ (cm.km))	PI_{E3} Exfiltration flow per unit sewer wet wall area	*Mean exfiltration flow per unit area of sewer wall. This indicator is related with the sewer wall area that is potentially subject to exfiltration. Literature values are between 0.08 and 1.20 (Amick and Burgess, 2000).*
$\frac{Q_{exf}}{\text{n° sewer length}}$	(m^3/s)	PI_{E4} Exfiltration flow per manhole	*Mean exfiltration flow per manhole unit. Manholes are an important source of exfiltration. This indicator gives an idea of the influence of the number of manholes on total exfiltration.*
$\frac{Q_{exf}}{\text{n° service connections}}$	(m^3/s)	PI_{E5} Exfiltration flow per service connection	*Mean exfiltration flow per service connection. Service connections are an important source of exfiltration. This indicator gives an idea of the influence of the number of service connections on total exfiltration*

For the global Torraccia catchment in Rome, the contribution of infiltration represents about 15% of the dry weather flow that will reach the WWTP (PI_{I2}). There are no impacts with regard to the hydraulic capacity of the system as PI_{I1} represents a value of only 0.06%. This means that infiltration has no relevant impacts in the system performance. For the Infernetto catchment, the contribution of infiltration represents about 48% of the dry weather flow that will reach the WWTP (PI_{I2}). This means that there are impacts in terms of operational costs.

Comparing the performance of the different catchments/sub-catchments, it is clear that the Yzeron catchment shows significant problems as a consequence of infiltration. In general, PIs for Yzeron are 22 times higher than for the Ecully and Torraccia cathments. In spite of the PI_{I2} indicator being 88%, the value for PI_{I1} is only 8%. This means that this Yzeron catchment presents severe infiltration problems and that impacts are higher in terms of costs although it seems that there are little or no associated hydraulic problems. On the contrary, Yzeron presents values for the PI_{I1} indicator of 45% and the PI_{I2} indicator is 59%. In this case, besides the impact on costs, there are also problems with hydraulic

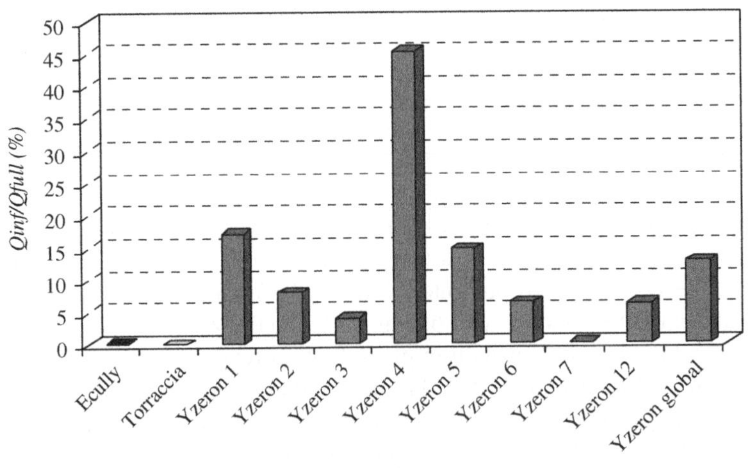

Figure 1: PI_{I1} Assessment: infiltration full capacity utilization.

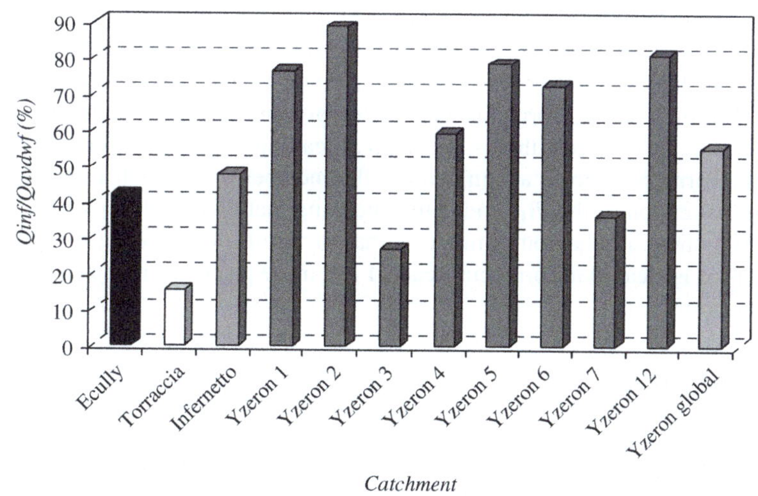

Figure 2: PI_{I2} Assessment: infiltration proportion of dry weather flow.

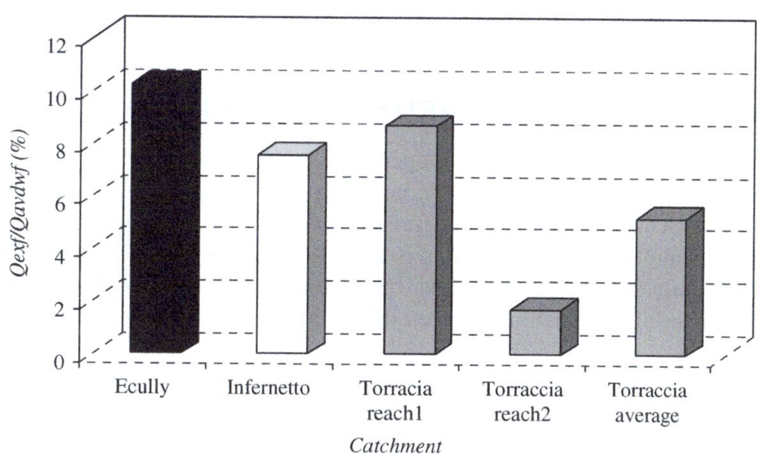

Figure 3: PI_{E1} Assessment: exfiltration proportion of domestic wastewater.

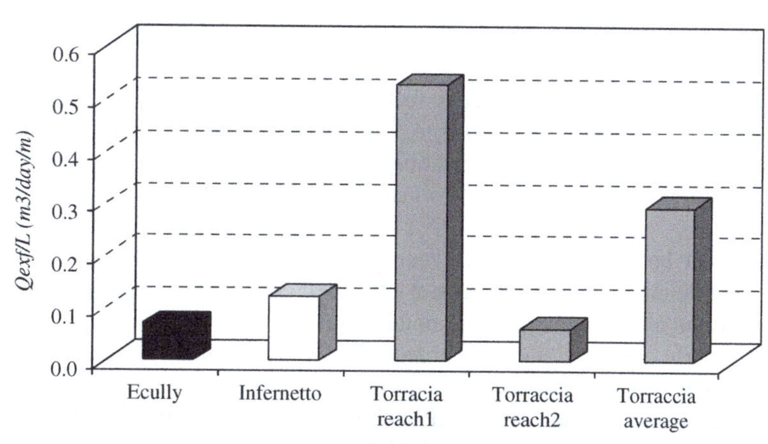

Figure 4: PI_{E2} Assessment: exfiltration flow per unit sewer length.

capacity. In general, the Yzeron catchment presents significantly higher values for the PI than Ecully and Torraccia. Thus Yzeron seems to have bigger problems related to infiltration occurrence than Ecully. Torraccia presents the lowest PI values and seems to have no problems related to infiltration while Infernetto seems to show some impacts of infiltration, for the assessed PI.

In terms of exfiltration, Ecully presents lower PI values than the other sewer systems, with the exception of the PI_{E1} indicator which represents 10% of dry weather flow lost; the highest value. This means that the unit rate of exfiltration along the sewers is not very high. However, if the sewer is long or the dry weather flow is low, this percentage increases. The average PI value for the Torraccia catchment presents higher values than that for Ecully, with the exception of the PI_{E1} indicator. In Torraccia this is about 5%, meaning that this percentage of dry weather flow is lost; half of the Ecully value. In Torraccia there is a significant difference between Reach 1 and Reach 2. PIs within Reach 1 being the highest, with the exception of the PI_{E1} indicator, meaning that this reach shows clear problems of exfiltration. At Infernetto, PI_{E1} and PI_{E2} show a behaviour similar to that of Ecully. Comparing the performance of the different catchments/sub-catchments, it is clear that Torraccia Reach 1 has more problems regarding exfiltration impacts and thus should have rehabilitation priority.

Conclusions on using PIs

The above methodology aims to support sewer system rehabilitation by using infiltration and exfiltration performance indicators (PIs), as a means of aggregating information on system characteristics and experimental data from monitoring or modelling, and translating these into performance values. The application of PIs for infiltration and exfiltration depends on available data, which means that on the one hand the approach can be applied at pipe or larger scales; on the other hand the quality and uncertainty of the PI results depend on the quality and uncertainty of data used. Any assessment using PIs must consider a set of PI and not just one, in order to give a global view having significant information and taking into account the different points of view that affect sewer system performance.

COST STRUCTURE AND COSTS OF APUSS METHODS

A questionnaire on sewer cost structures was developed and sent to selected APUSS associated end-users. The questionnaire was divided in two parts. Part 1 was concerned with general aspects of infiltration and exfiltration and part 2 was about technical and financial information concerning the sewer systems and their environment. An infiltration problem is in most of the cities already known and often is also normally accepted; only in a few cases does the operator take measures to reduce this problem. But, as expected, the technical data of the different sewer systems were strewn across a wide range. The transfer of information between different cities was not possible because of the large diversity of approaches and organisations.

However, a new cost-structure for sewer systems has been developed, dividing all arising costs into six groups. These are:

- costs for material,
- wages and salaries,
- miscellaneous operational costs,
- disposal of waste,
- depreciation.
- calculated interest.

This classification into six groups had been successfully tested in different benchmarking projects for wastewater treatment plants in Germany (e.g. Schulz *et al.*, 2002). All six groups can be divided into sub-groups if necessary. For example, the group "Materials" is divided into eleven sub-groups. In Table 3, the new developed (reduced) commercial data entry form for a sewer system is displayed, which constitutes the basis for both cost-benefit analysis and benchmark studies.

The application of the cost benefit analysis has been illustrated by means of four examples (rehabilitation vs. replacement in different installation depths). The necessary detailed cost data for the replacement and rehabilitation of a sewer pipe came from different investment projects undertaken by Emschergenossenschaft, Germany and from the literature. For comparison, yearly operation and investment costs were calculated, which revealed that the amortisation time of the planned measures was one of the main points in decision making.

For all measurement methods developed in the APUSS project, detailed ranges of costs (personal cost, cost for measuring equipment, and cost for consumption) have been evaluated by the APUSS partners (Table 4). They must be considered as costs for a "prototype" measurement which could be reduced when measurements are applied on a routine basis. With regard to the planning phase, more experience would speed up the preparatory analysis. With regard to the practical execution of the experiment, more experience leads to improved logistics and organisational aspects. This would reduce the overall personnel costs. Looking only at economical aspects, "classic" flow measurements appear more beneficial. However, the new tracer measurement methods provide values of infiltration and exfiltration having a better evaluation of uncertainty. The results obtained using these methods are thus regarded as being more reliable than those obtained with "classic" methods.

Table 3: Reduced commercial data entry form for a sewer system.

No.	**Indices** (Total costs for the year:______in €)	Unit	**Overall sewer system**	**Private sewerage system**			**Sewerage collection**					**Stormwater treatment**				**Miscella-neous**
			Total for cost group	Gully pots	Fat/light liquid seperator	in house wwtp	gravity sewer	pressure main	inverted siphon	pumping station	manhole	CSO	storage tank	infiltration facility	storage sewer	
54	**Materials**	€														
540	Power, water, natural gas, TP gas, transmitted heat	€														
541	other fuels (for motor-vehicles)	€														
544	Auxiliary and operational utilities	€														
547	Cost for purchases (service and material)	€														
5471	***of which*** *maintenance services*	€														
5472	***of which*** maintenance materials	€														
549	Miscellaneous	€														
55	**Wages and salaries**	€														
551	***of which*** maintenance and servicing	€														
552	***of which*** cleaning and inspection	€														
59	**Miscellaneous operational cost**	€														
90	**Disposal of waste**	€														
57	**Depreciation (standadized)**	€														
58	**Calculated interest**	€														
Total amount																

Table 4: Range of costs for experiment campaigns (estimation of each partner).

Methods developed within the APUSS project	**QUEST-C (exfiltration)**	**Pollutant time series method (infiltration)**	**Stable isotopes method (infiltration)**	**HC$^{*)}$ (in- and exfiltration)**
Measuring equipment	~ € 10 700 to € 13 200	~ € 24 500 to € 28 300	~ € 17 200	~ € 300$^{+)}$ to € 2 200
Including	Flow measuring unit Mixer	Flow measuring unit s::can probe	Flow measuring unit equipment for groundwater sampling	CCTV camera sealing balloons (only rental basis)
Personnel cost	~ € 1.500 to 2.000	~ € 1.500 to 5.000	~ € 3.000	~ € 400 to 800
Calculated hours	~ 70 h–80 h	~ 60 h–110 h	~ 120 h	~ 80 h–160 h
Calculated price per hour	€ 22–€ 27	€ 25–€ 45	€ 25	€ 5
Number of experiments/samples	1/4–10	1/20	1/44	2–3/-
Cost for consumption	~ € 300 to € 500	~ € 500 to € 1.000	~ € 2.700	~ € 100 to 200
Total cost	~ € 13.200 to € 15.000	~ € 27.500 to € 30.400	~ € 24.900	~ € 1.200 to € 2.800
Cost per experiment	~ € 1.800 to € 2.200	~ € 2.000 to € 5.500	~ € 5.700	~ € 600 to 900

Cost information gathered from APUSS partners
$^{*)}$ HC: Measurement of house connection for in- and exfiltration; $^{+)}$ Price for the measuring equipment is only rental basis calculated.

REFERENCES

Amick, R.S. and Burgess, E.H. (2000) *Exfiltration in Sewer Systems.* Report EPA/600/R-01/034. US Environmental Protection Agency, National Risk Management Research Laboratory. Cincinnati, Ohio, US.

ATV. (1999) *Standard A118 E: Hydraulic Dimensioning and Verification of Drainage Systems.* DWA. Hennef, Germany. 38p. ISBN 3-924063-49-4.

Belhadj, N., Joannis, C. and Raimbault, G. (1995) Modelling of rainfall induced infiltration into separate sewerage. *Water Science & Tech.*, **32**(1), 161–168.

Cardoso, M.A., Almeida, M.C. and Coelho, S.T. (2002) Evaluation of the infiltrartion impact on the performance of urban sewer systems. *Proc. 9th National Congress; Water, Wastewater and Solid Waste.* 16–19 September 2002. Braga, Portugal. 17p. (In Portuguese).

Cardoso, M.A., Coelho, S.T., Matos, J.S. and Matos, R.M. (1999) A new approach for diagnosis and rehabilitation of sewerage systems through the development of performance indicators. In Joliffe, I and Ball, J.E (eds): *Proc. 8th. Int. Conf. Urban Storm Drainage.* 30 August – 3 September 1999. Institution of Engineers Australia. Sydney, Australia. 610–617.

Cardoso, M.A. and Coelho, S.T. (2004) *Infiltration and Exfiltration Performance Indicators; Sewer Systems performance assessment Methodology, Formulation, Computational Application and Case Studies.* APUSS Deliverable 9.2. December 2004. Lisbon, Portugal.

Cardoso, M.A., Prigiobbe, V., Giulianelli, M., Baer, E., de Benedittis, J. and Coelho, S.T. (2005) Assesing the impact of infiltration and exfiltration in sewer systems using performance indicators: Case studies of the APUSS project. In Mikkelsen, P.S (ed.): *Proc. 10th Int. Conf. Urban Drainage* (*ICUD*), 21–26 August 2005. Copenhagen, Denmark. IWA Publishing, London. UK. ISBN 978 1843395 744.

de Benedittis, J. (2004) *Mesurage de l'infiltration et de l'Exfiltration Dans les Reseaux d'Assainissement.* Unpublished PhD Thesis. 16 April 2004. 331p. Insa de Lyon, Villeurbanne, France.

Ellis, J.B. (2001) Sewer infiltration/exfiltration and interactions with sewer flows and groundwater quality. *Proc. 2nd Int. Conf. Interactions Between Sewers, Treatment Plants and Receiving Waters in Urban Areas, INTERURBA II.* 19–22 February 2001. Lisbon, Portugal. 311–319.

Ellis, J.B., Revitt, D.M., Lister, P. and Willgress, C. (2002) The effects of sediments on sewer exfiltration. In Bertrand-Krajewski, J.-L. and Chebbo, G (eds): *Proc. 3rd. Int.Conf. Sewer Processes and Networks.* 15–17 April 2002. Paris, France. 147–154. IWA Publishing, London, UK. ISBN 978 1843394365.

Heaney, J.P., Pitt, R. and Field, R. (2001) *Innovative Urban Wet Weather Flow Management Systems.* Report EPA/600-R/029. Environmental Protection Agency, National Risk Management Research Laboratory. Cincinatti, Ohio. US.

NRC. (2003) *Infiltration/Inflow Control/Reduction for Wastewater Collection Systems: A Best Practice.* National Guide to Sustainable Municipal Infrastructure. Issue No.1.0. Federation of Canadian Municipalities. National Research Council of Canada. Canada.

Schultz, A., Egerland, B., Reicherter, E. and Obenaus, F. (2002) Elimination costs referring to different wastewater compounds. *Proc. 3rd IWA World Water Congress*. 7–12 April 2002. Melbourne, Australia. IWA Publishing, London, UK. ISBN 978 184339440.

Tchobanoglous, G., Burton, F.L. and Stensel, H.D. (2003) *Wastewater Engineering, Treatment and Reuse*. (Metcalf & Eddy) McGraw-Hill Science Engineering, 4th Edition. New York, US. ISBN 978 0070418783.

WRc. (2006) *Sewers for Adoption*. 6th Edition. Water Research Centre (WRc Publications). Swindon, Wilts.UK. ISBN 1898920573.

Chapter 13

APUSS: Perspectives and conclusions

J. Bryan Ellis, Jean-Luc Bertrand-Krajewski, Jörg Rieckermann and Oliver Kracht

THE APUSS CHALLENGE

The integrity of urban sewer systems in terms of infiltration gains and exfiltration losses represents a very important element in sustainable urban water resource management as well as having significant economic consequences for sewer system operators and wastewater utilities. This integrity is underwritten by the European Union Standard EN 752 which requires that sewer systems should be "watertight". The EU Waste Water Directive (91/271) also demands that large sewerage systems should comply with "leakage" requirements as laid down in Annex 1A. In addition, Article 4, Doc. 9085/99 DGI of the Water Framework Directive (WFD) emphasises that anthropogenic impacts on groundwater such as exfiltration, should be minimised or avoided wherever possible. Annex II, paragraphs 2.1 and 2.2 of the WFD further demands an identification of point and diffuse sources of surface and groundwater pollution together with an estimation of their environmental risks and mitigating measures. Thus there is a considerable groundswell of legislation which supports the principle of urban sewer integrity.

However, it is clear that this objective will not be readily achieved in practice given the condition, age, funding, socio-political and other constraints on the on-going maintenance and rehabilitation of urban sewer systems. It is estimated by French Water Agencies for example, that two Euros need to be spent for sewer upgrading and rehabilitation for every single Euro spent on sewage treatment, simply in order to maintain current levels of operational capability. Figure 1a illustrates major infiltration occurring into an ageing combined sewer pipe mainly associated with the corrosion and removal of the mortar infill between the brick lining, and is undoubtedly related to an elevated groundwater level which has provided a high hydraulic gradient which is "spouting" the clean groundwater into the sewered flow. Before the 1960s, many urban sewer pipes were fitted together with materials that tended to degrade and erode rapidly and leave the joints open. Further, the pipe material itself was often of inferior quality, leading to rapid degradation and subsequent leakage. It was not until the mid- to late 1980s that the quality of sewer joints and the range and type of materials known today was introduced. Figure 1b shows typical deformation of a small diameter ceramic pipe where exfiltration is occurring along the invert line, with the fracture in the soffit being a potential egress point for stormwater infiltration-inflow (I/I) as well as groundwater under elevated conditions. This type of "crushing" defect is frequently found in shallow trenched, pre-1934 small diameter pipes which are aligned parallel to the highway and is principally the result of surface over-loading as well as defective manufacture and/or jointing (Davies *et al.*, 2001). Irrespective of this, precautions are now taken to minimise leakage with both pipe materials and joints being required to meet specified quality standards, as must the sub-soil and trench conditions into which the sewer is laid. It is also common to test the sewer tightness, pressurising the system after installation. However, it is undoubtedly the case that infiltration has received considerably more attention than exfiltration, as the former impacts upon the operational

 Assessing infiltration and exfiltration on the Performance of Urban Sewer Systems (APUSS). Edited by Bryan Ellis and Jean-Luc Bertrand-Krajewski. ISBN: 9781843391494. Published by IWA Publishing, London, UK.

capacity and efficiency of the wastewater system (and consequent treatment capability), in a much more visible manner and with a more rapid response time.

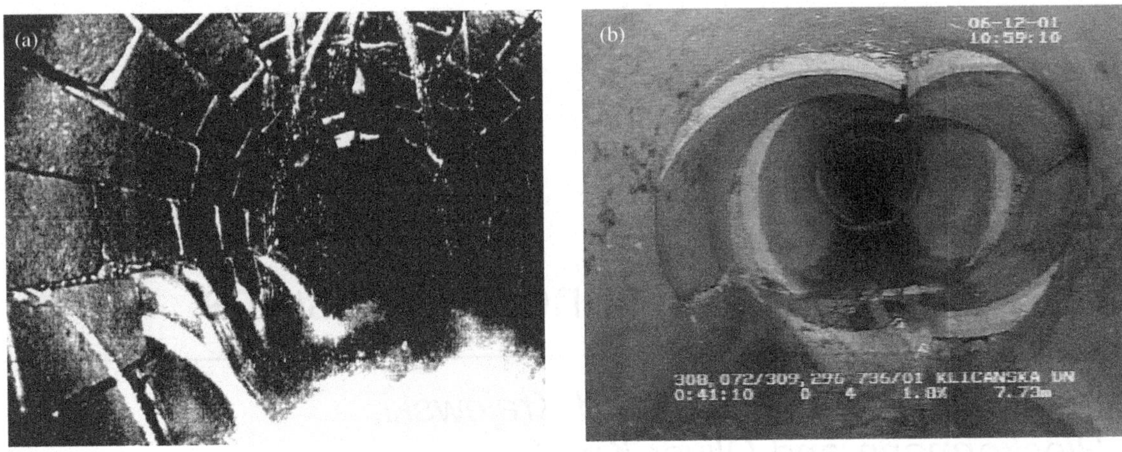

Figure 1: Sewer infiltration and exfiltration.

The incidence of poor sewer conditions is not that uncommon in most urban situations and for the 64,374 km sewer network of Thames Water for example, up to 25% is estimated by the Sewerage Rehabilitation Manual (WRC, 2001; SRM4) algorithm to fall into the worst Grades 4 and 5; the latter grading referring to a sewer pipe which is structurally insecure. Assuming that one-fifth of the sewer perimeter is obscured below the average "waterline" in the sewer, and with an assumed 10% of 2.5 m spaced joints being defective (with 10 mm joint openings) in a 1 km sewer length, a total exfiltration loss of between 26–260 $m^3\ km^{-1}\ year^{-1}$ might be expected on a worst-case scenario based on the APUSS sewer rig evidence (see Chapter 5). If this is scaled up to the 25% Grade 4 and 5 sewers of the total Thames Water sewer network, this would yield losses due to exfiltration of between 0.42M and 4.2M $m^3\ year^{-1}$. This APUSS estimate is significantly less than previous estimates for the Thames region of 109.5M m^3 based on a 3% exfiltration rate (Read, 1996), but is nevertheless still a concern for areas with abstractive groundwater beneath, or immediately adjacent to, cities in the light of the requirements of the WFD. Such conclusions of the APUSS project serve to confirm the findings of related EU 5th Framework projects such as AISUWRS (Wolf *et al.*, 2007a).

There may be a danger that experimental sewer rig studies tend to extrapolate results from worst-case scenarios without being clear that this is what is being done. This then tends to give a bias which may encourage critics who would wish there to be a problem. The working basis of such studies is often taken from a position that all defects in the sewer pipe are assumed to have occurred within the time frame of the experiment. Nevertheless, the APUSS sewer rig studies would indicate that exfiltration rates from leakages in sewer networks in general are less significant than previously thought (Vollertsen and Hvitved-Jacobsen., 2003), primarily due to the sealing effects of a clogging colmation layer and associated underlying transition zone as described in Chapter 6. However, as demonstrated by Ellis *et al.* (2009), this sealing is subject to episodic and sudden disturbances which are not restricted to the controls of sewer fill depths or hydraulic gradient, but which is more relatable to dynamic bed turbulence and colmation matrix properties as well as external groundwater gradients. Thus local disruption of the equilibrium sealing is to be expected in all urban sewers which is not necessarily related to flow depths in the sewer.

A standard US engineering textbook states that "*when designing for presently unsewered areas or relief of overtaxed existing sewers, allowance must be made for unavoidable infilt*ration "...and exfiltration (Metcalf and Eddy, 1991). A standard 10% allowance for infiltration (and infiltration-inflow, I/I) has long been included in the calculation of dry weather flow (DWF) by urban drainage engineers, although it is widely recognised that this design allowance may be as high as 75% or more, particularly where the pipe system lies below the watertable. Larger diameter sewer pipes do not provide the answer as they will have a larger circumference and joint section, potentially allowing more infiltration to enter the system. Standard pressure testing of sewer lines also includes allowance for some exfiltration. The American Society of Testing and Materials (ASTM, 2002) in their standard procedures for sewer leakage, accept a tolerance limit of 200 gallons per hour per inch diameter of sewer length (equivalent to 0.2 l/s loss for a 250 mm diameter pipe). Rather than accepting built-in design and operational allowances, a much sounder investment strategy would be to ensure sufficiently high standards of pipe manufacture, installation and testing as well as improved maintenance and repair schedules for existing drainage infrastructure.

APUSS OBJECTIVES

One principal problem associated with sewer infiltration and exfiltration is the identification and quantification of leakage for both specified sewer reaches and for the sewer network as a whole. Various studies have attempted to quantify exfiltration based on the degree and nature of urban groundwater pollution, but these suffer from difficulties of distinguishing between pollution from sewers and pollution from other sources and in the choice of distinctive marker species (Mull, 1996; Wolf *et al.*, 2007b). In addition, such approaches cannot identify which specific points or reaches of the sewer network are subject to leakage; they provide integrated catchment level values and the results are only rough estimations due to multifactor influences, including lateral flows and overlapping effects from leakage of septic tanks, supply pipes etc. Pressure testing and direct exfiltration measurements by means of double packer systems disturb the natural milieu in the sewer pipe and results in a diminished effectiveness of the clogging layer (Ellis *et al.*, 2009) and these methods normally derive an overestimate of the leakage rates.

Other approaches attempt to quantify sewer loss and gain as determined from indirect modelling or from water balance studies. These approaches are subject to considerable variability, particularly as the determination of individual parameter contributions to the water balance have significant uncertainties. There is general agreement that CCTV inspection is largely inconclusive in identifying if, where and to what degree, wastewater is leaking from sewers and groundwater gaining entry into sewers (Eisworth and Hotzl, 1994). Not all observable damages comprise exfiltration or infiltration routes and the occurrence of bed sediments and/or biofilm growths frequently obscure the invert and lower walls of a sewer pipe.

The dilemma of accurate and reliable identification and quantification of sewer leakage provided the core of the APUSS challenge and at which much of the work effort was directed. It was already recognised that the direct measurement of exfiltration for example, using tracer-based, in-sewer flow measurement is in principle a "better" approach (Jensen and Malden, 1996) However, determination of sewage loss gained by the use of artificial tracers has to be treated with care as boundary conditions can be dynamic i.e varying fill levels and discharge rate during the measuring period, inhomogeneous mixing of the tracer due to density differences etc., as well as to uncertainties resulting from the length of the investigated sewer reach. The APUSS objective was to develop and refine a generic trace methodology (based on chemicals and/or natural isotopes), to assess and quantify infiltration and exfiltration in sewer systems at different spatial scales from the sewer reach through the elementary sub-catchment up to the entire catchment scale. Such methods would need to cope with varying background conditions, variable flow rates and effluent composition, seasonal effects, dynamic groundwater levels etc. It was recognised at an early stage that such tracer methods would not be readily applicable to house connections particularly given difficulties of access and entry to such small diameter pipes, and following some initial exploratory runs, it was decided to adopt rather more conventional volumetric methods as described in Chapter 4. Accompanying tools and models would be required for large scale catchment extrapolation and screening application and for enduser decision support in conjunction with a GIS-based data information and display system. In addition, performance indicators (PIs) for the assessment of sewer condition and operation would need to accompany the tracer methodology together with costing implications and an analysis of drainage investment strategies based on a multi-criteria methodology. This MCA procedure would be employed to evaluate differing sewer management strategies based on the outcomes of infiltration and exfiltration analysis using the developed APUSS tracer methodologies.

APUSS ACHIEVEMENTS

As described in Chapter 1, the APUSS project defined a number of distinct aims and objectives having a mix of scientific, technological and management goals. In addition to major contributions in methodological procedures to improve urban drainage practice, the project has also delivered a much better and deeper understanding of the behaviour of sewer systems, particularly in relation to the dynamic processes controlling exfiltration and infiltration leakage. Supporting software and Standard Operating Procedures (SOPs) have been developed to provide drainage engineers and sewer managers with a full toolkit for operational implementation and modelling analysis.

For exfiltration measurements, two novel tracer methods have been developed, tested and validated under working field conditions in Switzerland, Italy, Germany, France and the UK with the collaboration of enduser groups within the European water industry. The QUEST procedure is a pulse injection method whilst QUEST-C is a continuous dosing method; both ionic tracers (NaCL, lithium, bromide) and fluorescent dyes (fluorescein, Rhodamine WT) have been tested in the tracer procedure (see chapter 2). However, in principle, any appropriate tracer could be used with the developed generic methods. Analytical procedures and data processing programs together with Standard Operating Procedures (SOPs) have been developed and tested by the APUSS partners and their associated city endusers. A significant effort has been devoted to the evaluation and the reduction of uncertainties and errors both in measurements and in the field data processing phase, A procedure for uncertainty analysis of exfiltration ratios has also been developed and validated for both methods and procedures in order to ensure that an optimal design and field implementation of the tracer experiment can be achieved by the enduser. Field testing of the QUEST-C method using a combination of tracers has clearly demonstrated that under well controlled conditions, the procedure has a near-zero bias and a precision in the exfiltration estimate of $\sim 2\%$ of the flow. The field testing in the Greater London area of the UK has indicated the potential viability of using Rhodamine WT in single-pulse QUEST studies, particularly in respect of short sewer lengths, whilst the simultaneous injection of multi-tracers under continuous dosing QUEST-C conditions facilitates realistic calculations of exfiltration rates for sewer travel times of about 20 minutes.

Two separate approaches have been developed for infiltration measurements and tested in working field conditions. The first method is based on oxygen and/or hydrogen isotopic ratios and has been tested most extensively in Rumlang (Switzerland), Lyon (France) and in Rome (Italy). More comprehensive investigations of the effects of heterogeneities and variabilities in the local groundwater as well as in the drinking water network were undertaken in Rumlang and Rome (see chapters 2 and 8). Such variations can obscure the natural tracer signal and this aspect remains crucial for the successful application of the method. The second method for infiltration determination is based on continuous in-situ flow and pollutant concentration time series. In particular, COD time series, measured by means of a UV-visible spectrophotometer, has been developed. Data analysis procedures have been established to evaluate infiltration rates from field data series and successfully tested in Switzerland, Italy and the UK (see chapter 9). For the tracer methods, computational tools were implemented in the platform R, which is freely available under the terms of GNU General Public License. Procedures for the evaluation of uncertainties in the pollutant time series methodology have been fully investigated and are documented in the corresponding SOP protocols which are available on the APUSS website.

Laboratory and field sewer rig studies have been carried out to establish the potential adsorption and interaction of tracers (especially Rhodamine WT) with wastewater solids and the effect of sewer solids and sediments on exfiltration. Field test runs in the UK supported by a consortium of UK wastewater utilities, using an experimental sewer reach in Dundee, Scotland, confirmed the utility and robustness of the tracer testing methodology, particularly in respect of Rhodamine WT for short sewer lengths (< 1 km). The laboratory and field sewer rig tests also demonstrated the effective sealing of open joints and fractures with sewer sediments and the concurrent growth of colmation layers, biofilm and sewer slimes. New definitions for the various clogging layers have been proposed (see Chapter 6) and the significance of minor hydraulic instabilities in the disruption of the ultimate equilibrium exfiltration state has been demonstrated. The work has established that it is growth and composition of the colmation layer and underlying transition zone which is primarily responsible for the low levels of exfiltration loss that have been noted in most urban drainage studies. However, the equilibrium state is easily disturbed and there can be significant short term leakages from, as well as infiltration inflows into, the sewer system irrespective of the existence of such clogging layers.

The developed tracer methods are generally applicable for the determination of exfiltration and infiltration in public sewer pipes and can derive leakage rates within $\pm 2\%$–3% when undertaken with care. However, as house connections (HCs) are also key components in I/E phenomena, volumetric methods have been employed and tested on experimental residential sites in Prague, Czech Republic. A basic working generic methodology has been established for application to single HCs, and a matrix procedure for extrapolation to sub-catchment and entire catchment-wide situations has been proposed (see Chapter 4). The matrix approach utilises conventional CCTV inspection records and a HC defects ranking procedure. A review (Chapter 3) of the definitions and objectives of HCs has identified the distinctive role of public and private connections in I/E quantification. Various approaches for the optimal design of measuring and monitoring surveys for HCs as well as procedures for installing and operating these crucial components of the sewer system are also presented in Chapter 3.

The new tracer methodology provides a generic working procedure for the *in-situ* monitoring and quantification of exfiltration and infiltration in urban sewer systems and can be utilised as a benchmark indicator. The commercial potential of the tracer procedures is substantial, particularly given the low uncertainties that can be achieved in applying the methodology. Endusers include wastewater utilities and sewer maintenance groups throughout the water sector who will now have access to new tools for accurately measuring losses and gains from and into sewer lines and open channels. It also provides a monitoring tool that can be applied in compliment with CCTV and is applicable across the entire range of operational and maintenance programmes of the wastewater sector within EU member states. Nevertheless, the field application is not straightforward and requires utmost care in order to obtain reliable results and places considerable demands on operational staff in terms of levels of training, analytical expertise and motivation. Detailed pre-investigation in the sewer reach of interest is necessary to select the most appropriate tracer method(s), and to establish flow and wastewater characteristics as these can heavily influence the outcome of the measurement result.

As only one data point is obtained per tracer experiment the test methods do not allow estimation of either the precise location or magnitude of leakage within a monitored pipe section. However, given a *priori* knowledge on water losses (obtainable from defect maps, pressure testing etc.), it is possible to convert the result of tracer experiments (as the loss ratio) into absolute exfiltration rates. Repetitive tracer applications can enhance the knowledge gap on operational sewer conditions and allow for a much more rational rehabilitation practice.

Conceptual models for infiltration at sub-catchment scale and for exfiltration at sewer reach scale have been developed and implemented within an integrated tool developed in the framework of the AquaBase GIS software. This displays a detailed map of the sewer system and the dry weather flow rates, the interpolated groundwater levels, and includes data and time series determined from the experimental tracer I/E results. Automatic calibration routines are included in order to calibrate the model parameters according to the data and time series. Uncertainty analysis procedures have also been implemented within the AquaBase routine and further models and tools established and tested within the context of a large scale application example in Prague. One must bear in mind in conceptual modelling that far too often estimates of sewer exfiltration losses are insufficiently critically evaluated. For example, theoretical models use perfect fluids, so there is nothing to clog holes and defects. In water balance studies, how is clean water leakage allowed for, and during dry periods are unbalanced returns due to irrigation or pool filling and evaporation allowed for? In the APUSS work, field experimental data series have been used to establish and test the leakage approach model for both infiltration and exfiltration, which were well reproduced together with their temporal variations. As I/E experiments and measurements are unlikely to cover all sewer pipes within a city, a statistical method based on the similarity approach has been developed in order to identify sub-catchment and groups of sewers which have similar characteristrics and which should potentially have similar orders of magnitude in I/E rates. This method showed interesting enduser results for the optimal positioning of flow gauges for the measurement of I/E rates and for the transfer of measurement results.

Performance indicators (PIs) for both infiltration and exfiltration have been established and a software program (the PIs Tool) has been adapted for the calculation of the PIs and their associated sewer asset information. Data can be transferred between the PIs tool and the AquaBase software. Data series from Lyon, France and Rome, Italy have been used to test and evaluate the proposed performance indicators. When I/E problems are detected and evaluated, endusers normally have various investment and rehabilitation strategies to solve them e.g sewer system rehabilitation, adaptation of the downstream wastewater treatment plant, storage tanks or any combination of these solutions. A generic methodology including modelling of the sewer system and of the wastewater treatment plant has been established to compare these investment options, and which accounts for different criteria such as environmental, operational, financial and other aspects which can effect the relative impacts of I/E. The multi-criteria method utilises ELECTRE III to compare and rank the various strategies and an application of the proposed methodology on a semi-virtual urban catchment in Ecully, near Lyon in central France has been undertaken.

Finally, a cost-structure for a typical urban sewer system has been developed which can be applied for general benchmark studies. The costs for the application of the APUSS experimental tracer methods have also been evaluated; they are undoubtedly more expensive than those associated with "classic" flow measurements, but the APUSS methods advantageously provide infiltration and exfiltration values with a much better accuracy and reproducibility together with a rigorous assessment of the uncertainties.

The APUSS research however, still leaves many questions unanswered and in particular that relating to the paucity of data availability regarding detailed sewer condition and the exchangeability of current sewer system data sets. The commercial data entry form developed in Chapter 12 provides a possible template which could serve as a benchmark approach as well as offering a costing methodology. Undoubtedly a standardised data format would be a very valuable tool for drainage engineers, and whilst some local municipalities have attempted such an approach, there are few if any national examples of such harmonisation. Such standardised sewer data formatting would not only support practitioner and sewer operator knowledge, but also support basic scientific knowledge on sewer behaviour.

APUSS AND SEWER MANAGEMENT

The APUSS project has demonstrated that accurate, robust and cost-effective I/E measurements can be achieved based on in-sewer tracer procedures. The question then is about the possible inclusion of such methodological procedures into regular management and operational processes. It may be that the exfiltration methods are only appropriate at the sewer reach (and possibly elementary 1st/2nd order sub-catchment) scale, with the infiltration procedures being most appropriately applied at the full catchment scale. A key point here is that relating to the sampling strategy and the uncertainties affecting the extrapolation process, and these issues are addressed in the modelling analysis described in Chapter 10. Figure 2 presents a staged decision schematic to assist sewer system operators and planners in developing an I/E control/reduction programme and where APUSS approaches and methods as developed in the various Work Areas (WA1 – WA4), might be applied in the management flow sequence. The staged flow sequences outlined in Figure 2 implies that I/E control should be approached as an adaptive management procedure which has no abrupt endpoint, but is considered as a rolling or continuous planning cycle. Continuous updating and field data from on-going monitoring and O&M procedures are fed back to improve sewer system analysis and condition forecasting.

As argued in Chapter 1, the process needs to address the whole collection system rather than just identified problem areas, and this is particularly the case for the control of infiltration/inflow (I/I), as has been proven in the companion CARE-S EU 5th Framework project (Saegrov, 2006). The root causes of the problems should be identified and remedial actions undertaken at the most appropriate locations; these may not necessarily be the same locations where the problems are evident in the pipe.

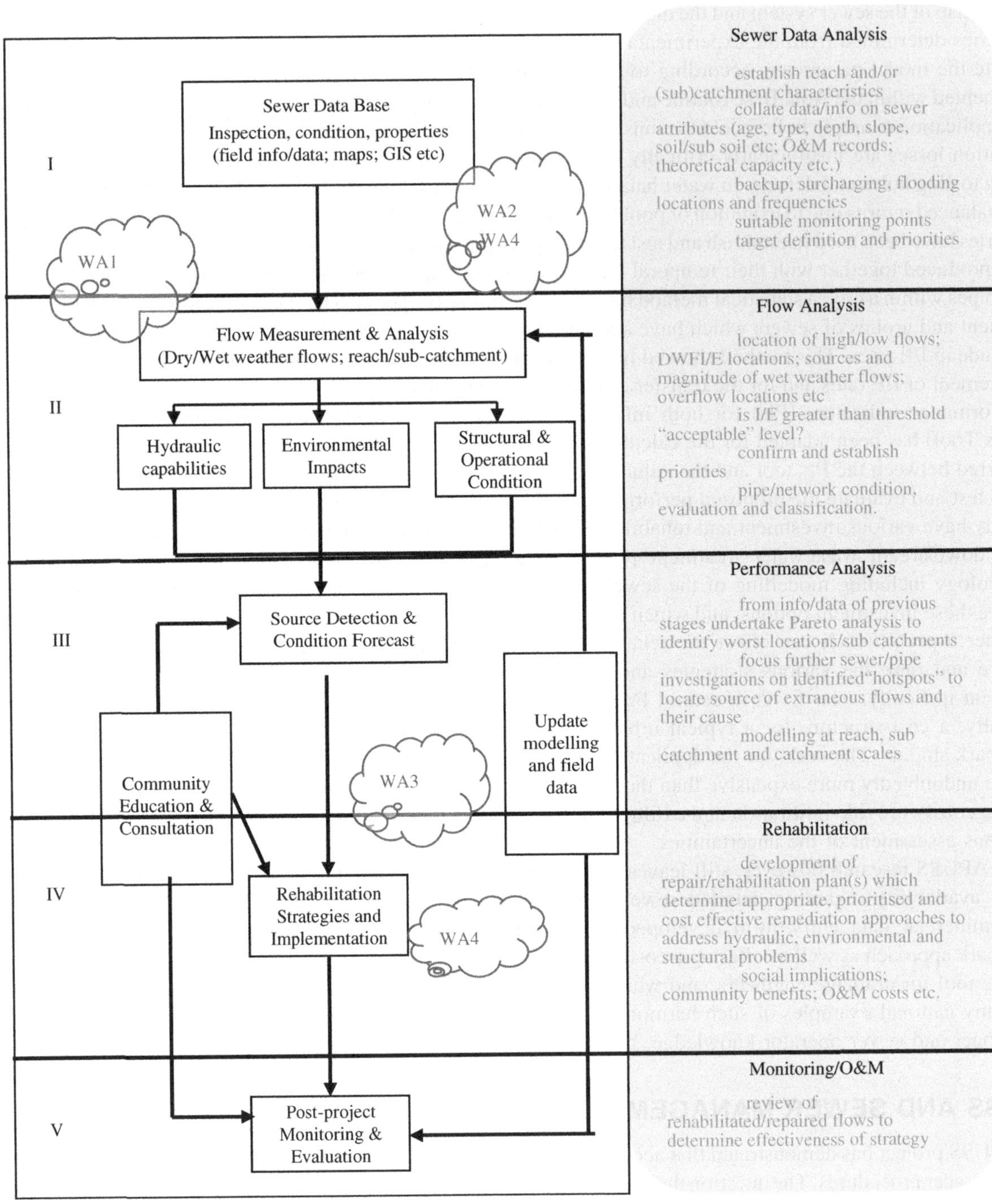

Figure 2: Sewer exfiltration and infiltration/inflow management.

Private building/house connections need to be considered in any I/E rehabilitation programme particularly given that they can contribute up to 70% of I/I inflows (see Chapter 3) and should be included in the Stage II and III surveys and analysis.

Computer modelling and up-scaling of reach/sub-catchment data obtained from the sewer monitoring Stage II as described in Chapters 10 and 11, will provide a conceptual overview of the likely distribution and magnitude of I/E for the whole catchment situation, which can then be used in the Stage IV rehabilitation analysis. The development of a computerised data-based management and decision-support system in the APUSS project (see Chapters 10 and 11) provides a community added-value through the provision of a surveillance data directory and visual display menus of benefit to endusers and for future Stage 1 and IV studies. Such computer-based outputs could form the basis for performance and cost-evaluation for different types of sewer system and wastewater treatment. The modelling and risk assessment analysis also provides a decision-making mechanism to support effective sewer management and control.

The rational approach outlined in Figure 2 is based on the assessment of actual rather than theoretical data related to the wastewater system and attempts to locate the cause of system problems and to ensure that there is a proper application of inspection, monitoring and construction procedures to protect the new/repaired system against future I/E problems.. It is also essential that source control and educational campaigns should be conducted in parallel with the development of remediation strategies (Stages III and IV) with bye-laws relating to private connections to ensure proper inspection during construction and rehabilitation. The educational programmes (Stages III and IV) should include not only raising the awareness of the local community of the adverse environmental and social impacts resulting from inefficient sewer system operation, but also serve to obtain public support for inputs to the rehabilitation plan e.g in identifying flooding events, locations and depths, infiltration overflows, reliability of existing sewer data etc. The general management programme outlined for the inspection, condition evaluation and remediation of sewer networks shown in Figure 2 follows the requirements of DIN EN752-5 and provides specific triggers for action, such as the levels of extraneous I/E flows.

The strategic elements incorporated in Stage IV should include:

- goals and targets for sewer asset management
- specific objective(s) for exfiltration and infiltration-inflow management
- adopted key sewer Performance Indicators (PIs) as identified in Chapter 12
- performance targets and policy strategies; these latter should be based on the identified key strategic issues and SWOT findings, including risk assessment in respect of I/E management and on the required management plan development level. The strategies should be supported by detailed action plans that might cover a period of between 3–5 years.
- performance measure outputs such as ratio of maximum day flow to average day flow or rainfall-dependent I/I (%); loss rates for selected sewer reaches (% flow); number of sewer surcharging/overflow complaints etc., as developed in Chapter 12. The US EPA for example, have identified an "acceptable" level of I/I for existing systems which is set at 450 l/capita/day although individual states have "acceptable" levels which vary considerably from this figure (Smith *et al.*, 1991).

As a minimum level of activity, it requires on-going monitoring and assessment of both dry and wet weather flows, for which the APUSS tracer methodology is well suited as a precursor to any further detailed CCTV or other survey (smoke/pressure testing etc). In addition, continuous updating of basic sewer data and plans should be undertaken based on post-project monitoring (Stage V). Post-project O&M programmes need to be pro-active, developed to prevent rather than being simply reactive to problems such as blockage, back-ups or surcharging. It should also be borne in mind that rehabilitation at one point or section in the sewer system will often result in transferring problems to another point/section, as for example, higher groundwater levels can be expected from the remediation work undertaken on the rehabilitated section.

The outcomes of the implementation of a properly effected sewer I/E rehabilitation/repair management strategy include:

- optimising system performance and efficiency
- minimisation of environmental impacts
- reducing hydraulic overloading on the collection system and treatment plant
- reduction in customer complaints
- deferral of asset augmentation and upgrading works and investment
- reduced energy and O&M costs
- better knowledge of the wastewater collection system and its condition and performance
- I/E monitoring and modelling show that potential sewer assets and community benefits can be achieved through the implementation of an appropriate management strategy.

Evaluation of expected benefits from an aggressive and appropriately executed I/E management programme suggest that cost-benefit ratios range from 1.5 to 4.4, with even a modest 30% peal I/I reduction for example, giving a saving investment of up to $200 M (€133 M). However, it must be realised that such benefits require substantial up-front investment and will be subject to considerable political as well as technical and financial scrutiny. The final long term outcomes however, rest on whether the implementation strategies lying with the local agencies are efficient enough to take advantage of the benefits. Under a worst case scenario, not only would the potential benefit be missed but customers might actually be faced with double payment; firstly for inefficient I/E control projects in the local agencies and then again for upgrading/expansion for regional conveyance and treatment capacity.

CONCLUSIONS

The APUSS project has correlated well with existing and proposed EU wastewater and environmental control policies and Directives being based on developing a better approach to the assessment and analysis of urban sewer performance in order to more accurately identify the sources and scale of exfiltration/infiltration leakage. Not only have innovative and robust tracer methods been developed, their cost-effective application at both pipe reach and sub-catchment scales have also be demonstrated in collaboration with endusers in the European water industry. Catchment-scale modelling extrapolating pipe reach and suc-catchment tracer-based data has been developed and evaluated. Key performance indicators (PIs) for sewer asset management have been developed and tested with endusers to support more cost-effective management and O&M procedures.

Combined with GIS-modelling, APUSS has delivered to endusers effective and innovative urban sewer performance tools and approaches for asset evaluation and decision-support. The added-value for European cities lies in the opportunities the APUSS approaches offer for improvement of their ageing and frequently inefficient collection systems. There are unfortunately only few opportunities for European urban sewer infrastructure to be re-developed from "scratch", and so dramatic changes in disposal practice are highly unlikely. Sewer asset management is much more likely to involve small step-changes through incremental improvements, with policies and approaches varying across the European community. Some wastewater utilities might be better placed and/or more inclined to adopt particular methodological changes than others. Nevertheless, despite conventional practice and traditional professional attitudes, European wastewater utilities have often been at the forefront of global technical and managerial innovation. Herein lies the advantage of the linkage and collaboration between researchers and endusers as demonstrated in the successful application of APUSS methods in the various European cities. The APUSS research offers the opportunity for justifying "evidence-based" intervention for sewer remediation in order to achieve "health and sustainability gains" for European urban citizens and supports basic quality-of-life environmental targets.

The APUSS project may be useful in informing future European approaches to standardisation protocols, in best practice recommendations or in supplementary mitigating measures within the WFD. The APUSS methods and tools offer urban decision-makers and wastewater authorities cost-effective approaches to their analysis of the operational condition and performance of urban sewer systems, to support in turn, the development of appropriate and sustainable mitigating and remedial management strategies.

The participation of APUSS within the CityNet cluster of six EU Research, Technology & Development (RTD) projects has served to address the wider issues of water and wastewater management in urban areas. This wider network consisting of 60 research institutions and over 56 endusers has made a significant contribution to European R&D capacity and implementation potential for future sustainable urban water management. This APUSS volume provides a further dissemination routeway for the outcomes of this fundamental and applied research, adding to the contribution of the CityNet consortium and to the scientific knowledge gained from common publication, workshop and conference proceedings.

REFERENCES

ASTM (2002) *Standard Test Methods for Hydrostatic Exfiltration and Infiltration Testing of Vitrified Clay Tile Pipes*. American Society of Testing and Materials. West Conshohocken, Pennsylvania, US.

Davies, J.P., Clarke, B.A., Whiter, J.T. and Cunningham, R.J. (2001) Factors influencing the structural deterioration and collapse of rigid sewer pipes. *Urban Water Journ.*, **3**, 73–89.

Eisworth, M. and Hotzl, H. (1994) Groundwater contamination by leaking sewerage systems. In *Proc. Conf. Water Down Under 94*. 21–25 November 1994. Adelaide. Institute of Engineers, Barton, Australia.

Ellis, J.B., Revitt, D.M., Vollertsen, J. and Blackwood, D.J. (2009) Sewer exfiltration and the colmation layer. *Water Science & Tech.*, **59**(11), 2273–2280.

Jensen, A. and Madsen, P.G. (1996) Exfiltration from sewers. *Stads og Havneingenioren*. **10**, 38–39.

Metcalf and Eddy. (1991) *Wastewater Engineering: Treatment, Disposal, Re-Use*. 3rd Edition (Revised by Tchobanoglous, G. and Burton, F.L.). McGraw-Hill International Inc., Singapore. ISBN 0-07-100824-1.

Mull, R. (1996) Water exchange between leaky sewers and aquifers. 695–700 in Sieker, F and Verworn, H-R (Edits): *Proc. 7th Int. Conf. Urban Storm Drainage*. 9–13 September 1996. Hannover, Germany. SuG-Verlagsgegellschaft, Hannover, Germany. ISBN 3-00-000860-8.

Read, G.F. (1996) *Sewers: Repair and Renovation*. Project Report 44, X171. Construction Industry Research & Information Association (CIRIA), London, UK.

Saegrov, S. (Edit). (2006) *CARE-S: Computer Aided Rehabilitation of Sewers and Storm Water Networks*. IWA Publishing, London. UK. ISBN 978 1843391159.

Smith, T.M., Bowkers, R.P. and Shah, H.J. (1991) *Handbook for Sewer System Infrastructure Analysis and Rehabilitation*. October 1991. EPA/625/6-91/030. Office of Water, US Environmental Protection Agency, US, 92 p.

Vollertsen, J. and Hvitved-Jacobsen, T. (2003) Exfiltration from gravity sewers: A pilot scale study. *Water Science & Tech.*, **47**(4), 69–76.

Wolf, L., Morris, B. and Burns, S. (2007a) *Urban Water Resources Toolbox: Integrating Groundwater into Urban Water Management*. IWA Publishing, London, UK. ISBN 978-1843391388.

Wolf, J., Klinger, J., Hotzl, H. and Mohrlok, U. (2007b) Quantifying mass fluxes from urban drainage systems to the urban soil-aquifer system. *Journ. Soils & Sediments*, **11**, 1–11.

WRC (2001) *Sewerage Rehabilitation Manual*. 4th Edition. Water Research Centre, Swindon, Wilts. UK. ISBN 978 1898920397. (see also http://srm.wrcplc.co.uk).

List of publications

PEER REVIEWED ARTICLES IN JOURNALS

Bertrand-Krajewski, J.-L., Barraud, S., Le Gauffre, P. and Baer, E. (2006) Infiltration in sewer systems: Multi-criteria comparison of investment/rehabilitation strategies. *Water Practice & Tech.*, **2**(1), doi 10.2166wpt2007.027.

Bertrand-Krajewski, J.-L., Barraud, S., LeGauffre, P. and Baer, E. (2007) Infiltration in sewer systems: A multi-criteria comparison of investment/rehabilitation strategies. *Water Practice & Tech.*, **2**(1), doi: 10. 2166/wpt2007.027.

Bertrand-Krajewski, J.-L., Ellis, J.B., Cardoso, M.A., Frehmann, T., Giulianelli, M., Gujer, W., Krebs, P., Pilska, Z., Pollert, J. and Pryl, K. (2006) Towards a better knowledge and management of infiltration and exfiltration in sewer systems: The APUSS project. *Water Practice & Tech.*, **1**(1), doi 10.2166/wpt.2006.021.

Blackwood, D.J., Ellis, J.B., Gilmour, D. and Revitt, D.M. (2005) Influencing exfiltration processes in sewers. *Water Science & Tech.*, **521**(2), 147–154.

Cardoso, M.A., Prigiobbe, V., Giulianelli, M., Baer, E., de Benedittis, J. and Coelho, S.T. (2005) Assessing the impact of infiltration and exfiltration in sewer systems using performance indicators: Case studies of the APUSS project. *Water Practice & Tech.*, **1**(1), doi 10.2166.

de Benedittis, J and Bertrand-Krakewski, J.-L. (2005) Infiltration in sewer systems: Comparison of measurement methods. *Water Science & Tech.*, **52**(3), 219–228.

de Benedittis, J. and Bertrand-Krajewski, J.-L. (2005) Measurement of infiltration rates in urban sewer systems by use of oxygen isotopes. *Water Science & Tech.*, **52**(3), 229–237.

de Benedittis, J. and Bertrand-Krajewski, J.-L. (2005) Mesurage de l'exfiltration en reseaux d'assainissement par tracage artificiel au NaCl. *La Houille Blanche*, **5**, 26–34 (In French).

Ellis, J.B. and Revitt, D.M. (2002) Sewer losses and interactions with groundwater quality. *Water Science & Tech.*, **45**(3), 195–202.

Ellis, J.B., Revitt, D.M., Lister, P., Willgress, C. and Buckley, A. (2003) Experimental studies of sewer exfiltration. *Water Science & Tech.*, **47**(4), 61–67.

Ellis, J.B., Revitt, D.M., Vollersten, J. and Blackwood, D.J. (2009) Sewer exfiltration and the colmation layer. *Water Science & Tech.*, **59**(11), 2273–2280.

Franz, T., Rutsch, M., Krebs, P., Frehmann, T., Bertrand-Krajewski, J.-L., Cardoso, M.A., Ellis, J.B., Giulianelli, M., Gujer, W., Pliska, Z., Pollert, J. and Pryl, K. (2007) Untersuchung und bewertung von in- und exfiltration undichter abwasserkanale-das EU Forschungsprojekt APUSS. *KA Abwasser Abfall*, **54**(4), 353–361 (In German).

Giulianelli, M. and Prigiobbe, V. (2004) Fenomeni di infiltrazione ed exfiltrazione nelle fognature urbane, bacino sperimentale a Roma. *L'Acqua*, **6**, 45–50 (In Italian).

Giulianelli, M., Priggiobbc, V. and Succhiarelli, C. (2004) Nuovi metodi per quantificare inmfiltrazioni ed exfiltrazione nelle fognature urbane. *Geologia dell'Ambiente*, **4**, 21–26 (In Italian).

Karpf, C. and Krebs, P. (2005) Application of the leakage model to assess exfiltration. *Water Science & Tech.*, **52**(5), 225–232.

Kracht, O. and Gujer, W. (2005) Quantification of infiltration into sewers based on time series of pollutant loads. *Water Science & Tech.*, **52**(3), 209–218.

Kracht, O., Gresch, M. and Gujer, W. (2007) A stable isotope approach for the quantification of sewer infiltration. *Environ. Science Technol.*, **41**, 5839–5845.

Kracht, O., Gresch, M. and Gujer, W. (2008) Innovative tracer methods for sewer infiltration modelling. *Urban Water Journ.*, **5**, 173–185.

Edited by Bryan Ellis and Jean-Luc Bertrand-Krajewski. ISBN: 9781843391494. Published by IWA Publishing, London, UK.

Prigiobbe, V. and Giulianelli, M. (2009) Quantification of sewer system infiltration using δ18O hydrograph separation. *Water Science & Tech.*, WST-WSTWS-EM08897R1 (in Press).
Revitt, D.M., Ellis, J.B. and Paterakis, N. (2006) Comparison of tracer techniques for monitoring sewer losses. *J. Environ. Monitoring*, **8**(5), 564–571.
Rieckermann, J., Borsuk, M., Reichart, P. and Gujer, W. (2005) A novel tracer method for quantifying sewer exfiltration. *Water Resources Research*, **41**(5), W05013. doi: 10. 1029/2004WR003699.
Rieckermann, J., Neumann, M., IOrt, C. and Gujer, W. (2005) Dispersion coefficients of sewers from tracer experiments. *Water Science & Tech.*, **52**(5), 123–133.
Rutsch, M., Franz, T. and Krebs, P. (2007) Transferability of exfiltration rates from sewer systems. *Journ. Soils and Sediments*, **7**(2), 69–74.
Rutsch, M., Mueller, L. and Krebs, P. (2005) Dynamics of rain-induced pollutographs of solubles in sewers. *Water Science & Tech.*, **52**(5), 169–177.
Rutsch, M., Rieckermann, J., Ellis, J.B., Cullmann, J., Vollersten, J. and Krebs, P. (2008) Towards a better understanding of sewer exfiltration. *Water Research*, **42**, 2385–2394.

REFEREED CONFERENCE PROCEEDINGS

Bertrand-Krajewski, J.-L., Barraud, S., Le Gauffre, P. and Baer, E. (2006) Infiltration in sewer systems: Multi-criteria comparison of investment/rehabilitation strategies. In *Proc. 2nd Int. Conf. Sewer Operation and Maintenance, SOM2006*. Ertl, T., Pressl, F., Kretschner, F. and Haberl, R. (eds). 26–28 October 2006, Vienna, Austria, SIG, Vienna, Austria, p. 8. ISBN 3-900962-67-7.
Bertrand-Krajewski, J.-L., Cardoso, M.A., Ellis, J.B., Frehmann, T., Giulianelli, M., Gujer, W., Krebs, P., Pliska, Z., Pollert, J. and Pryl, K. (2005) Towards a better knowledge and management of infiltration and exfilteration in sewer systems: The APUSS project. In *Proc. 10th Int. Conf. Urban Drainage*. Mikkelsen, P.S. (ed). 21–26 August 2005, Copenhagen, Denmark, p. 8, IWA Publishing, London, UK. ISBN 978 1843395 744.
Blackwood, D.J., Ellis, J.B., Revitt, D.M. and Gilmour, D. (2004) The influence of sewage solids and sewer sediments on rates of exfiltration from defects in sewers. In *Proc NOVATECH 04: Sustainable Technologies and Strategies for Urban Water Management*. Chocat, B. and Brelot, E. (eds). 6–10 June 2004, GRAIE, Insa de Lyon, Lyon, France, pp. 1609–1617. ISBN 2-9509337-5-0.
Blackwood, D.J., Gilmour, D., Ellis, J.B., Revitt, D.M. and Staines, A. (2005) Exfiltration from sewers: Is it a serious problem? In *Proc. 10th Int. Conf. Urban Drainage*. Mikkelsen, P.S. (ed). 21–26 August 2005, Copenhagen, Denmark, p. 8, IWA Publishing, London, UK. ISBN 978 1843395 744.
Cardoso, M.A., Almeida, M.C. and Coelho, S.T. (2002) Evaluation of the infiltration impact on the performance of urban sewer systems. In *Proc 9th National Congress: Water, Wastewater and Solid Waste, 9ENASB*. 16–19 September 2002, Braga, Portugal, p. 17 (In Portuguese).
Cardoso, M.A., Bertrand-Krajewski, J.-L., Ellis, J.B., Frehmann, T., Giulianelli, M., Gujer, W., Krebs, P., Pliska, Z., Pollert, J. and Pryl, K. (2006) Melhorar o conhecimento e a gestao da infiltracao e da exfiltracao em sistemas de drenagem urbana: O projecto APUSS. In *Proc. 2nd Encontro Nacional Saneamento Basico*. 24–27 October 2006, Cascais, Portugal, p. 12 (In Portuguese).
Cardoso, M.A., Coelho, S.T., Matos, J.S. and Matos, R.M. (1999) A new approach for diagnosis and rehabilitation of sewerage systems through the development of performance indicators. In *Proc. 8th Int. Conf. Urban Storm Drainage*. Joliffe, I. and Ball, J.E. (eds). 30 August–3 September 1999. Sydney, Australia, Institution of Engineers Australia, pp. 610–617.
Cardoso, M.A., Prigiobbe, V., Giulianelli, M., Baer, E., de Benedittis, J. and Coelho, S.T. (2005) Assessing the impact of infiltration and exfiltration in sewer systems using performance indicators: Case studies of the APUSS project. In *Proc. 10th Int. Conf. Urban Drainage*. Mikkelsen, P.S. (ed). 21–26 August 2005, Copenhagen, Denmark, p. 8, IWA Publishing, London, UK. ISBN 978 1843395 744.
de Benedittis, J. and Bertrand-Krajewski, J.-L. (2004) Measurement of infiltration rates in urban sewer systems by use of oxygen isotopes. In *Proc. 4th Int. Conf. Sewer Processes and Networks*. 22–24 November 2004, Madeira, Portugal, pp. 309–316. ISBN 1-84339-551-7.
de Benedittis, J. and Bertrand-Krajewski, J.-L. (2004) Mesurage de l'infiltration d'eaux souterraines dans les reseaux d'assainissemeent par tracage des isotopes stables del la molecule d'eau. In *Proc. Conf. AUGC Ville et Genie Civil*. 3–4 June 2004, Marne-la-Vallee, France, p. 8 (In French).
Ellis, J.B. (2001) Sewer infiltration/exfiltration and interactions with sewer flows and groundwater quality. In *Proc. 2nd Int. Conf. INTERURBAII: Interactions Between Sewers, Treatment Plants and Receiving Waters in Urban Areas*. 19–22 February 2001, Lisbon, Portugal, pp. 311–319.
Ellis, J.B., Revitt, D.M., Blackwood, D.J. and Gilmour, D. (2004) Leaky sewers: Assessing the hydrology and impact of exfiltration in urban sewers. In *Proc. Hydrology: Science & Practice for the 21st Century*. 12–16 July 2004, Imperial College, London, CD-ROM. Vol. II, British Hydrological Society, Inst. Civil Engineers, London. UK, pp. 266–271. ISBN 0-9539140-1-1.
Ellis, J.B., Revitt, D.M., Lister, P. and Willgress, C. (2002) The effects of sediments on sewer exfiltration. In *Proc. 3rd Int. Conf. Sewer Processes and Networks*. Bertrand-Krajewski, J.-L. and Chebbo, G. (eds). 15–17 April 2002, Paris, France, pp. 147–154. IWA Publishing, London, UK. ISBN 978 1843394365.
Ellis, J.B., Revitt, D.M., Vollertsen, J. and Blackwood, D.J. (2008) Factors influencing temporal exfiltration rates in sewer systems. In *Proc. 11th Int. Conf. Urban Drainage (ICUD)*. August 2008, Edinburgh, Scotland, CD-ROM. ISBN 978 1899796 212.
Franz, T. and Krebs, P. (2005) Statistical methods towards more efficient infiltration measurements. In *Proc. 10th Int. Conf. Urban Drainage*. Mikkelsen, P.S. (ed). 21–26 August 2005. Copenhagen, Denmark, p. 8, IWA Publishing, London, UK. ISBN 978 1843395 744.
Franz, T. and Krebs, P. (2007) Optimierung von messnetzen fur die bestimmung von fremdwasser. In *Proc. Dresdner Kolloquium zur Siedluingswasserwirtschaft 2007*. 4 October 2007, Dresdner, Germany, pp. 51–61. ISSN 1615–083-X (In German).
Franz, T. and Krebs, P. (2007) Transfer of infiltration measurement results to other sewers by means of discriminant analysis. In *Proc. 5th Int. Conf. Sewer Processes and Networks*. 28–31 August 2007, Delft, The Netherlands, pp. 245–253.

Giulianelli, M., Mazza, M., Prigiobbe, V. and Russo, F. (2003) Assesing exfiltration from a sewer by slug dosing of a chemical tracer (NaCl). In *Proc NATO Advanced Workshop: Enhancing the Urban Environment by Environmental Upgrading and Restoration.* Marsalek, J., Sztruher, D., Giulianelli, M. and Urbonas, B. (eds). 5–8 November 2003, Rome, Italy. Earth & Environment Series. Vol. 43, Springer, Berlin, Germany, pp. 351–362. ISBN 978-1-4020-2693-5.

Karpf, C. and Krebs, P. (2003) Bilanzierung von fremdwasser. In *Proc. ATV-DVWK Lanesgruppentagung Sachsen/Thuringen.* 20 May 2003, Suhl, Germany, pp. 29–44 (In German).

Karpf, C. and Krebs, P. (2004) Application of the leakage model to assess exfiltration. In *Proc. 6th Int. Conf. Urban Drainage Modelling*, 15–17 September 2004, Dresden, Germany, pp. 217–224. ISBN 1-94339-497-9.

Karpf, C. and Krebs, P. (2004) Sewers as drainage systems: Quantification of groundwater infiltration. In *Proc. NOVATECH 04: Sustainable Technologies and Strategies for Urban Water Management.* Chocat, B. and Brelot, E. (eds). 6–10 June 2004, GRAIE, Insa de Lyon, Lyon, France, pp. 969–976. ISBN 2-9509337-5-0.

Karpf, C. and Krebs, P. (2005) Assessment of extraneous water inflow in separated sewer networks. In *Proc. 10th Int. Conf. Urban Drainage.* Mikkelsen, P.S. (ed). 21–26 August 2005, Copenhagen, Denmark, p. 8, IWA Publishing, London, UK. ISBN 978 1843395 744.

Karpf, C., Franz, T. and Krebs, P. (2007) Fractionation of infiltration and inflow (I/I) in urban sewer systems with regression analysis. In *Proc. NOVATECH2007: Sustainable Techniques and Strategies in Urban Water Management.* June 2007. Lyon, France, GRAIE, Insa de Lyon, pp. 1227–1234. ISBN 2-9509337-7-7.

Karpf, C., Franz, T. and Krebs, P. (2007) Modelltechnische ansatze zur ermittlung von grundwasserinduzierten fremdwasserabflussen. In *Proc. Gewasserschutz-Wasser-Abwasser 207.* 14–16 March 2007, Essener Tagung, Aachen, Germany, pp. 1–11. ISSN 978–3938996–13–3.

Kracht, O. and Gujer, W. (2004) Quantification of infiltration into sewers based on time series of pollutant loads. In *Proc. 4th Int. Conf. Sewer Processes and Networks.* 22–24 November 2004, Madeira, Portugal, pp. 293–300. ISBN 1-84339-551-7.

Kracht, O. and Gujer, W. (2006) Tracer based quantification of sewer infiltration: Experiences using natural isotopes (δ18O, δ2H). In *Proc. 4th. Swiss Geoscience Meeting.* 24–25 November 2006, Bern, Switzerland.

Kracht, O., Gresch, M. and Gujer, W. (2005) Innovative tracer methods for infiltration monitoring. In *Proc. 10th Int. Conf. Urban Drainage.* Mikkelsen, P.S. (ed). 21–26 August 2005. Copenhagen, Denmark, p. 8, IWA Publishing, London, UK. ISBN 978 1843395 744.

Prigiobbe, V. and Giulianelli, M. (2004) Design of tests for quantifying sewer leakages by QUEST-C method. In *Proc. 6th Int. Conf. Urban Drainage Modelling.* 15–17 September 2004, Dresden, Germany, pp. 205–216. ISBN 1-94339-497-9.

Princ, I. and Kohout, D. (2003) Infiltration and exfiltration from house connections. In *Proc. Conf. Wastewater 2003.* 13–15 May 2003. Olomouc, Czech Republic, pp. 391–394. ISBN 80-239-0573-2.

Princ, I. and Kohout, D. (2003) Infiltration and exfiltration from house connections. In *Proc. Conf. Water Management Task 2003.* November 2003, Spinleruv Mlyn, Czech Republic, pp. 41–44 (In Czech).

Princ, I. and Kohout, D. (2003) Infiltration and exfiltration from house connections. In *Proc. Int. Conf. Hydrosphere 2003.* 2–3 October 2003.

Princ, I. and Kohout, D. (2004) Assessing infiltration to sewer systems in the locality Lazne Bohdanec. In *Proc. Int. Conf. Hydrosphere 2004.* December 2004, Brno, Czech Republic, pp. 275–278. ISBN 80-86020-41-4.

Princ, I., Kohout, D. and Tuma, R. (2002) Infiltration and exfiltration from house connections. In *Proc. Int. Conf. Hydrosphere 2002.* October 2002, Breclav, Czech Republic, p. 7 (In Czech). ISBN 80-86020-38-X.

Revitt, D.M., Ellis, J.B. and Paterakis, N. (2006) A comparison of pulse tracer dosing and continuous tracer dosing for the determination of sewer exfiltration. In *Proc. 2nd Int. Conf. Sewer Operation and Maintenance, SOM2006.* Ertl, T., Pressl, F., Kretschner, F. and Haberl, R. (eds). 26–28 October 2006, Vienna, Austria, SIG, Vienna, Austria, pp. 128–134. ISBN 3-900962-67-7.

Rieckermann, J. and Gujer, W. (2002) Quantifying exfiltration from leaky sewers with artificial tracers. In *Proc. Int. Conf. Sewer Operation and Maintenance, SOM02.* 26–28 November 2002, Bradford, UK, p. 8, CD-ROM. ISBN 1-85 143 213-2.

Rieckermann, J., Bares, V., Kracht, O., Braun. D and Gujer, W. (2003) Quantifying exfiltration with continuous dosing of artifiical tracers. In *Proc Int. Conf. Hydrosphere2003.* 2–3 October 2003. Brno, Czech Republic. p. 7. ISBN 80-86020-41-X.

Rieckermann, J., Borsuk, M.E. and Gujer, W. (2005) Using decision analysis to determine optimal experimental design for monitoring sewer exfiltration with tracers. In *Proc. 10th Int. Conf. Urban Drainage.* Mikkelsen, P.S. (ed). 21–26 August 2005, Copenhagen, Denmark, p. 8, IWA Publishing, London, UK. ISBN 978 1843395 744.

Rieckermann, J., Neumann, M., Ort, C. and Gujer, W. (2004) Dispersion coefficients of sewers from tracer experiments. In *Proc. 6th Int. Conf. Urban Drainage Modelling*, 15–17 September 2004, Dresden, Germany, pp. 417–426. ISBN 1-94339-497-9.

Rutsch, M., Mueller, I. and Krebs, P. (2004) Dynamics of rain-induced pollutographs of solubles in sewers. In *Proc. 6th Int. Conf. Urban Drainage Modelling*, 15–17 September 2004, Dresden, Germany, pp. 469–476. ISBN 1-94339-497-9.

Rutsch, M., Rieckermann, J. and Krebs, P. (2005) Quantification of sewer exfiltration: A review. In *Proc. 10th Int. Conf. Urban Drainage.* Mikkelsen, P.S. (ed). 21–26 August 2005, Copenhagen, Denmark, p. 8, IWA Publishing, London, UK. ISBN 978 1843395 744.

OTHER PUBLICATIONS

Bertrand-Krajewski, J.-L. (2003) Assessing infiltration and exfiltration on the performance of urban sewer systems. In *Proc. 2nd IMAGE-TRAIN Cluster Meeting, Sustainable Management of Soil and Groundwater Resources in Urban Areas.* 2–4 October 2002, Krakow, Poland. *Umweltbundesamt, Tagengsberichte*, **33**, 57–61. ISBN 3-85457-694-3.

Blackwood, D.J., Ellis, J.B., Gilmour, D., Revitt, D.M. and Staines, A. (2004) *Final Report: Exfiltration Research Project, Riverside Dundee.* On behalf of UK Water Industry Consortium, p. 102. Urban Pollution Research Centre, Middlesex University, London and Urban Water Technology Centre, University of Abertay, Dundee, Scotland, p. 102.

de Benedittis, J. and Bertrand-Krajewski, J.-L. (2003) Measurement of infiltration rates in sewer systems: Use of natural water isotopes. In *Int. Seminar In-Sewer Processes, SewNet.* 10–11 April 2003, p. 41, Delft, The Netherlands.

de Benedittis, J. and Bertrand-Krajewski, J.-L. (2003) Measurement of infiltration rates in urban sewer systems: Use of oxygen isotopes. In *Proc. 18th European Junior Scientists Workshop*. 7–11 November 2003, p. 8, Almograve, Portugal.

Franz, T., Karpf, C. and Krebs, P. (2006) Grossraeumige modellierung von infiltration und exfiltration. In *Proc. DWA Gemeinschaftstagung Undichte Kanaele:Kein Risiko?* 11–12 October 2006, pp. 131–142, Frankfurt, Germany. ISBN 3-939057-60-6 (In German).

Franz, T., Rutsch, M. and Krebs, P. (2004) Field data to estimate infiltration into sewer systems. In *Proc. 19th European Junior Scientist Workshop, Process Data and Integrated Urban Water Modelling*. 11–14 March 2004, p. 8, Lyon, France.

Franz, T., Rutsch, M., Krebs, P., Frehmann, T., Bertrand-Krajewski, J.-L., Cardoso, M.A., Ellis, J.B., Giolianelli, M., Gujer, W., Pliska, Z., Pollert, J. and Pryl, K. (2007) Untersuchung und bewertung von in- und exfiltration undichter abwasserkanale: Das EU Forschungsprojekt APUSS. *KA Abwasser Abfall*, **54**(4), 353–361 (In German).

Giulianelli, M., Paoluzzi, M. and Prigiobbe, V. (2002) GIS for infiltration and exfiltration management on urban drainage networks in Rome. In *Proc. 16th European Junior Scientist Workshop: RTC and Measurements in Urban Drainage Networks*. 7–10 November 2002, Milo, Italy, pp. 107–115.

Kohout, D. (2002) Exfiltration from house connections. In *Proc. Water Management Task 2002*. November 2002, Spindleruv Mlyn, Czech Republic, p. 4. ISBN 80-01-02644-2 (In Czech).

Kohout, D. (2003) Infiltration and exfiltration form house connections. In *Proc. 18th European Junior Scientists Workshop*. 7–11 November 2003, p. 8, Almograve, Portugal.

Kracht, O., Gresch, M., de Benedittis, J., Prigiobbe, V. and Gujer, W. (2003) Stable isotopes of water as a natural tracer for infiltration into urban sewer systems. *Geophysical Research Abtsracts*, 5, 07852. ISSN 1029–7006.

Kracht, O., Rieckermann, J. and Gujer, W. (2007) Projet APUSS: Quantification de l'in- et l'efiltration des canalisations. In *Proc. Seminar: Infiltration des Eaux Pluviales et Protection des Eaux Souterraines. Assoc. Suisse des Professionnels de la Protection Des Eaux (VSA)*. 7 March 2007. Ecole d'ingenieurs et d'architectes de Fribourg. Switzerland (In French).

Princ, I. (2002) Infiltration from house connections. In *Proc. Water Management Task 2002*. November 2002, Spindleruv Mlyn, Czech Republic, p. 4. ISBN 80-01-02644-2 (In Czech).

Princ, I., Kohout, D. and Tuma, R. (2003) Infiltrace a exfiltrace do'z domovnich pripojek. In *Proc. Conf. SOVAK: Casopis Oboru Vodovodu a Kanalizaci*. 12 January 2003, Prague, Czech Republic (In Czech).

Rieckermann, J. and Gujer, W. (2002) Quantifying exfiltration from sewer systems with a novel tracer method. In *Proc. 16th European Junior Scientist Workshop: RTC and Measurements in Urban Drainage Networks*. 7–10 November 2002, Milo, Italy, pp. 71–81.

Rieckermann, J., Bares, V., Braun, D., Kracht, O., Prigiobbe, V. and Gujer, W. (2004) Assessing exfiltration from sewers with dynamic analysis of tracer experiments. In *Proc. 19th European Junior Scientist Workshop, Process Data and Integrated Urban Water Modelling*. 11–14 March 2004, p. 8, Lyon, France.

Rieckermann, J., Sydler, D., Borsuk, M.E., Gujer, W. and Reichert, P. (2006) Using tracer experiments and prior knowledge for the optimal diagnosis of sewer leakage. 131–142 in: *DWA Gemeinschaftstaging Undichte Kanale: (K)ein Risiko?* 11–12 October 2006. Frankfurt, Germany. ISBN 2-939057-60-6.

Rutsch, M. and Uibrig, H. (2003) Classification system to estimate the leakage of sewers. In *Proc. 17th European Junior Scientists Workshop, Rehabilitation Management of Urban Infrastructure Networks*. 3–7 September 2003, p. 8, Neunzehnhain, Germany.

Schilling, W., Bertrand-Krajewski, J.-L., Eiswirth, M., Krebs, P., Saegrov, S. and Thevenot, D. (2002) Integrated urban water management approaches: CityNet. In *Proc. Int. Conf. AQUAECO: Science in Support of European Water Policies: Sustainability of Aquatic Ecosystems*. 26–28 November 2002, p. 4, Stresa, Italy.

Stedman, L. and Bertrand-Krajewski, J.-L. (2004) Seeking certainty on sewer assessment. *Water 21*. August 2004, pp. 56–57, IWA Publishing, London. UK.

Uibrig, H., Karpf, C., Rutsch, M. and Krebs, P. (2002) Possibilities of assessing extraneous water. In *Proc. Seminar, Rehabilitation and Maintaining Sewer Networks*. 21–22 March 2002, p. 12, Rostock, Germany (In German).

DOCTORAL THESES

Cardoso, M.A. (2007) *Performance Assessment of Urban Drainage Systems*. Technical University of Lisbon, Portugal.

de Benedittis, J. (2004) *Mesurage de l'Infiltration et de l'Exfiltration Dans les Reseaux d'Assainissement*. Insa de Lyon, Villeurbanne, France. 16 April 2004, p. 331 (In French).

Franz, T. (2007) *Spatial Classification Methods for Efficient Infiltration Measurements and Transfer of Measuring Results*. Institut fur Siedlungs und Industriewasserwirtschaft. University of Technology, Dresden, Germany. ISSN 1615–083X.

Kracht, O. (2007) *Tracer-Based Hydrograph Separation Methods for Sewer Systems*. ETH, Swiss Federal Institute of Technology, Zurich, Switzerland.

Prigiobbe, V. (2005) *Analysis and Application of Novel Method for Quantifying Infiltration and Exfiltration in Urban Sewer Systems*. Universita Degli Studi di Roma, Tor Vergata. Rome, Italy, 5 May 2005.

Rieckermann, J. (2006) *Quantification of Exfiltration from Sewers With Tracers*. ETH, Swiss Federal Institute of Technology, Zurich, Switzerland.

Rutsch, M. (2007) *Assessment of Sewer Leakage by Means of Exfiltration Measurements and Modelling Tests*. Institut fur Siedlungs und Industriewasserwirtschaft. University of Technology, Dresden, Germany. ISSN 1615–083X.

APUSS PROJECT DELIVERABLES

All internal project reports produced in support of APUSS are included in the deliverables listing available on the official APUSS website (*http://www.insa-lyon.fr/Laboratoires/URGC-HU/apuss*). This deliverable listing includes the Standard Operating Procedures (SOPs) produced in support of the tracer methods and uncertainty analysis described in Chapters 2 and 4, the sediment-tracer studies described in Chapter 5, the conceptual catchment modelling, costing structure and performance indicators and the GIS AquaBase software described in Chapters 10, 11 and 12.

IWA Publishing's authorised EU representative for General Product Safety Regulations is Diane D'Arras, 15 rue Duret, 75116 Paris, France, e-mail: safety@iwap.co.uk.

Printed and bound by CPI Group (UK) Ltd, Croydon, CR0 4YY
03/10/2026
02227463-0005